SMART DUST: SENSOR NETWORK APPLICATIONS, ARCHITECTURE, AND DESIGN

SMART DUST:
SENSOR NETWORK APPLICATIONS, ARCHITECTURE, AND DESIGN

IMAD MAHGOUB
MOHAMMAD ILYAS

Taylor & Francis
Taylor & Francis Group
Boca Raton London New York

A CRC title, part of the Taylor & Francis imprint, a member of the Taylor & Francis Group, the academic division of T&F Informa plc.

The material was previously published in *Handbook of Sensor Networks: Compact Wireless and Wired Sensing Systems*.

Published in 2006 by
CRC Press
Taylor & Francis Group
6000 Broken Sound Parkway NW, Suite 300
Boca Raton, FL 33487-2742

Printed in the United States of America on acid-free paper
10 9 8 7 6 5 4 3 2 1

International Standard Book Number-10: 0-8493-7037-X (Hardcover)
International Standard Book Number-13: 978-0-8493-7037-3 (Hardcover)
Library of Congress Card Number 2005022133

Library of Congress Cataloging-in-Publication Data

Smart dust : sensor network applications, architecture, and design / editors Imad Mahgoub, and Mohammad Ilyas.
p. cm.
Includes bibliographical references and index.
ISBN 0-8493-7037-X (9780849370373 : alk. paper)
1. Sensor networks. I. Ilyas, Mohammad, 1953- II. Mahgoub, Imad.

TK7872.D48S63 2006
681'.2--dc22 2005022133

informa
Taylor & Francis Group
is the Academic Division of Informa plc.

Visit the Taylor & Francis Web site at
http://www.taylorandfrancis.com

and the CRC Press Web site at
http://www.crcpress.com

Preface

Advances in wireless communications and microelectronic mechanical system technologies have enabled the development of networks of a large number of small inexpensive, low-power multifunctional sensors. These networks nicknamed "Smart Dust" present a very interesting and challenging area and have tremendous potential applications.

Wireless sensor networks consist of a large number of sensor nodes that may be randomly and densely deployed. Sensor nodes are small electronic components capable of sensing many types of information from the environment including temperature, light, humidity, radiation, the presence or nature of biological organisms, geological features, seismic vibrations, specific types of computer data, and more. Recent advancements have made it possible to make these components small, powerful, and energy efficient, and they can now be manufactured cost-effectively in quantity for specialized telecommunication applications. The sensor nodes are very small in size and are capable of gathering, processing, and communicating information to other nodes and to the outside world.

This handbook is expected to capture the current state of sensor networks, and specifically address the architecture, applications, and design of such networks. This handbook has a total of 17 chapters written by experts from around the world.

The targeted audience for this handbook includes professionals who are designers and planners for emerging telecommunication networks, researchers (faculty members and graduate students), and those who would like to learn about this field.

Although this handbook is not precisely a textbook, it can certainly be used as a textbook for graduate courses and research-oriented courses that deal with wireless sensor networks. Any comments from the readers will be highly appreciated.

Many people have contributed to this handbook in their unique ways. The first and the foremost group that deserves immense gratitude is the group of highly-talented and skilled researchers who have contributed to this handbook. All of them have been extremely cooperative and professional. It has also been a pleasure to work with Nora Konopka, Helena Redshaw, and Allison Taub of Taylor & Francis, and we are extremely gratified for their support and professionalism. Our families have extended their unconditional love and strong support throughout this project and they all deserve very special thanks.

Imad Mahgoub and Mohammad Ilyas
Boca Raton, Florida

Editors

Imad Mahgoub, Ph.D., received his B.Sc. degree in electrical engineering from the University of Khartoum, Khartoum, Sudan, in 1978. From 1978 to 1981, he worked for the Sudan Shipping Line Company, Port Sudan, Sudan, as an electrical and electronics engineer. He received his M.S. in applied mathematics in 1983 and his M.S. in electrical and computer engineering in 1986, both from North Carolina State University. In 1989, he received his Ph.D. in computer engineering from The Pennsylvania State University.

Since August 1989, Dr. Mahgoub has been with the College of Engineering at Florida Atlantic University, Boca Raton, Florida, where he is currently professor of computer science and engineering. He is the director of the Computer Science and Engineering Department Mobile Computing Laboratory at Florida Atlantic University.

Dr. Mahgoub has conducted successful research in various areas, including mobile computing; interconnection networks; performance evaluation of computer systems; and advanced computer architecture. He has published more than 80 research articles and supervised three Ph.D. dissertations and 22 M.S. theses to completion. He has served as a consultant to industry. Dr. Mahgoub served as a member of the executive committee/program committee of the 1998, 1999, and 2000 IEEE International Performance, Computing and Communications Conferences. He has served on the program committees of several international conferences and symposia. He was the vice chair of the 2003, 2004, and 2005 International Symposium on Performance Evaluation of Computer and Telecommunication Systems. Dr. Mahgoub is a senior member of IEEE and a member of ACM.

Mohammad Ilyas, Ph.D., received his B.Sc. degree in electrical engineering from the University of Engineering and Technology, Lahore, Pakistan, in 1976. From March 1977 to September 1978, he worked for the Water and Power Development Authority in Pakistan. In 1978, he was awarded a scholarship for his graduate studies and he completed his M.S. degree in electrical and electronic engineering in June 1980 at Shiraz University, Shiraz, Iran. In September 1980, he joined the doctoral program at Queen's University in Kingston, Ontario, Canada; he completed his Ph.D. degree in 1983. Dr. Ilyas's doctoral research was about switching and flow control techniques in computer communication networks. Since September 1983, he has been with the College of Engineering at Florida Atlantic University, Boca Raton, Florida, where he is currently associate dean for graduate studies and research. From 1994 to 2000, he was chair of the department. During the 1993–1994 academic year, he was on sabbatical leave with the Department of Computer Engineering, King Saud University, Riyadh, Saudi Arabia.

Dr. Ilyas has conducted successful research in various areas, including traffic management and congestion control in broadband/high-speed communication networks; traffic characterization; wireless communication networks; performance modeling; and simulation. He has published one book, three handbooks, and more than 140 research articles. He has supervised 10 Ph.D. dissertations and more than 35 M.S. theses to completion. Dr. Ilyas has been a consultant to several national and international organizations; a senior member of IEEE, he is an active participant in several IEEE technical committees and activities.

Contributors

Özgür B. Akan
Georgia Institute of Technology
Atlanta, Georgia

Cristian Borcea
Rutgers University
Piscataway, New Jersey

Athanassios Boulis
University of California at Los Angeles
Los Angeles, California

Erdal Cayirci
Istanbul Technical University
Istanbul, Turkey

Anantha Chandrakasan
Engim, Inc.
Acton, Massachusetts

Duminda Dewasurendra
Virginia Polytechnic Institute and State University
Blacksburg, Virginia

Jessica Feng
University of California at Los Angeles
Los Angeles, California

Vicente González–Millán
University of Valencia
Valencia, Spain

Joel I. Goodman
MIT Lincoln Laboratory
Lexington, Massachusetts

Martin Haenggi
University of Notre Dame
Notre Dame, Indiana

Hossam Hassanein
Queen's University
Kingston, Ontario, Canada

Chi-Fu Huang
National Chiao-Tung University
Hsin-Chu, Taiwan

Liviu Iftode
Rutgers University
Piscataway, New Jersey

Chaiporn Jaikaeo
University of Delaware
Newark, Delaware

Porlin Kang
Rutgers University
Piscataway, New Jersey

Zdravko Karakehayov
Technical University of Sofia
Sofia, Bulgaria

Farinaz Koushanfar
University of California at Berkeley
Berkeley, California

Sheng-Po Kuo
National Chiao-Tung University
Hsin-Chu, Taiwan

Antonio A.F. Loureiro
Federal University of Minas Gerais
Belo Horizonte, Brazil

David R. Martinez
MIT Lincoln Laboratory
Lexington, Massachusetts

Amitabh Mishra
Virginia Polytechnic Institute and State University
Blacksburg, Virginia

José Marcos Nogueira
Federal University of Minas Gerais
Belo Horizonte, Brazil

Miodrag Potkonjak
University of California at Los Angeles
Los Angeles, California

Albert I. Reuther
MIT Lincoln Laboratory
Lexington, Massachusetts

Linnyer Beatrys Ruiz
Pontifical Catholic University of Paraná
Curitiba, Brazil
and Federal University of Minas Gerais
Belo Horizonte, Brazil

Ayad Salhieh
Wayne State University
Detroit, Michigan

Enrique Sanchis-Peris
University of Valencia
Valencia, Spain

Loren Schwiebert
Wayne State University
Detroit, Michigan

Chien-Chung Shen
University of Delaware
Newark, Delaware

Amit Sinha
Engim, Inc.
Acton, Massachusetts

Sasha Slijepcevic
University of California at Los Angeles
Los Angeles, California

Chavalit Srisathapornphat
University of Delaware
Newark, Delaware

Weilian Su
Georgia Institute of Technology
Atlanta, Georgia

Yu-Chee Tseng
National Chiao-Tung University
Hsin-Chu, Taiwan

Quanhong Wang
Queen's University
Kingston, Ontario, Canada

Brett Warneke
Dust Networks
Berkeley, California

Jennifer L. Wong
University of California at Los Angeles
Los Angeles, California

Kenan Xu
Queen's University
Kingston, Ontario, Canada

Contents

1 Opportunities and Challenges in Wireless Sensor Networks *Martin Haenggi*

1.1 Introduction **1**-1
1.2 Opportunities **1**-2
1.3 Technical Challenges **1**-4
1.4 Concluding Remarks **1**-11

2 Next-Generation Technologies to Enable Sensor Networks *Joel I. Goodman, Albert I. Reuther, David R. Martinez*

2.1 Introduction **2**-1
2.2 Goals for Real-Time Distributed Network Computing for Sensor Data Fusion **2**-5
2.3 The Convergence of Networking and Real-Time Computing **2**-6
2.4 Middleware **2**-11
2.5 Network Resource Management **2**-11
2.6 Experimental Results **2**-16

3 Sensor Network Management *Linnyer Beatrys Ruiz, José Marcos Nogueira, Antonio A. F. Loureiro*

3.1 Introduction **3**-1
3.2 Management Challenges **3**-2
3.3 Management Dimensions **3**-3
3.4 MANNA as an Integrating Architecture **3**-15
3.5 Putting It All Together **3**-25
3.6 Conclusion **3**-25

4 Models for Programmability in Sensor Networks *Athanassios Boulis*

4.1 Introduction **4**-1
4.2 Differences between Sensor Networks and Traditional Data Networks **4**-2
4.3 Aspects of Efficient Sensor Network Applications **4**-2
4.4 Need for Sensor Network Programmability **4**-3
4.5 Major Models for System-Level Programmability **4**-4
4.6 Frameworks for System-Level Programmability **4**-6
4.7 Conclusions **4**-12

5 Miniaturizing Sensor Networks with MEMS *Brett Warneke*

5.1 Introduction 5-1
5.2 MEMS Basics 5-2
5.3 Sensors 5-4
5.4 Communication 5-5
5.5 Micropower Sources 5-10
5.6 Packaging 5-12
5.7 Systems 5-13
5.8 Conclusion 5-15

6 Sensor Network Architecture and Applications *Chien-Chung Shen, Chaiporn Jaikaeo, Chavalit Srisathapornphat*

6.1 Introduction 6-1
6.2 Sensor Network Applications 6-1
6.3 Functional Architecture for Sensor Networks 6-3
6.4 Sample Implementation Architectures 6-4
6.5 Summary 6-12

7 A Practical Perspective on Wireless Sensor Networks *Quanhong Wang, Hossam Hassanein, Kenan Xu*

7.1 Introduction 7-1
7.2 WSN Applications 7-2
7.3 Classification of WSNs 7-6
7.4 Characteristics, Technical Challenges, and Design Directions 7-7
7.5 Technical Approaches 7-11
7.6 Conclusions and Considerations for Future Research 7-22

8 Sensor Network Architecture *Jessica Feng, Farinaz Koushanfar, Miodrag Potkonjak*

8.1 Overview 8-1
8.2 Motivation and Objectives 8-1
8.3 SNs — Global View and Requirements 8-3
8.4 Individual Components of SN Nodes 8-4
8.5 Sensor Network Node 8-8
8.6 Wireless SNs as Embedded Systems 8-13
8.7 Summary 8-16

9 Power-Efficient Topologies for Wireless Sensor Networks *Ayad Salhieh, Loren Schwiebert*

9.1 Motivation 9-1
9.2 Background 9-2
9.3 Issues for Topology Design 9-3
9.4 Assumptions 9-8

9.5 Analysis of Power Usage **9**-10
9.6 Directional Source-Aware Routing Protocol (DSAP) **9**-13
9.7 DSAP Analysis **9**-15
9.8 Summary **9**-19

10 Overview of Communication Protocols for Sensor Networks *Weilian Su, Erdal Cayirci, Özgür B. Akan*
10.1 Introduction **10**-1
10.2 Applications/Application Layer Protocols **10**-2
10.3 Localization Protocols **10**-4
10.4 Time Synchronization Protocols **10**-5
10.5 Transport Layer Protocols **10**-7
10.6 Network Layer Protocols **10**-9
10.7 Data Link Layer Protocols **10**-11
10.8 Conclusion **10**-14

11 Positioning and Location Tracking in Wireless Sensor Networks *Yu-Chee Tseng, Chi-Fu Huang, Sheng-Po Kuo*
11.1 Introduction **11**-1
11.2 Fundamentals **11**-2
11.3 Positioning and Location Tracking Algorithms **11**-4
11.4 Experimental Location Systems **11**-10
11.5 Conclusions **11**-12

12 Comparison of Data Processing Techniques in Sensor Networks *Vicente González-Millán, Enrique Sanchis-Peris*
12.1 Sensor Networks: Organization and Processing **12**-1
12.2 Architectures for Sensor Integration **12**-3
12.3 Example of Architecture Evaluation in High-Energy Physics **12**-18

13 Cooperative Computing in Sensor Networks *Liviu Iftode, Cristian Borcea, Porlin Kang*
13.1 Introduction **13**-1
13.2 The Cooperative Computing Model **13**-3
13.3 Node Architecture **13**-4
13.4 Smart Messages **13**-5
13.5 Programming Interface **13**-7
13.6 Prototype Implementation and Evaluation **13**-8
13.7 Applications **13**-12
13.8 Simulation Results **13**-14
13.9 Related Work **13**-15
13.10 Conclusions **13**-18

14 Dynamic Power Management in Sensor Networks *Amit Sinha, Anantha Chandrakasan*
14.1 Introduction **14**-1
14.2 Idle Power Management **14**-2
14.3 Active Power Management **14**-5
14.4 System Implementation **14**-6
14.5 Results **14**-12

15 Design Challenges in Energy-Efficient Medium Access Control for Wireless Sensor Networks *Duminda Dewasurendra, Amitabh Mishra*
15.1 Introduction **15**-1
15.2 Unique Characteristics of Wireless Sensor Networks **15**-2
15.3 MAC Protocols for Wireless Ad Hoc Networks **15**-4
15.4 Design Challenges for Wireless Sensor Networks **15**-10
15.5 Medium Access Protocols for Wireless Sensor Networks **15**-13
15.6 Open Issues **15**-22
15.7 Conclusions **15**-24

16 Security and Privacy Protection in Wireless Sensor Networks *Sasha Slijepcevic, Jennifer L. Wong, Miodrag Potkonjak*
16.1 Introduction **16**-1
16.2 Unique Security Challenges in Sensor Networks and Enabling Mechanisms **16**-2
16.3 Security Architectures **16**-4
16.4 Privacy Protection **16**-11
16.5 Conclusion **16**-15

17 Low-Power Design for Smart Dust Networks *Zdravko Karakehayov*
17.1 Introduction **17**-1
17.2 Location **17**-1
17.3 Sensing **17**-2
17.4 Computation **17**-2
17.5 Hardware–Software Interaction **17**-5
17.6 Communication **17**-7
17.7 Orientation **17**-10
17.8 Conclusion **17**-10

Index **I**-1

1

Opportunities and Challenges in Wireless Sensor Networks

1.1 Introduction 1-1
1.2 Opportunities 1-2
Growing Research and Commercial Interest • Applications
1.3 Technical Challenges 1-4
Performance Metrics • Power Supply • Design of Energy-Efficient Protocols • Capacity/Throughput • Routing • Channel Access and Scheduling • Modeling • Connectivity • Quality of Service • Security • Implementation • Other Issues
1.4 Concluding Remarks 1-11

Martin Haenggi
University of Notre Dame

1.1 Introduction

Due to advances in wireless communications and electronics over the last few years, the development of networks of low-cost, low-power, multifunctional sensors has received increasing attention. These sensors are small in size and able to sense, process data, and communicate with each other, typically over an RF (radio frequency) channel. A sensor network is designed to detect events or phenomena, collect and process data, and transmit sensed information to interested users. Basic features of sensor networks are:

- Self-organizing capabilities
- Short-range broadcast communication and multihop routing
- Dense deployment and cooperative effort of sensor nodes
- Frequently changing topology due to fading and node failures
- Limitations in energy, transmit power, memory, and computing power

These characteristics, particularly the last three, make sensor networks different from other wireless ad hoc or mesh networks.

Clearly, the idea of mesh networking is not new; it has been suggested for some time for wireless Internet access or voice communication. Similarly, small computers and sensors are not innovative per se. However, combining small sensors, low-power computers, and radios makes for a new technological platform that has numerous important uses and applications, as will be discussed in the next section.

1.2 Opportunities

1.2.1 Growing Research and Commercial Interest

Research and commercial interest in the area of wireless sensor networks are currently growing exponentially, which is manifested in many ways:

- The number of Web pages (Google: 26,000 hits for sensor networks; 8000 for wireless sensor networks in August 2003)
- The increasing number of
  - Dedicated annual workshops, such as IPSN (information processing in sensor networks); SenSys; EWSN (European workshop on wireless sensor networks); SNPA (sensor network protocols and applications); and WSNA (wireless sensor networks and applications)
  - Conference sessions on sensor networks in the communications and mobile computing communities (ISIT, ICC, Globecom, INFOCOM, VTC, MobiCom, MobiHoc)
  - Research projects funded by NSF (apart from ongoing programs, a new specific effort now focuses on sensors and sensor networks) and DARPA through its SensIT (sensor information technology), NEST (networked embedded software technology), MSET (multisensor exploitation), UGS (unattended ground sensors), NETEX (networking in extreme environments), ISP (integrated sensing and processing), and communicator programs

Special issues and sections in renowned journals are common, e.g., in the *IEEE Proceedings* [1] and signal processing, communications, and networking magazines. Commercial interest is reflected in investments by established companies as well as start-ups that offer general and specific hardware and software solutions.

Compared to the use of a few expensive (but highly accurate) sensors, the strategy of deploying a large number of inexpensive sensors has significant advantages, at smaller or comparable total system cost: much higher spatial resolution; higher robustness against failures through distributed operation; uniform coverage; small obtrusiveness; ease of deployment; reduced energy consumption; and, consequently, increased system lifetime. The main point is to position sensors close to the source of a potential problem phenomenon, where the acquired data are likely to have the greatest benefit or impact.

Pure sensing in a fine-grained manner may revolutionize the way in which complex physical systems are understood. The addition of actuators, however, opens a completely new dimension by permitting management and manipulation of the environment at a scale that offers enormous opportunities for almost every scientific discipline. Indeed, Business 2.0 (http://www.business2.com/) lists sensor robots as one of "six technologies that will change the world," and *Technology Review* at MIT and Globalfuture identify WSNs as one of the "10 emerging technologies that will change the world" (http://www.globalfuture.com/mit-trends2003.htm). The combination of sensor network technology with MEMS and nanotechnology will greatly reduce the size of the nodes and enhance the capabilities of the network.

The remainder of this chapter lists and briefly describes a number of applications for wireless sensor networks, grouped into different categories. However, because the number of areas of application is growing rapidly, every attempt at compiling an exhaustive list is bound to fail.

1.2.2 Applications

1.2.2.1 General Engineering

- *Automotive telematics.* Cars, which comprise a network of dozens of sensors and actuators, are networked into a system of systems to improve the safety and efficiency of traffic.
- *Fingertip accelerometer virtual keyboards.* These devices may replace the conventional input devices for PCs and musical instruments.
- *Sensing and maintenance in industrial plants.* Complex industrial robots are equipped with up to 200 sensors that are usually connected by cables to a main computer. Because cables are

expensive and subject to wear and tear caused by the robot's movement, companies are replacing them by wireless connections. By mounting small coils on the sensor nodes, the principle of induction is exploited to solve the power supply problem.

- *Aircraft drag reduction.* Engineers can achieve this by combining flow sensors and blowing/sucking actuators mounted on the wings of an airplane.
- *Smart office spaces.* Areas are equipped with light, temperature, and movement sensors, microphones for voice activation, and pressure sensors in chairs. Air flow and temperature can be regulated locally for one room rather than centrally.
- *Tracking of goods in retail stores.* Tagging facilitates the store and warehouse management.
- *Tracking of containers and boxes.* Shipping companies are assisted in keeping track of their goods, at least until they move out of range of other goods.
- *Social studies.* Equipping human beings with sensor nodes permits interesting studies of human interaction and social behavior.
- Commercial and residential security.

1.2.2.2 Agriculture and Environmental Monitoring

- *Precision agriculture.* Crop and livestock management and precise control of fertilizer concentrations are possible.
- *Planetary exploration.* Exploration and surveillance in inhospitable environments such as remote geographic regions or toxic locations can take place.
- *Geophysical monitoring.* Seismic activity can be detected at a much finer scale using a network of sensors equipped with accelerometers.
- *Monitoring of freshwater quality.* The field of hydrochemistry has a compelling need for sensor networks because of the complex spatiotemporal variability in hydrologic, chemical, and ecological parameters and the difficulty of labor-intensive sampling, particularly in remote locations or under adverse conditions. In addition, buoys along the coast could alert surfers, swimmers, and fishermen to dangerous levels of bacteria.
- *Zebranet.* The Zebranet project at Princeton aims at tracking the movement of zebras in Africa.
- *Habitat monitoring.* Researchers at UC Berkeley and the College of the Atlantic in Bar Harbor deployed sensors on Great Duck Island in Maine to measure humidity, pressure, temperature, infrared radiation, total solar radiation, and photosynthetically active radiation (see http://www.greatduckisland.net/).
- *Disaster detection.* Forest fire and floods can be detected early and causes can be localized precisely by densely deployed sensor networks.
- *Contaminant transport.* The assessment of exposure levels requires high spatial and temporal sampling rates, which can be provided by WSNs.

1.2.2.3 Civil Engineering

- *Monitoring of structures.* Sensors will be placed in bridges to detect and warn of structural weakness and in water reservoirs to spot hazardous materials. The reaction of tall buildings to wind and earthquakes can be studied and material fatigue can be monitored closely.
- *Urban planning.* Urban planners will track groundwater patterns and how much carbon dioxide cities are expelling, enabling them to make better land-use decisions.
- *Disaster recovery.* Buildings razed by an earthquake may be infiltrated with sensor robots to locate signs of life.

1.2.2.4 Military Applications

- *Asset monitoring and management.* Commanders can monitor the status and locations of troops, weapons, and supplies to improve military command, control, communications, and computing (C4).

- *Surveillance and battle-space monitoring.* Vibration and magnetic sensors can report vehicle and personnel movement, permitting close surveillance of opposing forces.
- *Urban warfare.* Sensors are deployed in buildings that have been cleared to prevent reoccupation; movements of friend and foe are displayed in PDA-like devices carried by soldiers. Snipers can be localized by the collaborative effort of multiple acoustic sensors.
- *Protection.* Sensitive objects such as atomic plants, bridges, retaining walls, oil and gas pipelines, communication towers, ammunition depots, and military headquarters can be protected by intelligent sensor fields able to discriminate between different classes of intruders. Biological and chemical attacks can be detected early or even prevented by a sensor network acting as a warning system.
- *Self-healing minefields.* The self-healing minefield system is designed to achieve an increased resistance to dismounted and mounted breaching by adding a novel dimension to the minefield. Instead of a static complex obstacle, the self-healing minefield is an intelligent, dynamic obstacle that senses relative positions and responds to an enemy's breaching attempt by physical reorganization.

1.2.2.5 Health Monitoring and Surgery

- *Medical sensing.* Physiological data such as body temperature, blood pressure, and pulse are sensed and automatically transmitted to a computer or physician, where it can be used for health status monitoring and medical exploration. Wireless sensing bandages may warn of infection. Tiny sensors in the blood stream, possibly powered by a weak external electromagnetic field, can continuously analyze the blood and prevent coagulation and thrombosis.
- *Microsurgery.* A swarm of MEMS-based robots may collaborate to perform microscopic and minimally invasive surgery.

The opportunities for wireless sensor networks are ubiquitous. However, a number of formidable challenges must be solved before these exciting applications may become reality.

1.3 Technical Challenges

Populating the world with networks of sensors requires a fundamental understanding of techniques for connecting and managing sensor nodes with a communication network in scalable and resource-efficient ways. Clearly, sensor networks belong to the class of ad hoc networks, but they have specific characteristics that are not present in general ad hoc networks.

Ad hoc and sensor networks share a number of challenges such as energy constraints and routing. On the other hand, general ad hoc networks most likely induce traffic patterns different from sensor networks, have other lifetime requirements, and are often considered to consist of *mobile* nodes [2–4]. In WSNs, most nodes are static; however, the network of basic sensor nodes may be overlaid by more powerful mobile sensors (robots) that, guided by the basic sensors, can move to interesting areas or even track intruders in the case of military applications.

Network nodes are equipped with wireless transmitters and receivers using antennas that may be omnidirectional (isotropic radiation), highly directional (point-to-point), possibly steerable, or some combination thereof. At a given point in time, depending on the nodes' positions and their transmitter and receiver coverage patterns, transmission power levels, and cochannel interference levels, a wireless connectivity exists in the form of a random, multihop graph between the nodes. This ad hoc topology may change with time as the nodes move or adjust their transmission and reception parameters.

Because the most challenging issue in sensor networks is *limited and unrechargeable* energy provision, many research efforts aim at improving the energy efficiency from different aspects. In sensor networks, energy is consumed mainly for three purposes: *data transmission*, *signal processing*, and *hardware operation* [5]. It is desirable to develop energy-efficient processing techniques that minimize power requirements across all levels of the protocol stack and, at the same time, minimize message passing for network control and coordination.

1.3.1 Performance Metrics

To discuss the issues in more detail, it is necessary to examine a list of metrics that determine the performance of a sensor network:

- *Energy efficiency/system lifetime.* The sensors are battery operated, rendering energy a very scarce resource that must be wisely managed in order to extend the lifetime of the network [6].
- *Latency.* Many sensor applications require delay-guaranteed service. Protocols must ensure that sensed data will be delivered to the user within a certain delay. Prominent examples in this class of networks are certainly the sensor-actuator networks.
- *Accuracy.* Obtaining accurate information is the primary objective; accuracy can be improved through joint detection and estimation. Rate distortion theory is a possible tool to assess accuracy.
- *Fault tolerance.* Robustness to sensor and link failures must be achieved through redundancy and collaborative processing and communication.
- *Scalability.* Because a sensor network may contain thousands of nodes, scalability is a critical factor that guarantees that the network performance does not significantly degrade as the network size (or node density) increases.
- *Transport capacity/throughput.* Because most sensor data must be delivered to a single base station or fusion center, a *critical area* in the sensor network exists (the gray area in Figure 1.1.), whose sensor nodes must relay the data generated by virtually all nodes in the network. Thus, the traffic load at those critical nodes is heavy, even when the average traffic rate is low. Apparently, this area has a paramount influence on system lifetime, packet end-to-end delay, and scalability.

Because of the interdependence of energy consumption, delay, and throughput, all these issues and metrics are tightly coupled. Thus, the design of a WSN necessarily consists of the resolution of numerous trade-offs, which also reflects in the network protocol stack, in which a cross-layer approach is needed instead of the traditional layer-by-layer protocol design.

1.3.2 Power Supply

The most difficult constraints in the design of WSNs are those regarding the minimum energy consumption necessary to drive the circuits and possible microelectromechanical devices (MEMS) [5, 7, 8]. The energy problem is aggravated if actuators are present that may be substantially hungrier for power than the sensors. When miniaturizing the node, the energy density of the power supply is the primary issue. Current technology yields batteries with approximately 1 J/mm^3 of energy, while capacitors can achieve as much as 1 mJ/mm^3. If a node is designed to have a relatively short life span, for example, a few months, a battery is a logical solution. However, for nodes that can generate sensor readings for long periods of time, a charging

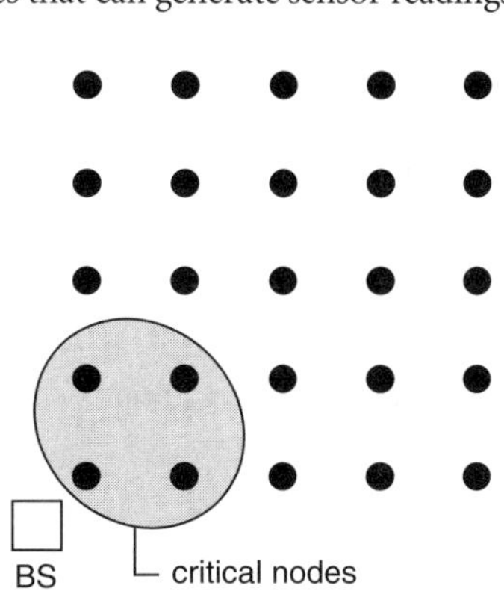

FIGURE 1.1 Sensor network with base station (or fusion center). The gray-shaded area indicates the critical area whose nodes must relay all the packets.

method for the supply is preferable. Currently, research groups are investigating the use of solar cells to charge capacitors with photocurrents from the ambient light sources. Solar flux can yield power densities of approximately 1 mW/mm^2. The energy efficiency of a solar cell ranges from 10 to 30% in current technologies, giving 300 μW in full sunlight in the best-case scenario for a 1-mm^2 solar cell operating at 1 V. Series-stacked solar cells will need to be utilized in order to provide appropriate voltages.

Sensor acquisition can be achieved at 1 nJ per sample, and modern processors can perform computations as low as 1 nJ per instruction. For wireless communications, the primary candidate technologies are based on RF and optical transmission techniques, each of which has its advantages and disadvantages. RF presents a problem because the nodes may offer very limited space for antennas, thereby demanding very short-wavelength (i.e., high-frequency) transmission, which suffers from high attenuation. Thus, communication in that regime is not currently compatible with low-power operation. Current RF transmission techniques (e.g., Bluetooth [9]) consume about 100 nJ per bit for a distance of 10 to 100 m, making communication very expensive compared to acquisition and processing.

An alternative is to employ free-space optical transmission. If a line-of-sight path is available, a well-designed free-space optical link requires significantly lower energy than its RF counterpart, currently about 1 nJ per bit. The reason for this power advantage is that optical transceivers require only simple baseband analog and digital circuitry and no modulators, active filters, and demodulators. Furthermore, the extremely short wavelength of visible light makes it possible for a millimeter-scale device to emit a narrow beam, corresponding to an antenna gain of roughly five to six orders of magnitude compared to an isotropic radiator. However, a major disadvantage is that the beam needs to be pointed very precisely at the receiver, which may be prohibitively difficult to achieve.

In WSNs, where sensor sampling, processing, data transmission, and, possibly, actuation are involved, the trade-off between these tasks plays an important role in power usage. Balancing these parameters will be the focus of the design process of WSNs.

1.3.3 Design of Energy-Efficient Protocols

It is well acknowledged that *clustering* is an efficient way to save energy for static sensor networks [10–13]. Clustering has three significant differences from conventional clustering schemes. First, data compression in the form of distributed source coding is applied within a cluster to reduce the number of packets to be transmitted [14, 15]. Second, the *data-centric* property makes an identity (e.g., an address) for a sensor node obsolete. In fact, the user is often interested in phenomena occurring in a specified area [16], rather than in an individual sensor node. Third, randomized rotation of cluster heads helps ensure a balanced energy consumption [11].

Another strategy to increase energy efficiency is to use *broadcast and multicast trees* [6, 17, 18], which take advantage of the *broadcast property* of omnidirectional antennas. The disadvantage is that the high computational complexity may offset the achievable benefit. For sensor networks, this *one-to-many* communication scheme is less important; however, because all data must be delivered to a single destination, the traffic scheme (for application traffic) is the opposite, i.e., *many to one*. In this case, clearly the *wireless multicast advantage* offers less benefit, unless path diversity or cooperative diversity schemes are implemented [19, 20].

The exploitation of *sleep modes* [21, 22] is imperative to prevent sensor nodes from wasting energy in receiving packets unintended for them. Combined with efficient medium access protocols, the "sleeping" approach could reach optimal energy efficiency without degradation in throughput (but at some penalty in delay).

1.3.4 Capacity/Throughput

Two parameters describe the network's capability to carry traffic: *transport capacity* and *throughput*. The former is a distance-weighted sum capacity that permits evaluation of network performance. Throughput is a traditional measure of how much traffic can be delivered by the network [23–30]. In a packet network, the (network-layer) throughput may be defined as the expected number of successful packet transmissions of a given node per timeslot.

The capacity of wireless networks in general is an active area of research in the information theory community. The results obtained mostly take the form of scaling laws or "order-of" results; the prefactors are difficult to determine analytically. Important results include the scaling law for point-to-point coding, which shows that the throughput decreases with $1/\sqrt{N}$ for a network with N nodes [23]. Newer results [28] permit network coding, which yields a slightly more optimistic scaling behavior, although at high complexity. Grossglauser and Tse [26] have shown that mobility may keep the per-node capacity constant as the network grows, but that benefit comes at the cost of unbounded delay.

The throughput is related to (error-free) transmission rate of each transmitter, which, in turn, is upper bounded by the channel capacity. From the pure information theoretic point of view, the capacity is computed based on the ergodic channel assumption, i.e., the code words are long compared to the coherence time of the channel. This Shannon-type capacity is also called *throughput capacity* [31]. However, in practical networks, particularly with delay-constrained applications, this capacity cannot provide a helpful indication of the channel's ability to transmit with a small probability of error.

Moreover, in the multiple-access system, the corresponding power allocation strategies for maximum achievable capacity always favor the "good" channels, thus leading to unfairness among the nodes. Therefore, for delay-constrained applications, the channel is usually assumed to be nonergodic and the capacity is a random variable, instead of a constant in the classical definition by Shannon. For a delay-bound D, the channel is often assumed to be block fading with block length D, and a *composite channel* model is appropriate when specifying the capacity. Correspondingly, given the noise power, the channel state (a random variable in the case of fading channels), and power allocation, new definitions for *delay-constrained* systems have been proposed [32–35].

1.3.5 Routing

In ad hoc networks, routing protocols are expected to implement three main functions: *determining and detecting network topology changes* (e.g., breakdown of nodes and link failures); *maintaining network connectivity*; and *calculating and finding proper routes.* In sensor networks, up-to-date, less effort has been given to routing protocols, even though it is clear that ad hoc routing protocols (such as *destination-sequenced distance vector* (DSDV), *temporally-ordered routing algorithm* (TORA), *dynamic source routing* (DSR), and *ad hoc on-demand distance vector* (AODV) [4, 36–39]) are not suited well for sensor networks since the main type of traffic in WSNs is "many to one" because all nodes typically report to a single base station or fusion center. Nonetheless, some merits of these protocols relate to the features of sensor networks, like *multihop communication* and *QoS routing* [39]. Routing may be associated with data compression [15] to enhance the scalability of the network.

1.3.6 Channel Access and Scheduling

In WSNs, scheduling must be studied at two levels: the *system level* and the *node level.* At the node level, a scheduler determines which flow among all multiplexing flows will be eligible to transmit next (the same concept as in traditional wired scheduling); at the system level, a scheme determines which nodes will be transmitting. System-level scheduling is essentially a medium access (MAC) problem, with the goal of minimum collisions and maximum spatial reuse — a topic receiving great attention from the research community because it is tightly coupled with energy efficiency and throughput.

Most of the current wireless scheduling algorithms aim at improved *fairness, delay, robustness* (with respect to network topology changes) and *energy efficiency* [62, 64, 65, 66]. Some also propose a distributed implementation, in contrast to the centralized implementation in wired or cellular networks, which originated from general fair queuing. Also, wireless (or sensor) counterparts of other wired scheduling classes, like *priority scheduling* [67, 68] and *earliest deadline first (EDF)* [69], confirm that prioritization is necessary to achieve *delay balancing* and *energy balancing.*

The main problem in WSNs is that all the sensor data must be forwarded to a base station via multihop routing. Consequently, the traffic pattern is highly nonuniform, putting a high burden on the sensor nodes close to the base station (the critical nodes in Figure 1.1). The scheduling algorithm and routing protocols must aim at *energy and delay balancing*, ensuring that packets originating close and far away from the base station experience a comparable delay, and that the critical nodes do not die prematurely due to the heavy relay traffic [40].

At this point, due to the complexity of scheduling algorithms and the wireless environment, most performance measures are given through simulation rather than analytically. Moreover, *medium access* and *scheduling* are usually considered separately. When discussing scheduling, the system is assumed to have a single user; whereas in the MAC layer, all flows multiplexing at the node are treated in the same way, i.e., a default FIFO buffer is assumed to schedule flows. It is necessary to consider them jointly to optimize performance figures such as delay, throughput, and packet loss probability.

Because of the bursty nature of the network traffic, random access methods are commonly employed in WSNs, with or without carrier sense mechanisms. For illustrative purposes, consider the simplest sensible MAC scheme possible: all nodes are transmitting packets independently in every timeslot with the same transmit probability p at equal transmitting power levels; the next-hop receiver of every packet is one of its neighbors. The packets are of equal length and fit into one timeslot. This MAC scheme was considered in Silvester and Kleinrock [41], Hu [42], and Haenggi [43]. The resulting (per-node) throughput turns out to be a polynomial in p of order N, where N is the number of nodes in the network.

A typical throughput polynomial is shown in Figure 1.2. At $p = 0$, the derivative is 1, indicating that, for small p, the throughput equals p. This is intuitive because there are few collisions for small p and the throughput $g(p)$ is approximately linear. The region in which the packet loss probability is less than 10% can be denoted as the *collisionless* region. It ranges from 0 to about $p_{max}/8$. The next region, up to p_{max}, is the practical region in which energy consumption (transmission attempts) is traded off against throughput; it is therefore called the *trade-off* region. The difference $p - g(p)$ is the *interference loss*. For small networks, all N nodes interfere with each other because spatial reuse is not possible: If more than one node is transmitting, a collision occurs and all packets are lost. Thus, the (per-node) throughput is $p(1-p)^{N-1}$, and the optimum transmit probability is $1/N$. The maximum throughput is $(1 - 1/N)^{N-1}/N$. With increasing N, the throughput approaches $1/(eN)$, as pointed out in Silvester and Kleinrock [41] and LaMaire et al. [44]. Therefore the difference $p_{max} - 1/N$ is the *spatial reuse gain* (see Figure 1.2). This simple example illustrates the concepts of collisions, energy-throughput trade-offs, and spatial reuse, which are present in every MAC scheme.

1.3.7 Modeling

The bases for analysis and simulations and analytical approaches are accurate and tractable models. Comprehensive network models should include the number of nodes and their relative distribution; their degree and type of mobility; the characteristics of the wireless link; the volume of traffic injected by the sources and the lifespan of their interaction; and detailed energy consumption models.

1.3.7.1 Wireless Link

An attenuation proportional to d^α, where d is the distance between two nodes and α is the so-called path loss exponent, is widely accepted as a model for path loss. Alpha ranges from 2 to 4 or even 5 [45], depending on the channel characteristics (environment, antenna position, frequency). This path loss model, together with the fact that packets are successfully transmitted if the signal-to-noise-and-interference ratio (SNIR) is bigger than some threshold [8], results in a deterministic model often used for analysis of multihop packet networks [23, 26, 41, 42, 46–48]. Thus, the radius for a successful transmission has a deterministic value, irrespective of the condition of the wireless channel. If only interferers within a certain distance of the receiver are considered, this "physical model" [23] turns into a "disk model."

The stochastic nature of the fading channel and thus the fact that the SINR is a random variable are mostly neglected. However, the volatility of the channel cannot be ignored in wireless networks [5, 8];

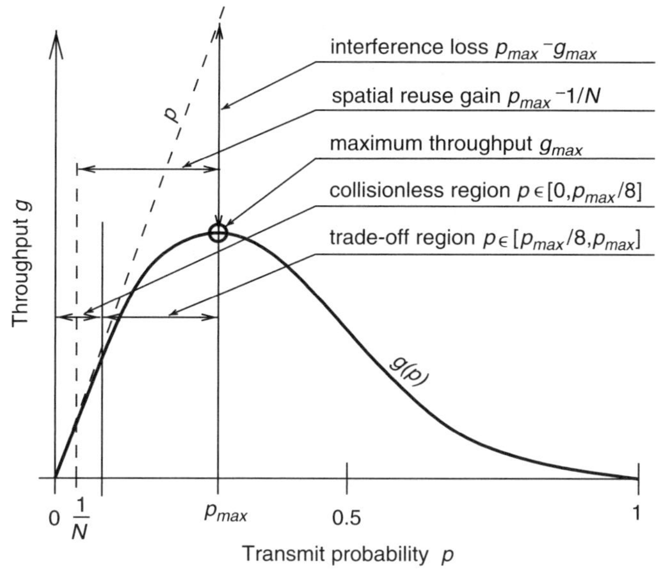

FIGURE 1.2 Generic throughput polynomial for a simple random MAC scheme.

Sousa and Silvester have also pointed out the inaccuracy of disk models [49] and it is easily demonstrated experimentally [50, 51]. In addition, this "prevalent all-or-nothing model" [52] leads to the assumption that a transmission over a multihop path fails completely or is 100% successful, ignoring the fact that end-to-end packet loss probabilities increase with the number of hops. Although fading has been considered in the context of packet networks [53, 54], its impact on the throughput of multihop networks and protocols at the MAC and higher layers is largely an open problem.

A more accurate channel model will have an impact on most of the metrics listed in Section 1.3.1. In the case of Rayleigh fading, first results show that the energy benefits of routing over many short hops may vanish completely, in particular if latency is taken into account [20, 55, 56]. The Rayleigh fading model not only is more accurate than the disk model, but also has the additional advantage of permitting separation of noise effects and interference effects due to the exponential distribution of the received power. As a consequence, the performance analysis can conveniently be split into the analysis of a zero-interference (noise-analysis) and a zero-noise (interference-analysis) network.

1.3.7.2 Energy Consumption

To model energy consumption, four basic different states of a node can be identified: transmission, reception, listening, and sleeping. They consist of the following tasks:

- *Acquisition:* sensing, A/D conversion, preprocessing, and perhaps storing
- *Transmission:* processing for address determination, packetization, encoding, framing, and maybe queuing; supply for the baseband and RF circuitry (The nonlinearity of the power amplifier must be taken into account because the power consumption is most likely not proportional to the transmit power [56].)
- *Reception:* Low-noise amplifier, downconverter oscillator, filtering, detection, decoding, error detection, and address check; reception even if a node is not the intended receiver
- *Listening:* Similar to reception except that the signal processing chain stops at the detection
- *Sleeping:* Power supply to stay alive

Reception and transmission comprise all the processing required for physical communication and networking protocols. For the physical layer, the energy consumption depends mostly on the circuitry, the error correction schemes, and the implementation of the receiver [57]. At the higher layers, the choice of protocols (e.g., routing, ARQ schemes, size of packet headers, number of beacons and other infrastructure packets) determines the energy efficiency.

1.3.7.3 Node Distribution and Mobility

Regular grids (square, triangle, hexagon) and uniformly random distributions are widely used analytically tractable models. The latter can be problematic because nodes can be arbitrarily close, leading to unrealistic received power levels if the path attenuation is assumed to be proportional to d^{α}. Regular grids overlaid with Gaussian variations in the positions may be more accurate. Generic mobility models for WSNs are difficult to define because they are highly application specific, so this issue must be studied on a case-by-case basis.

1.3.7.4 Traffic

Often, simulation work is based on constant bitrate traffic for convenience, but this is most probably not the typical traffic class. Models for bursty many-to-one traffic are needed, but they certainly depend strongly on the application.

1.3.8 Connectivity

Network connectivity is an important issue because it is crucial for most applications that the network is not partitioned into disjoint parts. If the nodes' positions are modeled as a Poisson point process in two dimensions (which, for all practical purposes, corresponds to a uniformly random distribution), the problem of connectivity has been studied using the tool of *continuum percolation theory* [58, 59]. For large networks, the phenomenon of a sharp phase transition can be observed: the probability that the network *percolates* jumps abruptly from almost 0 to almost 1 as soon as the density of the network is bigger than some critical value. Most such results are based on the geometric disk abstraction. It is conjectured, though, that other connectivity functions lead to better connectivity, i.e., the disk is apparently the hardest shape to connect [60]. A practical consequence of this conjecture is that fading results in improved connectivity. Recent work [61] also discusses the impact of interference. The simplifying assumptions necessary to achieve these results leave many open problems.

1.3.9 Quality of Service

Quality of service refers to the capability of a network to deliver data reliably and timely. A high *quantity of service*, i.e., throughput or transport capacity, is generally not sufficient to satisfy an application's delay requirements. Consequently, the *speed of propagation* of information may be as crucial as the throughput. Accordingly, in addition to network capacity, an important issue in many WSNs is that of quality-of-service (QoS) guarantees. Previous QoS-related work in wireless networks mostly focused on delay (see, for example, Lu et al. [62], Ju and Li [63], and Liu et al. [64]). QoS, in a broader sense, consists of the triple (R, P_e, D), where R denotes throughput; P_e denotes reliability as measured by, for example, bit error probability or packet loss probability; and D denotes delay. For a given R, the reliability of a connection as a function of the delay will follow the general curve shown in Figure 1.3.

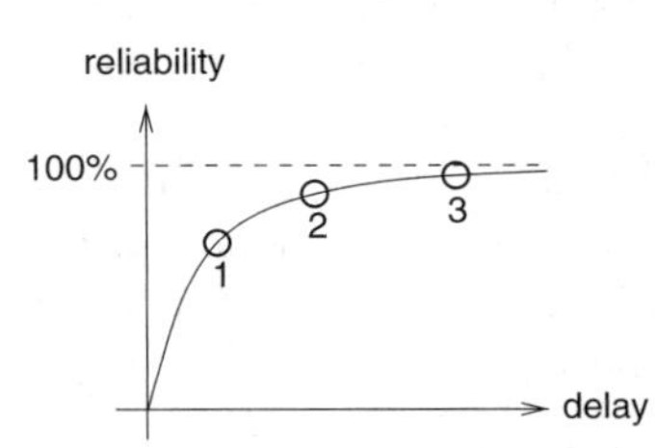

FIGURE 1.3 Reliability as a function of the delay. The circles indicate the QoS requirements of different possible traffic classes.

Note that capacity is only one point on the reliability-delay curve and therefore not always a relevant performance measure. For example, in certain sensing and control applications, the value of information quickly degrades as the latency increases. Because QoS is affected by design choices at the physical, medium-access, and network layers, an integrated approach to managing QoS is necessary.

1.3.10 Security

Depending on the application, security can be critical. The network should enable intrusion detection and tolerance as well as robust operation in the case of failure because, often, the sensor nodes are not protected against physical mishandling or attacks. Eavesdropping, jamming, and listen-and-retransmit attacks can hamper or prevent the operation; therefore, access control, message integrity, and confidentiality must be guaranteed.

1.3.11 Implementation

Companies such as Crossbow, Ember, Sensoria, and Millenial are building small sensor nodes with wireless capabilities. However, a per-node cost of $100 to $200 (not including sophisticated sensors) is prohibitive for large networks. Nodes must become an order of magnitude cheaper in order to render applications with a large number of nodes affordable. With the current pace of progress in VLSI and MEMS technology, this is bound to happen in the next few years. The fusion of MEMS and electronics onto a single chip, however, still poses difficulties. Miniaturization will make steady progress, except for two crucial components: the antenna and the battery, where it will be very challenging to find innovative solutions. Furthermore, the impact of the hardware on optimum protocol design is largely an open topic. The characteristics of the power amplifier, for example, greatly influence the energy efficiency of routing algorithms [56].

1.3.12 Other Issues

- *Distributed signal processing*. Most tasks require the combined effort of multiple network nodes, which requires protocols that provide coordination, efficient local exchange of information, and, possibly, hierarchical operation.
- *Synchronization and localization*. The notion of time is critical. Coordinated sensing and actuating in the physical world require a sense of global time that must be paired with relative or absolute knowledge of nodes' locations.
- *Wireless reprogramming*. A deployed WSN may need to be reprogrammed or updated. So far, no networking protocols are available to carry out such a task reliably in a multihop network. The main difficulty is the acknowledgment of packets in such a joint multihop/multicast communication.

1.4 Concluding Remarks

Wireless sensor networks have numerous exciting applications in virtually all fields of science and engineering, including health care, industry, military, security, environmental science, geology, agriculture, and social studies. In particular, the combination with macroscopic or MEMS-based actuators is intriguing because it permits manipulation of the environment in an unprecedented manner. Researchers and operators currently face a number of critical issues that need be resolved before these applications become reality. Wireless networking and distributed data processing of embedded sensing/actuating nodes under tight energy constraints demand new approaches to protocol design and hardware/software integration.

References

1. Sensor networks and applications, *IEEE Proc.*, 8, Aug. 2003.
2. Internet Engineering Task Force, Mobile ad-hoc networks (MANET). See http://www.ietf.org/html.charters/manet-charter.html.
3. Z.J. Haas et al., Eds., Wireless ad hoc networks, *IEEE J. Selected Areas Commun.*, 17, Aug. 1999. Special ed.
4. C.E. Perkins, Ed., *Ad Hoc Networking*. Addison Wesley, Reading, MA, 2000.
5. A.J. Goldsmith and S.B. Wicker, Design challenges for energy-constrained ad hoc wireless networks, *IEEE Wireless Commun.*, 9, 8–27, Aug. 2002.
6. A. Ephremides, Energy concerns in wireless networks, *IEEE Mag. Wireless Commun.*, 9, 48–59, Aug. 2002.
7. V. Rodoplu and T.H. Meng, Minimum energy mobile wireless networks, *IEEE J. Selected Areas Commun.*, 17(8), 1333–1344, 1999.
8. A. Ephremides, Energy concerns in wireless networks, *IEEE Wireless Commun.*, 9, 48–59, Aug. 2002.
9. Bluetooth wireless technology. Official Bluetooth site: http://www.bluetooth.com.
10. S. Tilak, N.B. Abu–Ghazaleh, and W. Heinzelman, A taxonomy of wireless micro-sensor network models, *ACM Mobile Computing Commun. Rev.*, 6(2), 28–36, 2002.
11. W.B. Heinzelman, A.P. Chandrakasan, and H. Balakrishnan, An application-specific protocol architecture for wireless microsensor networks, *IEEE Trans. Wireless Commun.*, 1, 660–670, Oct. 2002.
12. J. Kulik, W. Heinzelman, and H. Balakrishnan, Negotiation-based protocols for disseminating information in wireless sensor networks, *Wireless Networks*, 8, 169–185, March–May 2002.
13. A.B. McDonald and T.F. Znati, A mobility-based framework for adaptive clustering in wireless ad-hoc networks, *IEEE J. Selected Areas Commun.*, 17, 1466–1487, Aug. 1999.
14. S.S. Pradhan, J. Kusuma, and K. Ramchandran, Distributed compression in a dense microsensor network, *IEEE Signal Process. Mag.*, 19, 51–60, Mar. 2002.
15. A. Scaglione and S. Servetto, On the interdependence of routing and data compression in multi-hop sensor networks, in *Proc. ACM Int. Conf. Mobile Comp. Networks (MobiCom'02)*, Atlanta, GA, 140–147, Sept. 2002.
16. C. Intanagowiwat, R. Govindan, and D. Estrin, Directed diffusion: a scalable and robust communication paradigm for sensor networks, in *ACM Int. Conf. Mobile Computing Networking (MobiCom'00)*, Boston, MA, 56–67, Aug. 2000.
17. J.E. Wieselthier, G.D. Nguyen, and A. Ephremides, On the construction of energy-efficient broadcast and multicast trees in wireless networks, in *IEEE INFOCOM*, Tel Aviv, Israel, 585–594, Mar. 2000.
18. J.E. Wieselthier, G.D. Nguyen, and A. Ephremides, An insensitivity property of energy-limited wireless networks for session-based multicasting, in *IEEE ISIT*, Washington, D.C., June 2001.
19. J. Laneman, D. Tse, and G. Wornell, Cooperative diversity in wireless networks: efficient protocols and outage behavior, *IEEE Trans. Inf. Theory*. Accepted for publication. Available at: http://www.nd.edu/jnl/pubs/it2002.pdf.
20. M. Haenggi, A formalism for the analysis and design of time and path diversity schemes in wireless sensor networks, in *2nd Int. Workshop Inf. Process. Sensor Networks (IPSN'03)*, Palo Alto, CA, 417–431, Apr. 2003. Available at http://www.nd.edu/mhaenggi/ipsn03.pdf.
21. C.S. Raghavendra and S. Singh, PAMAS — power aware multi-access protocol with signaling for ad hoc networks, 1999. *ACM Computer Commun. Rev.* Available at: http://citeseer.nj.nec.com/460902.html.
22. C.-K. Toh, Maximum battery life routing to support ubiquitous mobile computing in wireless ad hoc networks, *IEEE Commun. Mag.*, 39, 138–147, June 2001.
23. P. Gupta and P.R. Kumar, The capacity of wireless networks, *IEEE Trans. Inf. Theory*, 46, 388–404, Mar. 2000.
24. P. Gupta and P.R. Kumar, Towards an information theory of large networks: an achievable rate region, in *IEEE Int. Symp. Inf. Theory*, Washington, D.C., 159, 2001.

25. L.-L. Xie and P.R. Kumar, A network information theory for wireless communication: scaling laws and optimal operation, Apr. 2002. submitted to *IEEE Trans. Inf. Theory.* Available at: http://black1.csl.uiuc.edu/prkumar/publications.html.
26. M. Grossglauser and D. Tse, Mobility increases the capacity of ad-hoc wireless networks, in *IEEE INFOCOM*, Anchorage, AL, 2001.
27. D. Tse and S. Hanly, Effective bandwidths in wireless networks with multiuser receivers, in *IEEE INFOCOM*, 35–42, 1998.
28. M. Gastpar and M. Vetterli, On the capacity of wireless networks: the relay case, in *IEEE INFOCOM*, New York, 2002.
29. G. Mergen and L. Tong, On the capacity of regular wireless networks with transceiver multipacket communication, in *IEEE Int. Symp. Inf. Theory*, Lausanne, Switzerland, 350, 2002.
30. S. Toumpis and A. Goldsmith, Capacity regions for wireless ad hoc networks, *IEEE Trans. Wireless Commun.*, 2, 736–748, July 2003.
31. D.N.C. Tse and S.V. Hanly, Multiaccess fading channels — part I: polymatroid structure, optimal resource allocation and throughput capacities, *IEEE Trans. Inf. Theory*, 44(7), 2796–2815, 1998.
32. S.V. Hanly and D.N.C. Tse, Multiaccess fading channels — part II: delay-limited capacities, *IEEE Trans. Inf. Theory*, 44(7), 2816–2831, 1998.
33. R. Negi and J.M. Cioffi, Delay-constrained capacity with causal feedback, *IEEE Trans. Inf. Theory*, 48, 2478–2494, Sept. 2002.
34. R.A. Berry and R.G. Gallager, Communication over fading channels with delay constraints, *IEEE Trans. Inf. Theory*, 48, 1135–1149, May 2002.
35. D. Tuninetti, On multiple-access block-fading channels, Mar. 2002. Ph.D. thesis, Institut EURECOM. Available at: http://www.eurecom.fr/tuninett/publication.html.
36. J. Broch, D. Maltz, D. Johnson, Y. Hu, and J. Jetcheva, A performance comparison of multi-hop wireless ad hoc network routing protocols, in *ACM Int. Conf. Mobile Computing Networking (MobiCom)*, Dallas, TX, 85–97, Oct. 1998.
37. P. Johansson, T. Larsson, and N. Hedman, Scenario-based performance analysis of routing protocols for mobile ad-hoc networks, in *ACM MobiCom*, Seattle, WA, Aug. 1999.
38. S.R. Das, C.E. Perkins, and E.M. Royer, Performance comparison of two on-demand routing protocols for ad hoc networks, in *IEEE INFOCOM*, Mar. 2000.
39. C.R. Lin and J.-S. Liu, QoS Routing in ad hoc wireless networks, *IEEE J. Selected Areas Commun.*, 17, 1426–1438, Aug. 1999.
40. M. Haenggi, Energy-balancing strategies for wireless sensor networks, in *IEEE Int. Symp. Circuits Syst. (ISCAS'03)*, Bangkok, Thailand, May 2003. Available at http://www.nd.edu/mhaenggi/iscas03.pdf.
41. J.A. Silvester and L. Kleinrock, On the capacity of multihop slotted ALOHA networks with regular structure, *IEEE Trans. Commun.*, COM-31, 974–982, Aug. 1983.
42. L. Hu, Topology control for multihop packet networks, *IEEE Trans. Commun.*, 41(10), 1474–1481, 1993.
43. M. Haenggi, Probabilistic analysis of a simple MAC scheme for ad hoc wireless networks, in *IEEE CAS Workshop on Wireless Communications and Networking*, Pasadena, CA, Sept. 2002.
44. R.O. LaMaire, A. Krishna, and H. Ahmadi, Analysis of a wireless MAC protocol with client–server traffic and capture, *IEEE J. Selected Areas Commun.*, 12(8), 1299–1313, 1994.
45. T.S. Rappaport, *Wireless Communications — Principles and Practice*, 2nd ed. Prentice Hall, Englewood Cliffs, NJ.
46. H. Takagi and L. Kleinrock, Optimal transmission ranges for randomly distributed packet radio terminals, *IEEE Trans. Commun.*, COM-32, 246–257, Mar. 1984.
47. J.L. Wang and J.A. Silvester, Maximum number of independent paths and radio connectivity, *IEEE Trans. Commun.*, 41, 1482–1493, Oct. 1993.
48. C. Schurgers, V. Tsiatsis, S. Ganeriwal, and M. Srivastava, Optimizing sensor networks in the energy–latency–density design space, *IEEE Trans. Mobile Computing*, 1(1), 70–80, 2002.

49. E.S. Sousa and J.A. Silvester, Optimum transmission ranges in a direct-sequence spread-spectrum multihop packet radio network, *IEEE J. Selected Areas Commun.*, 8, 762–771, June 1990.
50. D.A. Maltz, J. Broch, and D.B. Johnson, Lessons from a full-scale multihop wireless ad hoc network testbed, *IEEE Personal Commun.*, 8, 8–15, Feb. 2001.
51. D. Ganesan, B. Krishnamachari, A. Woo, D. Culler, D. Estrin, and S. Wicker, An empirical study of epidemic algorithms in large scale multihop wireless networks, 2002. Intel Research Report IRB-TR-02-003. Available at www.intel-research.net/Publications/Berkeley/05022002170319.pdf.
52. T.J. Shepard, A channel access scheme for large dense packet radio networks, in *ACM SIGCOMM*, Stanford, CA, Aug. 1996. Available at: http://www.acm.org/sigcomm/sigcomm96/papers/shepard.ps.
53. M. Zorzi and S. Pupolin, Optimum transmission ranges in multihop packet radio networks in the presence of fading, *IEEE Trans. Commun.*, 43, 2201–2205, July 1995.
54. Y.Y. Kim and S. Li, Modeling multipath fading channel dynamics for packet data performance analysis, *Wireless Networks*, 6, 481–492, 2000.
55. M. Haenggi, On routing in random rayleigh fading networks, *IEEE Trans. Wireless Commun.*, 2003. Submitted for publication. Available at http://www.nd.edu/mhaenggi/routing.pdf.
56. M. Haenggi, The impact of power amplifier characteristics on routing in random wireless networks, in *IEEE Global Commun. Conf. (GLOBECOM'03)*, San Francisco, CA, Dec. 2003. Available at http://www.nd.edu/mhaenggi/globecom03.pdf.
57. H. Meyr, M. Moenecleay, and S.A. Fechtel, *Digital Communication Receivers: Synchronization, Channel Estimation, and Signal Processing*. Wiley Interscience, 1998.
58. R. Meester and R. Roy, *Continuum Percolation*. Cambridge University Press, New York, 1996.
59. B. Bollobás, *Random Graphs*, 2nd ed. Cambridge University Press, New York, 2001.
60. L. Booth, J. Bruck, M. Cook, and M. Franceschetti, Ad hoc wireless networks with noisy links, in *IEEE Int. Symp. Inf. Theory*, Yokohama, Japan, 2003.
61. O. Dousse, F. Baccelli, and P. Thiran, Impact of interferences on connectivity in ad-hoc networks, in *IEEE INFOCOM*, San Francisco, CA, 2003.
62. S. Lu, V. Bharghavan, and R. Srikant, Fair scheduling in wireless packet networks, *IEEE/ACM Trans. Networking*, 7, 473–489, Aug. 1999.
63. J.-H. Ju and V.O.K. Li, TDMA scheduling design of multihop packet radio networks based on Latin squares, *IEEE J. Selected Areas Commun.*, 1345–1352, Aug. 1999.
64. H. Luo, S. Lu, and V. Bharghavan, A new model for packet scheduling in multihop wireless networks, in *ACM Int. Conf. Mobile Computing Networking (MobiCom'00)*, Boston, MA, 76–86, 2000.
65. H. Luo, P. Medvedev, J. Cheng, and S. Lu, A self-coordinating approach to distributed fair queueing in *ad hoc* wireless networks, *IEEE INFOCOM*, Anchorage, Apr. 2001.
66. A.E. Gamal, C. Nair, B. Prabhakar, E. Uysal-Biyikoglu, and S. Zahedi, Energy-efficient scheduling of packet transmissions over wireless networks, *IEEE INFOCOM*, New York, 2002, pp. 1773–1782.
67. S. Bhatnagar, B. Deb, and B. Nath, Service differentiation in sensor networks, *Fourth International Symposium on Wireless Personal Multimedia Communications*, Sept. 2001.
68. V. Kanodia, C. Li, A. Sabharwal, B. Sadeghi, and E. Knightly, Distributed multi-hop scheduling and medium access with delay and throughput constraints, *ACM MobiCom*, Rome, July 2001.
69. A. Striegel and G. Manimaran, Best-effort scheduling of (m, k)-firm real-time streams in multihop networks, *Workshop of Parallel and Distributed Real-Time Systems (WPDRTS) at IPDPS 2000*, Apr. 2000.

2

Next-Generation Technologies to Enable Sensor Networks*

2.1 Introduction 2-1
Geolocation and Identification of Mobile Targets • Long-Term Architecture
2.2 Goals for Real-Time Distributed Network Computing for Sensor Data Fusion 2-5
2.3 The Convergence of Networking and Real-Time Computing 2-6
Guaranteeing Network Resources • Guaranteeing Storage Buffer Resources • Guaranteeing Computational Resources
2.4 Middleware 2-11
Control and Command of System • Parallel Processing
2.5 Network Resource Management 2-11
Graph Generator • Metrics Object • Graph Search • NRM Agents • Sensor Interface • Mapping Database • Topology Database • NRM Federation • NRM Fault Tolerance
2.6 Experimental Results 2-16

Joel I. Goodman
MIT Lincoln Laboratory

Albert I. Reuther
MIT Lincoln Laboratory

David R. Martinez
MIT Lincoln Laboratory

2.1 Introduction

Several important technical advances make extracting more information from intelligence, surveillance, and reconnaissance (ISR) sensors very affordable and practical. As shown in Figure 2.1, for the radar application the most significant advancement is expected to come from employing collaborative and network centric sensor netting. One important application of this capability is to achieve ultrawideband multifrequency and multiaspect imaging by fusing the data from multiple sensors. In some cases, it is highly desirable to exploit multimodalities, in addition to multifrequency and multiaspect imaging.

Key enablers to fuse data from disparate sensors are the advent of high-speed fiber and wireless networks and the leveraging of distributed computing. ISR sensors need to perform enough on-board computation to match the available bandwidth; however, after some initial preprocessing, the data will be distributed across the network to be fused with other sensor data so as to maximize the information content. For example, on an experimental basis, MIT Lincoln Laboratory has demonstrated a virtual radar with ultrawideband frequency [1]. Two radars, located at the Lincoln Space Surveillance Complex

*This work is sponsored by the United States Air Force under Air Force contract F19628-00-C-002. Opinions, interpretations, conclusions, and recommendations are those of the authors and are not necessarily endorsed by the U.S. government.

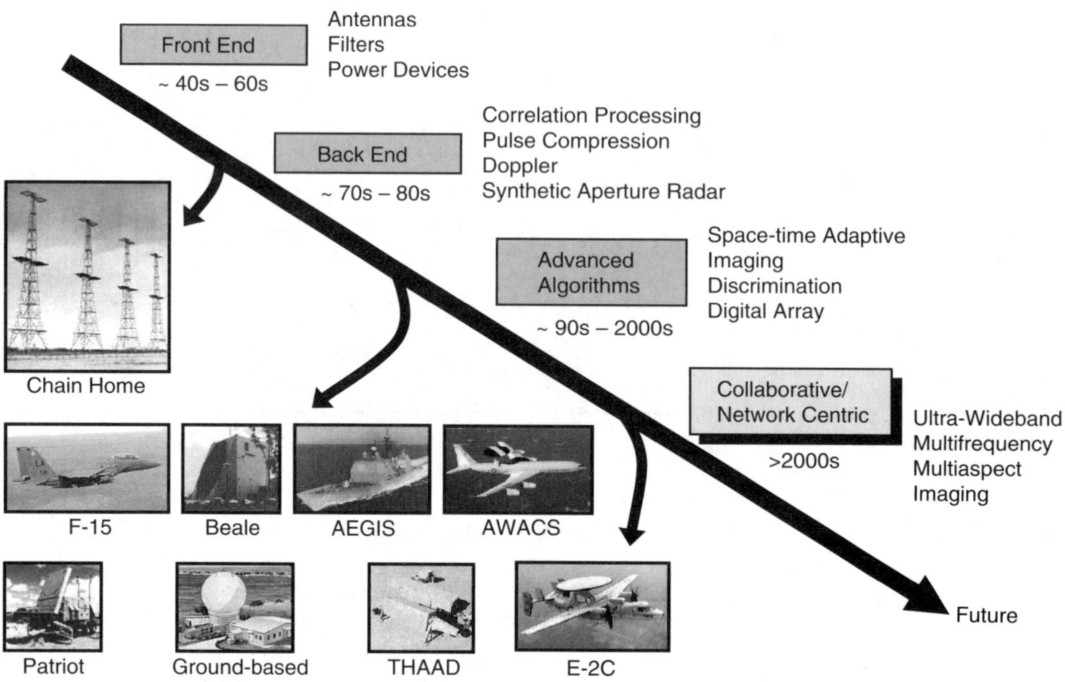

FIGURE 2.1 Radar technology evolution.

in Westford, Massachusetts, were employed; each of the two independent radars transmitted the data via a high-speed fiber network. The total bandwidth transmitted via fiber exceeded 1 Gbits/sec (billion bits per second). One radar was operating at X-band with 1-MHz bandwidth, and the second was operating at Ku-band with a 2-MHz bandwidth. A synthetic radar with an instantaneous bandwidth of 8 MHz was achieved after employing advanced ultrawideband signal processing [2].

These capabilities are now being extended to include high-speed wireless and fiber networking with distributed computing. As the Internet protocol (IP) technologies continue to advance in the commercial sector, the military can begin to leverage IP formatted sensor data to be compatible with commercial high-speed routers and switches. Sensor data from theater can be posted to high-speed networks, wireless and fiber, to request computing services as they become available on this network. The sensor data are processed in a distributed fashion across the network, thereby providing a larger pool of resources in real time to meet stringent latency requirements. The availability of distributed processing in a grid-computing architecture offers a high degree of robustness throughout the network. One important application to benefit from these advances is the ability to geolocate and identify mobile targets accurately from multiaspect sensor data.

2.1.1 Geolocation and Identification of Mobile Targets

Accurately geolocating and identifying mobile targets depends on the extraction of information from different sensor data. Typically, data from a single sensor are not sufficient to achieve a high probability of correct classification and still maintain a low probability of false alarm. This goal is challenging because mobile targets typically move at a wide range of speeds, tend to move and stop often, and can be easily mistaken for a civilian target. While the target is moving the sensor of choice is the ground moving target indication (GMTI). If the target stops, the same sensor or a different sensor working cooperatively must employ synthetic aperture radar (SAR). Before it can be declared foe, the target must often be confirmed with electro-optical or infrared (EO/IR) images. The goal of future networked systems is to have multiple sensors providing the necessary multimodality data to maximize the chances of accurately declaring a target.

A typical sensing sequence starts by a wide area surveillance platform, such as the Global Hawk unmanned aerial vehicle (UAV), covering several square kilometers until a target exceeds a detection threshold. The wide area surveillance will typically employ GMTI and SAR strip maps. Once a target has been detected, the on-board or off-board processing starts a track file to track the target carefully, using spot GMTI and spot SAR over a much smaller region than that initially covered when performing wide area surveillance. It is important to recognize that a sensor system is not merely tracking a single target; several target tracks can be going on in parallel. Therefore, future networked sensor architectures rely on sharing the information to maximize the available resources.

To date, the most advanced capability demonstrated is based on passing target detections among several sensors using the Navy cooperative engagement capability (CEC) system. Multisensor tracks are formed from the detection inputs arriving at a central location. Although this capability has provided a significant advancement, not all the information available from multimodality sensors has been exploited. The limitation is with the communication and available distributed computing. Multimodality sensor data together with multiple look angles can substantially improve the probability of correct classification vs. false alarm density. In addition to multiple modalities and multiple looks on the target, it is also desirable to send complex (amplitude and phase) radar GMTI data and SAR images to permit the use of high-definition vector imaging (HDVI) [3]. This technique permits much higher resolution on the target by suppressing noise around it, thereby enhancing the target image at the expense of using complex video data and much higher computational rates.

Another important tool to improve the probability of correct classification with minimal false alarm is high-range resolution (HRR) profiles. With this tool, the sensor bandwidth or, equivalently, the size of the resolution cell must be small resulting in a large data rate. However, it has been demonstrated that HRR can provide a significant improvement [4]. Therefore, next generation sensors depend on available communication pipes with enough bandwidth to share the individual sensor information effectively across the network. Once the data are posted on the network, the computational resources must exist to maintain low latencies from the time data become available to the time a target geoposition and identification are derived. The next subsection discusses the long-term architecture to implement netting of multiple sensor data efficiently.

2.1.2 Long-Term Architecture

In the future it will be desirable to minimize the infrastructure (foot print) forwardly deployed in the battlefield. It is most desirable to leverage high-speed satellite communication links to bring sensor data back to a combined air operations center (CAOC) established in the continental United States (CONUS).

The technology enablers for the long-term architecture shown in Figure 2.2 are high-speed, IP-based wireless and fiber communication networks, together with distributed grid computing. The in-theater commander's ability to task his organic resources to perform reconnaissance and surveillance of the opposing forces, and then to relay that information back to CONUS, allows significant reduction in the complexity, level, and cost of in-theater resources. Furthermore, this approach leverages the diverse analysis resources in CONUS, including highly trained personnel to support the rapid, accurate identification and localization of targets necessary to enable the time-critical engagement of surface mobile threats.

Space, air, and surface sensors will be deployed quickly to the battlefield. As shown in Figure 2.3, the stage in the processing chain at which the sensor data are tapped off to be sent via the network will dictate the amount of data transferred. For example, in a few applications one needs to send the data directly out of the analog-to-digital converters (A/D) to exploit coherent data combining from multiple sensors. Most commonly, it is preferable to perform on-board signal preprocessing to minimize the amount of data transferred. However, one must still be able to preserve content in the transferred data that is required to exploit features in the data not available from processing a signal sensor end to end. For example, one might be interested in transmitting wide area surveillance (WAS) data from SAR with high resolution to be followed by multiaspect SAR processing (shown in Figure 2.3 as application B). The data volume will be larger than the second example shown in Figure 2.3 as application A, in which

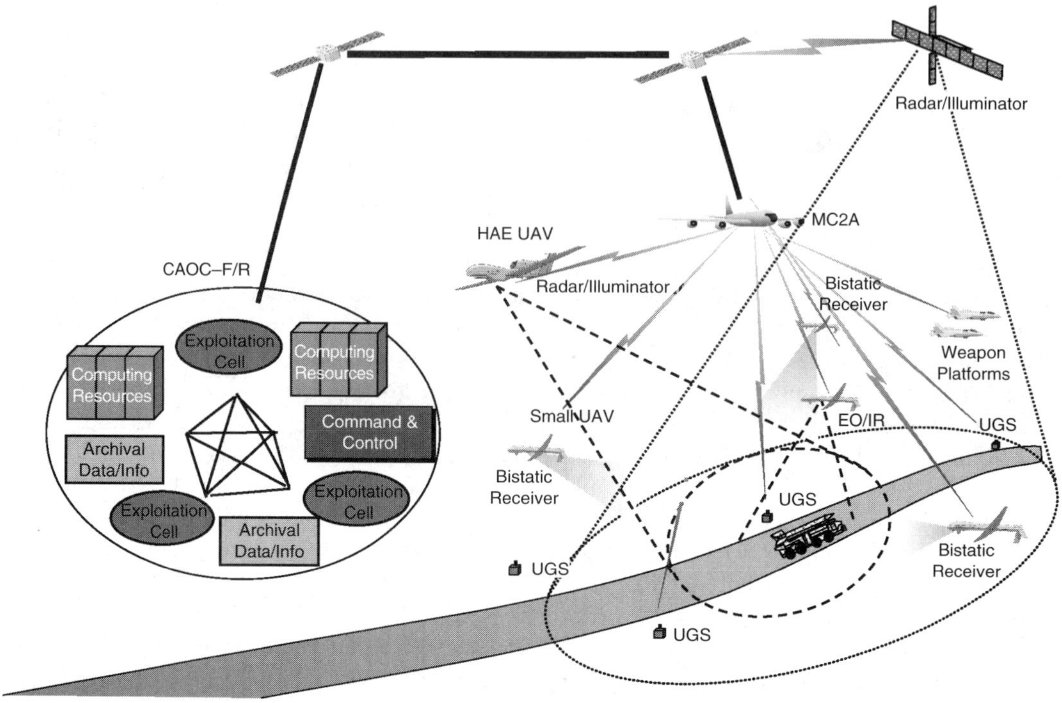

FIGURE 2.2 Postulated long-term architecture.

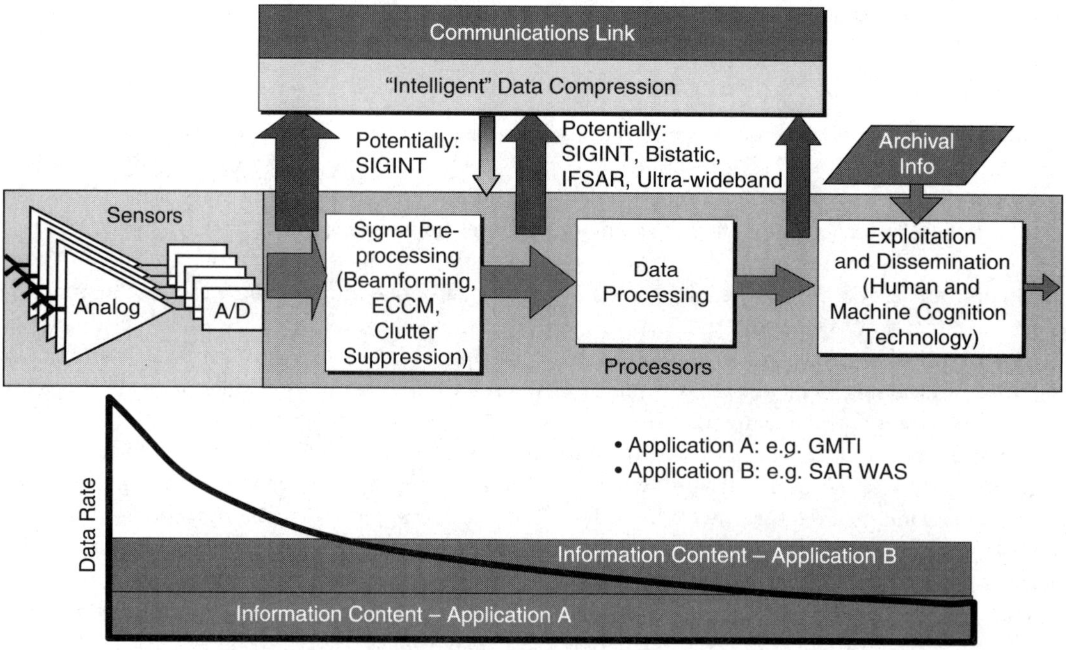

FIGURE 2.3 Sensor signal processing flow.

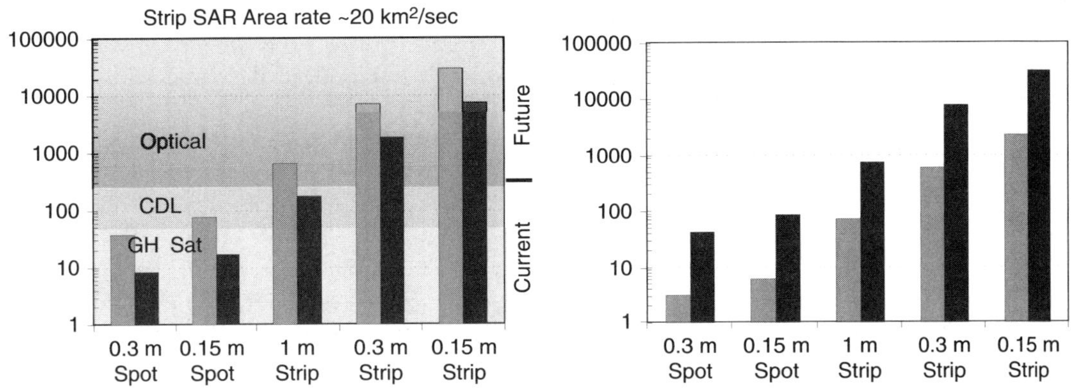

FIGURE 2.4 SAR data rate and computational throughput trade.

most of the GMTI processing is done on board. In any of these applications, it is paramount that "intelligent" data compression be done on board before data transmission to send only the necessary parts of the data requiring additional processing off board.

Each sensor will be capable of generating on-board processed data greater than 100 Mbits/sec (million bits per second). Figure 2.4 shows the trade-off between communication link data rates vs. on-board computation throughputs for different postulated levels of image resolution (for spot or strip map SAR modes). For example, for an assumed 1-m strip map SAR, one can send complex video radar data to then perform super-resolution processing off board. This approach would require sending between 100 to 1000 Mbits/sec. Another option is to perform the super-resolution processing on board, requiring between 100 billion floating-point operations per second (GFLOPS) to 1 trillion floating-point operations per second (TFLOPS).

Specialized military equipment, such as the common data link (CDL), can achieve data rates reaching 274 Mb/sec. If higher communication capacity were available, one would much prefer to send the large data volume for further processing off board to leverage information content available from multiple sensor data. As communication rates improve in the forthcoming years, it will not matter to the in-theater commander if the data are processed off board with the benefit of allowing exploitation of multiple sensor data at much rawer levels than is possible to date.

2.2 Goals for Real-Time Distributed Network Computing for Sensor Data Fusion

Several advantages can be gained by utilizing real-time distributed network computing to enable greater sensor data fusion processing. Distributed network computing potentially reduces the cost of the signal processing systems and the sensor platform because each individual sensor platform no longer needs as much processing capability as a stove-piped stand-alone system (although each platform may need higher bandwidth communications capabilities). Also, fault tolerance of the processing systems is increased because the processing and network systems are shared between sensors, thereby increasing the pool of available signal processors for all of the sensors. Furthermore, the granularity of managed resources is smaller; individual processors and network resources are managed as independent entities rather than managing an entire parallel computer and network as independent entities. This affords more flexible configuration and management of the resources.

To enable collaborative network processing of sensor signals, three technological areas are required to evolve and achieve maturity:

- Guaranteed *communication, storage buffer,* and *computation resources* must keep up with the high-throughput streams of data coming from the sensors. If any stage of the processing falls behind due to a network problem or interruption in the processor, buffering the data will become a problem quickly as increasing volumes of data must be stored to accommodate the delayed processors. Section 2.3 addresses technological possibilities to mitigate these resource availability issues.

- *Middleware* in the network of processors must be developed to accommodate a heterogeneous mix of computer and network resources. This middleware consists of a task control interface, which facilitates the communication between network resource management agents and entities, and an application programming interface for programming applications executed on the collaborative network processors. Section 2.4 will address these middleware interfaces.
- A *network resource manager* (NRM) system is necessary for orchestrating the execution of the application components on the computation and communication resources available in the collaborative network. Section 2.5 will discuss the components and functionality of the NRM.

2.3 The Convergence of Networking and Real-Time Computing

To date, networking of sensors has been demonstrated primarily using localized- and limited-capacity data links. As a result, the data available on the network from each sensor node typically represent the product of extensive prior processing of the radar data carried at the individual sensor. For example, the Navy CEC system, a relatively advanced current system, uses detection reports from independent sensors in the network to build composite tracks of targets. Access to raw (or possibly minimally preprocessed) multisensor data opens the opportunity for more effective exploitation of these data through integrated sensor data processing. The future network-centric ISR architecture will likely employ worldwide wideband communication networks to interconnect sensors with distributed processing and fusion sites. The resulting distributed database will provide a common operational picture for deployed forces. The sensor data will return to a CONUS entry point and pass over a wideband fiber network to the various processing centers where the sensor data will be fused. The data link from the theater to CONUS is expected to be optical to achieve very high link capacity [5].

This section discusses technologies that will guarantee that wireless and terrestrial network resources, storage buffer resources, and computational resources are available for sensor signal processing.

2.3.1 Guaranteeing Network Resources

Sensor data will traverse wireless and terrestrial (e.g., optical, twisted-copper) networks in which bit errors, packet loss, and delay could adversely affect the quality and timeliness of the ultimate result. The goal then is to choose a network and processing architecture to ameliorate the deleterious effects of data loss and network delay in the data fusion process. Due to the costs associated with developing, deploying, and maintaining a fixed terrestrial infrastructure, as well as inventing wholly new modulation protocols and standards for wireless and terrestrial signaling, it is cost-effective and expedient for military technology to ride the "commercial wave" of technical investment and progress in communication technologies.

With a fixed network infrastructure consisting primarily of commercial components, combating data loss and delay in terrestrial networks involves choosing the right protocols so that the network can enforce quality of service (QoS) demands; in wireless networks, this involves aggressive coding, modulation, and "lightweight" flow control for efficient bandwidth utilization. With sufficient complexity and bandwidth, it is possible with today's IP-based protocols to differentiate high-priority data to impart the mandated QoS for time-critical applications.

2.3.1.1 Terrestrial Networks

Reserving bandwidth on an IP-based network that is uniformly recognized across administrative domains involves employing protocols like RSVP-TE [6] or CR-LDP [7]. Although having sufficient communication bandwidth is an important aspect of processing sensor data in real time on a distributed network of resources, it does not guarantee real-time performance. For example, time-critical applications mapped onto networked resources should not have processing interrupted to service unmanaged traffic or be subject to a computational resource's resident operating system switching contexts to a lower priority task. For data that originate from sensors at very high streaming rates, a storage solution, as discussed in Section 2.3.2, is needed that is capable of recording sensor data in real time as well as robust in the face of network resource failures; this insures that a high-priority application can continue processing in the presence of malfunctioning or compromised networked equipment. However, adding a buffering storage solution only alleviates part of the problem; it does not mitigate the underlying problem of losing packets during network equipment failures or periods of network traffic that exceed network capacities.

For an IP-based network, one solution to this problem is to use remote agents deployed on primary compute resources or networked terminals located at switches that can dynamically filter unmanaged traffic. This is implemented by programming computer hardware specifically tasked with packet filtering (e.g., next generation gigabit Ethernet card) or dynamically reconfiguring the switch that directly connects to the compute resource in question by supplying an access control list (ACL) to block all packets except those associated with time-critical targeting. The formation of these exclusive networks using agents has been dubbed *dynamic private networks* (DPNs) — in effect, mechanisms for virtually overlaying a circuit switch onto a packet-switched network.

2.3.1.2 Wireless Networks

Unlike terrestrial networks, flow control and routing in mobile wireless sensor networks must contend with potentially long point-to-point propagation delays (e.g., satellite to ground) as well as a constantly changing topology. In a traditional terrestrial network employing link-state routing (e.g., OSPF), each node maintains a consistent view of a (primarily) fixed network topology so that a shortest path algorithm [8] can be used to find desirable routes from source to destination. This requires that nodes gather network connectivity information from other routers.

If OSPF were employed in a mobile wireless network, the overhead of exchanging network connectivity information about a transient topology could potentially consume the majority of the available bandwidth [9]. Routing protocols have been specifically designed to address the concerns of mobile networks [10]; these protocols fall into two general categories: proactive and reactive. Proactive routing protocols keep track of routes to all destinations, while reactive protocols acquire routes on demand. Unlike OSPF, proactive protocols do not need a consistent view of connectivity; that is, they trade optimal routes for feasible routes to reduce communication overhead. Reactive routes suffer a high initial overhead in establishing a route; however, the overall overhead of maintaining network connectivity is substantially reduced. The category of routing used is highly dependent upon how the sensors communicate with one another over the network.

Traditional flow control mechanisms over terrestrial networks that deliver reliable transport (e.g., TCP) may be inappropriate for wireless networks because, unlike wireless networks, terrestrial networks generally have a very low bit error rate (BER) on the order of 10^{-10}, so errors are primarily due to packet loss. Packet loss occurs in heavily congested networks when an ingress or egress queue of a switch or router begins to fill, requiring that some packets in the queue be discarded [11]. This condition is detected when acknowledgments from the destination node are not received by the source, prompting the source's flow control to throttle back the packet transmit rate [12].

In a wireless network in which BERs are four to five orders of magnitude higher than those of terrestrial networks, packet loss due to bit errors can be mistakenly associated with network congestion, and source flow control will mistakenly reduce the transmit rate of outgoing packets. Furthermore, when the source and destination are far apart, such as the communication between a satellite and ground terminal, where propagation delays can be on the order of 240 ms, delayed acknowledgments from the destination result in source flow control inefficiently using the available bandwidth. This is due to source flow control incrementally increasing the transmit rate as destination acknowledgments are received even though the entire frame of packets may have already been transmitted before the first packet reaches the receiver [13]. Therefore, to use bandwidth efficiently in a wireless network for reliable transport, flow control must be capable of differentiating BER from packet loss and account for long-haul packet transport by more efficiently using the available bandwidth. Some work in this area is reflected in RFC 2488 [14], as well as proposals for an explicit congestion warning, where, for example, the destination site would respond to packet errors with an acknowledgment that it received the source packets with a corruption notification.

At the physical layer, high data rates for a given BER have been realized by employing low-density parity check codes, such as turbo codes, in conjunction with bandwidth efficient modulation to achieve spectral efficiencies to within 0.7 dB of the Shannon limit [15]. Furthermore, extremely high spectral efficiencies have been demonstrated using multiple input, multiple output (MIMO) antenna systems whose theoretical channel capacity increases linearly with the number of transmit/receive antenna pairs [16]. Although turbo codes are advantageous as a forward error correction mechanism in wireless systems when trying to maximize throughput, MIMO systems achieve high spectral efficiencies only when operating in rich scattering environments [17]. In environments in which little scattering occurs, such as in some air-to-air communication links, MIMO systems offer very little improvement in spectral efficiency.

2.3.2 Guaranteeing Storage Buffer Resources

For a variety of reasons, it may be very desirable to record streaming sensor data directly to storage media while simultaneously sending the data on for immediate processing. For sensor signal processing applications, this enables multimodality data fusion of archived data with real-time (perishable) data from in-theatre sensors for improved target identification and visualization [18]. Storage media could also be used for rate conversion in cases in which the transmission rate exceeds the processing rate and for time-delay buffering for real-time robust fault tolerance (discussed in the next section). The storage media buffer reuse is deterministic and periodic so that management of the buffer is straightforward.

A number of possible solutions exist:

- *Directly attached storage* is a set of hard disks connected to a computer via SCSI or IDE/EIDE/ATA; however, this technology does not scale well to the volume of streaming sensor data.
- *Storage area networks* are hard disk storage cabinets attached to a computer with a fast data link like Fibre Channel. The computer attached to the storage cabinet enjoys very fast access to data, but because the data must travel through that computer, which presents a single point of failure, to get to other computers on the network, this option is not a desirable solution.
- *Network-attached storage* connects the hard disk storage cabinet directly to the network as a file server. However, this technology offers only midrange performance, a single point of failure, and relatively high cost.

A visionary architecture in which data storage centers operate in parallel at a wide-area network (WAN) and local area network (LAN) level is described in Cooley et al. [19]. In this architecture, developed by MIT Lincoln Laboratory, high-rate streaming sensor data are stored in parallel across a partitioned network of storage arrays, which affords a highly scalable, low-cost solution that is relatively insensitive to communications or storage equipment failure. This system employs a novel and computationally efficient encoding and decoding algorithm using low-density parity check codes [20] for erasure recovery. Initial system performance measures indicate the erasure coding method described in Cooley et al. [19] has a significantly higher throughput and greater reliability when compared to Reed–Solomon, Tornado [21], and Luby [20] codes. This system offers a promising low-cost solution that scales in capability with the performance gains of commodity equipment.

2.3.3 Guaranteeing Computational Resources

The exponential growth in computing technology has contributed to making viable the implementation of advanced sensor processing in cost-effective hardware with form factors commensurate with the needs of military users. For example, several generations of embedded signal processors are shown in Figure 2.5.

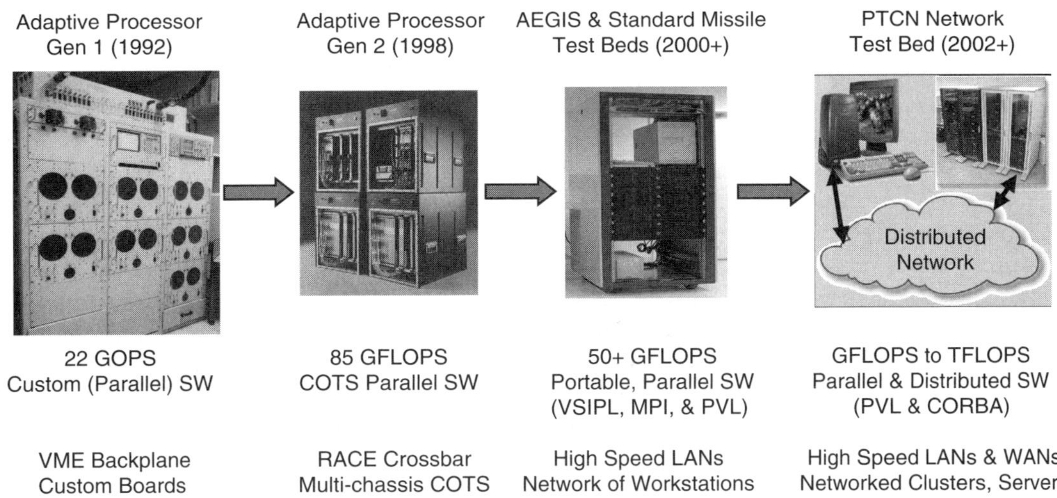

FIGURE 2.5 Embedded signal processor evolution.

In the early 1990s, embedded signal processors were built using custom hardware and software. In the late 1990s, a move occurred from custom hardware to COTS processor systems running vendor-specific software together with application-specific parallel software tuned to each specific application. Most recently, the military embedded community is beginning to demonstrate requisite performance employing parallel and portable software running on COTS hardware.

Continuing technology advances in computation and communication will permit future signal processors to be built from commodity hardware distributed across a high-speed network and employing distributed, parallel, and portable software. These computing architectures will deliver 10^9 to 10^{12} floating point operations per second (GFLOPs to TFLOPs) in computational throughput. The distributed nature of the software will apply to on-board sensor processing as well as off-board processing. Clearly, on-board embedded processor systems will need to meet the stringent platform requirements in size, weight, and power.

Wireless and terrestrial network resources are not the only areas in which delays, failures, and errors must be avoided to process sensor data in a timely fashion. The system design must also guarantee that the marshaled compute nodes will keep up with the required computational throughput of streaming data at every stage of the processing chain. This guarantee encompasses two important facets: (1) keeping the processors from being interrupted while they are processing tasks and (2) implementing fail-over that is tolerant of fault.

2.3.3.1 Avoiding Processor Interruption

It is easy to take for granted that laptop and desktop computers will process commands as fast as the hardware and software are capable of doing so. A fact not generally known is that general computers are interrupted by system task processes and the processes of other applications (one's own and possibly from others working in the background on one's system). System task processes include keyboard and mouse input; communications on the Ethernet; system I/O; file system maintenance; log file entries; etc. When the computer interrupts an application to attend to such tasks, the execution of the application is temporarily suspended until the interrupting task has finished execution. However, because such interruptions often only consume a few milliseconds of processing time, they are virtually imperceptible to the user [22].

Nevertheless, the interruptions are detrimental to the execution of real-time applications. Any delay in processing these streams of data will instigate a need for buffering the data that will grow to insurmountable size as the delays escalate. A solution for these interrupt issues is to use a real-time operating system on the computation processors.

Simply put, real-time operating systems (RTOS) give priority to computational tasks. They usually do not offer as many operating system features (virtual memory, threaded processing, etc.) because of the interrupting processing nature of these features [22]. However, an RTOS can ensure that real-time critical tasks have guaranteed success in meeting streamed processing deadlines. An RTOS does not need to be run on typical embedded processors; it can also be deployed on Intel and AMD Pentium-class or Motorola G-series processor systems. This includes Beowulf clusters of standard desktop personal computers and commodity servers. This is an important benefit, providing a wide range of candidate heterogeneous computing resources.

A great deal of press has been generated in the past several years about real-time operating systems; however, the distinction between soft real-time and hard real-time operating systems is seldom discussed. Hard real-time systems guarantee the completion of tasks in a deterministic time period, while soft real-time systems give priority to critical tasks over other tasks but do not guarantee the completion of tasks in a deterministic time period [22]. Examples of hard real-time operating systems are VxWorks (Wind River Systems, Inc. [23]); RTLinux/Pro (FSMLabs, Inc. [24]); and pSOS (Wind River Systems, Inc. [23]), as well as dedicated massively parallel embedded operating systems like MC/OS (Mercury Computer Systems, Inc. [25]). Examples of soft real-time operating systems are Microsoft Pocket PC; Palm OS; certain real-time Linux releases [24, 26]; and others.

2.3.3.2 Working through System Faults

When fault tolerance in massively parallel computers is addressed, usually the solution is parallel redundant systems for fail-over. If a power supply or fan fails, another power supply or fan that is redundant in the system takes over the workload of the failed device. If a hard disk drive fails on a redundant array of independent disks (RAID) system, it can be hot swapped with a new drive and the contents of the drive rebuilt from the contents of the other drives along with checksum error correction code information. However, if an individual processor fails on a parallel computer, it is considered a failure of the entire parallel computer, and an identical backup computer is used as a fail-over. This backup system is then used as the primary computer, while the failed parallel computer is repaired to become the backup for the new primary eventually.

If, however, it were possible to isolate the failed processor and remap and rebind the processes on other processors in that computer — in real time — it would then be possible to have only a number of redundant processors in the system rather than entire redundant parallel computers. There are two strategies for determining the remapping as well as two strategies for handling the remapping and rebinding; each has its advantages and disadvantages.

To discuss these fail-over strategies, it is necessary to define the concepts of tasks and mappings. A signal processing application can be separated into a series of pipelined stages or tasks that are executed as part of the given application. A mapping is the task-parallel assignment of a task to a set of computer and network resources. In terms of determining the fail-over remapping, it is possible to choose a single remapping for each task or to choose a completely unique secondary path — a new mapping for each task that uses a set of processors mutually exclusive from the processors in the primary mapping path. If task backup mappings are chosen for each task, the fail-over will complete faster than a full processing chain fail-over; however, the rebinding fail-over for a failed task mapping is more difficult because the mappings from the task before and the task after the failed task mapping must be reconfigured to send data to and receive data from the new mapping. Conversely, if a completely unique secondary path is chosen as a fail-over, then fail-over completion will have a longer latency than performing a single task fail-over. However, the fail-over mechanics are simpler because the completely unique secondary path could be fully initialized and ready to receive the stream of data in the event of a failure in the primary mapping path.

In terms of handling the remapping and rebinding of tasks, it is possible to choose the fail-over mappings when the application is initially launched or immediately after a fault occurs. In either case, greater latency is incurred at launch time or after the occurrence of a fault. For these advanced options, support for this fault tolerance comes mainly from the middleware support, which is discussed in the next section, and from the NRM discussed in Section 2.5.

2.4 Middleware

Middleware not only provides a standard interface for communications between network resources and sensors for plug-and-play operation, but also enables the rapid implementation of high-performance embedded signal processing.

2.4.1 Control and Command of System

Because many systems use a diverse set of hardware, operating systems, programming languages, and communication protocols for processing sensor data, the manpower and time-to-deployment associated with integration have a significant cost. A middleware component providing a uniform interface that abstracts the lower-level system implementation details from the application interface is the common object request broker architecture (CORBA) [27]. CORBA is a specification and implementation that defines a standard interface between a client and server. CORBA leverages an interface definition language (IDL) that can be compiled and linked with an object's implementation and its clients. Thus, the CORBA standard enables client and server communications that are independent of the host hardware platforms, programming language, operating systems, and so on. CORBA has specifications and implementations to interface with popular communication protocols such as TCP/IP. However, this architecture has an open specification, general interORB protocol (GIOP) that enables developers to define and plug in platform-specific communication protocols for unique hardware and software interfaces that meet application-specific performance criteria.

For real-time and parallel embedded computing, it is necessary to interface with real-time operating systems, define end-to-end QoS parameters, and enact efficient data reorganization and queuing at communication interfaces. CORBA has recently included specifications for real-time performance and parallel processing, with the expectation that emerging implementations and specification addendums will produce efficient implementations. This will enable CORBA to move out of the command and control domain and be included as a middleware component involved in real-time and parallel processing of time-critical sensor data.

2.4.2 Parallel Processing

The ability to choose one of many potential parallel configurations enables numerous applications to share the same set of resources with various performance requirements. What is needed is a method to decouple the mapping, that is, the parallel instantiation of an application on target hardware, from generic serial application development. Automating the mapping process is the only feasible way of exploring the large parameter space of parallel configurations in a timely and cost-effective manner.

MIT Lincoln Laboratory has developed a C++-based library known as the parallel vector library (PVL) [28]. This library contains objects with parameterized methods deeply rooted in linear algebraic expressions commonly found in sensor signal processing. The parameters are used to direct the object instance to process data as one constituent part of a parallel whole. The parameters that organize objects in parallel configurations are run-time parameters so that new parallel configurations can be instantiated without having to recompile a suite of software. The technology of PVL is currently being incorporated into the parallel vector, signal, and image processing library for C++ (parallel VSIPL++) standard library [29].

2.5 Network Resource Management

Given the stated goals for distributed network computing for sensor fusion as outlined in Section 2.3, the associated network communication, storage, and processing challenges in Section 2.3, and the desire for standard interfaces and libraries to enable application parallelism and plug-and-play integration in Section 2.4, an integrated solution is needed that bridges network communications, distributed storage, distributed processing, and middleware. Clearly, it is possible for a development team to implement a

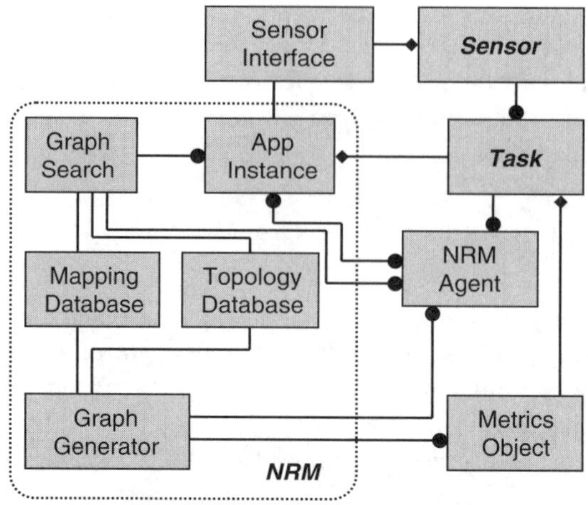

FIGURE 2.6 Object model for network resource manager (NRM).

"point" solution, but this is inherently not scalable and very difficult to maintain. Therefore an additional goal is to fully automate the process of configuring network communication, storage, and computational resources to process data for sensor fusion applications in real time, provide robust fault tolerance in the face of network resource failures, and impart this service in a highly dynamic network in the face of competing interests.

To address these needs, the network resource manager (NRM) was developed. The novelty and potency of the NRM is its capability of taking a sensor signal processing application designed and tested on single target processing element (PE) and mapping it in a task- and a data-parallel fashion across a network of computational resources to achieve real-time performance [30]. Figure 2.6 is an object-oriented model of the components that constitute the NRM. A high-level overview of the NRM follows, and details will be provided in the following subsections. The task of building a model from which the NRM launches parallel applications is broken into three distinct phases:

1. Map generation involves breaking an application into various task- and data-parallel components.
2. Map timing collects performance metric information associated with the components (or tasks) running on host resources. Using the performance metrics, the NRM creates a weighted graph-theoretic view of various permutations of an application mapped in parallel across networked resources.
3. Map selection finds the path through the graph that best meets system and application performance requirements.

The graph generator and graph search objects will heavily leverage PVL (discussed earlier) objects in the instantiation of task- and data-parallel configurations of applications on host resources. It should be noted, however, that the NRM's capabilities are fully general and independent from those of PVL and could work with other applications that are not developed using PVL to instantiate task- and data parallelism.

2.5.1 Graph Generator

As noted previously, PVL uses run-time parameters to generate new parallel configurations. This enables the NRM to launch applications in arbitrary parallel configurations using software developed for a single target PE without having to recompile the application software suite. The central challenge is to select a subset of the potentially astronomical number of permutations of parallel configurations as candidate parallel mappings. It is expected that the NRM will receive guidance in the form of performance and resource utilization bounds to help it avoid choosing undesirable configurations. It will also be given a

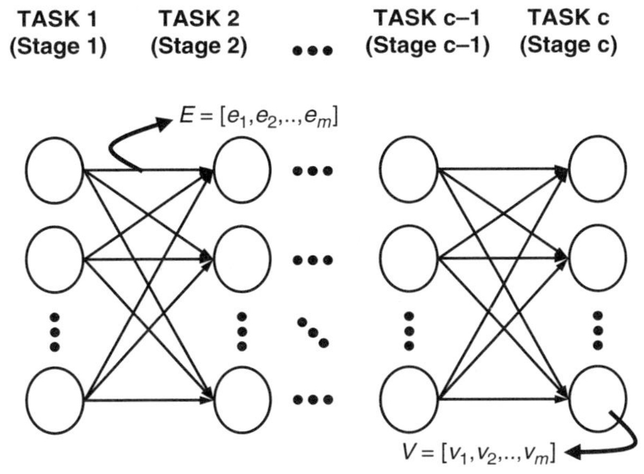

FIGURE 2.7 Sample graph with edge and vertex weights.

series of constituent tasks that comprise an application, so that its primary objective is to choose candidate data-parallel configurations for each of the individual tasks. Using a graph-theoretic model, the application space may be broken up as shown in Figure 2.7.

Each column in the graph is populated with vertices; each vertex corresponds to a mapping of the task corresponding to the given column to a potentially unique set of computational resources in the system. Each vertex has edges entering and exiting: entering edges correspond to communications with preceding tasks and exiting edges correspond to communications with succeeding tasks. Sensor signal processing applications may be represented as a stream signal processing flow, in which data move in one direction from task to task as they are processed. In this graph-theoretic model, task parallelism is represented along the horizontal axis of the graph, i.e., pipelined, overlapping execution intervals, while data parallelism is represented by the mapping of each task in the application onto one or more parallel computational resources of each vertex. The graph-theoretic representation of data- and task-parallel applications and the corresponding flow of communication enable the graph generator of the NRM to capture the potentially astronomical number of combinations of application-to-resource mappings in a concise and efficient fashion.

Finally, the graph generator is also responsible for launching the executable for each task mapping (vertex) on target resources so that performance metrics can be collected as discussed in the next subsection.

2.5.2 Metrics Object

The metrics object (MO) is responsible for collecting performance metrics of tasks launched by the graph generator. The MO works closely with the graph generator to weight the graph. Each of the resources that hosts a task is time synchronized; metric agents (see NRM agents in Subsection 2.5.4) on each of the resources will provide the MO measurements for it to formulate the following performance parameters associated with graph weights: throughput; latency; RAM memory; and PE utilization. The MO will calculate another metric known as processor cost, which is a ratio of compute horsepower used in the mapping to the overall processing horsepower available in the network.

Link utilization percentages within each mapping are also measured, as well as intertask utilization percentages. Map generation uses task column pairs to gather performance metrics in order to reduce the effort and time involved drastically. This is possible because the graph search algorithm will use a running tabulation of resource utilization percentages to ensure that simple linear superposition of path weights hold, given that these percentages remain under a given threshold. This is explained further in the next subsection. Once above the threshold, weight modifiers will be applied to subsequent stages during search. Finally, the metrics object will calculate a *network cost*, analogous to processor cost, which is a ratio of communications bandwidth used by a mapping pair with respect to the overall bandwidth available in the network.

2.5.3 Graph Search

The NRM must choose a path through the graph that determines the task mappings with which an application is launched on network resources. The choice of a path by the NRM is constrained by the time to result and the mandate to use a minimum set of networked resources. The data rate of the sensor data stream will drive required throughput for each task column in the graph; overall latency, which represents the total pipeline delay, is defined as the time period after which all data have been transmitted that a result is generated. To minimize any one application's impact on resource consumption, the path through the graph could be chosen to minimize the overall usage of computational or communication resources. This choice will depend upon whether an application is launched in a network that is compute resource or communication bandwidth limited.

The graph search problem may be formalized as a discrete and constrained optimization problem: given a set of hard constraints, minimize (or maximize) a given objective function. As described in the metrics object subsection, the NRM may choose constraints and an objective function from the set of weights shown in Table 2.1.

Scalar weights are singular — that is, only one is associated with a given vertex or edge; vector weights may include many elements in an edge or vertex association. Because each vertex and edge may represent the combination of many PE and network communication elements associated with a mapping pair, processor and network utilization may constitute weight vectors with many elements.

Although all weights tabulated previously may be chosen as constraints, memory, throughput, and network and PE utilization are not parameters that can be chosen as an objective function to optimize. This is because throughput is only a function of data rate; maximizing throughput has no impact on performance. Utilization also has no impact on performance and is only a measure of the validity of the solution. That is, subsequent stages in the graph may include resources from earlier stages, so keeping a running tabulation of utilization gives an indication of the onset of usage exceeding capacity and thereby degrading performance.

Network utilization and cost, PE utilization and cost, and memory are weights derived and constrained by the NRM, while data rate (throughput) and latency are application dependent and imposed by the sensor. The objective function that the NRM uses is chosen based on the desire to minimize an application's impact on resource usage or minimize the latency associated with an application's execution. For example, in a bandwidth-limited network, the graph search problem may be formulated as follows. While meeting application latency and throughput constraints, using less than 80% of the bandwidth available in the chosen network conduits and PEs and less than 100% of the available local PE-RAM memory, and using only a fraction of the overall processing bandwidth available network wide, select a parallel configuration for the

TABLE 2.1 Graph Weights Associated with Individual Edges and Vertices, and Corresponding Sizes (Types)

Weight	Type
Latency	Scalar
Throughput	Scalar
PE utilization	Vector
Processor cost	Scalar
Network utilization	Vector
Network cost	Scalar
Memory	Scalar

application and the associated host resources using the smallest fraction of overall network bandwidth available. Even for moderately sized graphs (e.g., 1000 vertices by 10 stages), this is a complex combinatorial optimization problem; the general problem is NP complete. The authors have developed an iterative heuristic algorithm that has shown favorable performance for this class of problem in the quality of the solution and time to solution compared to other popular combinatorial optimization algorithms [31].

2.5.4 NRM Agents

The NRM agents are information and service links between the NRM and each of the resources. Agents must first register and be authenticated (e.g., using Kerberos [32]) before an NRM will invoke their services. This registration includes a characterization of the resource capabilities and services. When registered, the NRM will use these remotely deployed agents on computational resources to download and launch parameterized executables and modify the access control list (ACL) of switches and routers under its control in the formation of DPNs. Agents also provide a mechanism for centralized software maintenance and configuration by acting as transaction managers in the download and installation of applications, databases, middleware, etc. As stated earlier, the agents also provide a measurement object that is instantiated by applications to provide the NRM's MO with performance metrics during graph generation. Finally, agents give the NRM a view of the network state, periodically sending diagnostic messages indicating its operational status.

2.5.5 Sensor Interface

Sensors can be thought of as resources much like computational and communication resources, which are served by the NRM agents; thus, the sensor interface can be thought of as another type of NRM agent. Because many different sensor platforms could be served by an NRM-managed resource network, the sensor interface provides a common, abstract mechanism for communication between the NRM and the sensor platforms.

Sensors will request services through the sensor interface from the NRM using a well-defined middleware interface such as CORBA. This request for services involves requesting the proper application for the data stream that the sensor will be delivering to the network of resources as well as a request for the required metric constraints, such as throughput and latency (discussed in Subsection 2.5.2), needed to process the sensor data stream effectively. The determination of required constraints could involve negotiations between the sensor and the NRM through the sensor interface. The NRM uses the sensor interface to direct the sensor platform to start sending a data stream once the NRM has marshaled the resources that the sensor will need to satisfy the request. Finally, the sensor interface also facilitates communications between the sensor platform and the NRM regarding flow control, application shutdown, etc.

2.5.6 Mapping Database

This mapping database is populated with data structures generated by the graph generator and metrics object; it represents the weighted graph-theoretic characterization of the various parallel permutations of an application that is mapped to networked resources. Graph search uses the mapping database to reconstitute a weighted graph for each application for which it is asked to find resources and the degree and form of parallelism needed to meet real-time constraints.

2.5.7 Topology Database

The topology database stores the current state of each of the resources; the graph generator and graph search use this database. Graph generator uses the topology database to determine which resources are available and most appropriate for candidate task-application mappings. Graph search uses this database to verify that resources are functional before a set of resources is chosen to host an application, as well as for generating and modifying weights associated with resource utilization. The topology database is generated during the discovery phase when the NRM first comes online (e.g., see Breitbart et al. [33] and Astic and Foster [34]). Alternatively, an administrator could choose to generate a topology database for the NRM that enumerates connectivity and capability among all computation and storage resources under its control. Agent reports (or lack thereof) will affect state changes in this database indicating whether the resource is online or offline.

2.5.8 NRM Federation

In a large network with a sizeable number of resources, using a single NRM may not be the most effective solution. In such a scenario, multiple NRMs are organized in a bilevel hierarchy; wide-area network (WAN) NRMs interface with sensors and administer backbone communication resources, underneath which local-area network (LAN) NRMs administer and allocate compute resources for regional compute centers (RCCs). The primary responsibility of a WAN NRM is to choose a location on the network at which distributed computing is conducted for each application and to allocate WAN bandwidth for data flow between sensors and LAN resources. The objective of the WAN NRM is to load balance WAN traffic and computational load, taking into account the relative overall processing capability of each RCC. Each LAN NRM advertises its current processing capability using standardized metrics.

Each NRM is a federated collection, using a voting mechanism to elect an executor independently at the LAN and WAN levels. Each federation monitors the health of its executor by inspecting periodic diagnostic reports that the executor broadcasts. In response to an executor's diagnostic report (or lack thereof), the federation may choose to relieve the current executor of its responsibility and elect a new one. This prevents any one NRM failure from rendering resources unusable or disabling a sensor from contracting for network services.

Earlier paragraphs have detailed the LAN NRMs graph-theoretic representation of network resources, as well as its construction, weighting, and search criteria. The WAN NRM graph-theoretic representation and weighting are somewhat different from that of a LAN NRM; however, its construction and search criteria are formulated in an identical manner. The vertices in a WAN graph represent RCCs and each column corresponds to an application, while the concatenation of applications across the columns in a WAN NRM graph spans a mission. This is in contrast to a LAN NRM, in which the concatenation of tasks in its graph spans an application.

2.5.9 NRM Fault Tolerance

The absence of a heartbeat or the delivery of an error report by an agent alerts the NRM to a system fault. The NRM's fault tolerance policy is application dependent and is derived from a mandate by the developer and/or client. The policy is a trade-off between resource usage and seamless fail-over and includes redundant processing, surgical replacement, or restart of the application. Redundant processing is the most robust fail-over mechanism; the NRM simply assigns duplicate sets of resources to process the same data. If one set of resources fails, results are obtained from one of the duplicate sets. Redundant processing has the highest resource cost of all fault tolerant policies.

Conversely, the NRM may choose to replace the failed component dynamically so that processing is able to continue. In this case, the NRM may have allocated distributed network storage to act as a time-delay buffer in the event of resource failure. This would enable the application, if so instrumented, to pick up processing at the point at which the failure occurred. Finally, the NRM could simply choose to halt execution of the application and start over with a new set of processing resources, although a certain amount of data and the corresponding results may be lost irrevocably.

2.6 Experimental Results

A proof-of-concept experiment has been conducted at MIT Lincoln Laboratory in which the NRM allocates distributed networked resources for a sensor data fusion application in various scenarios [35].

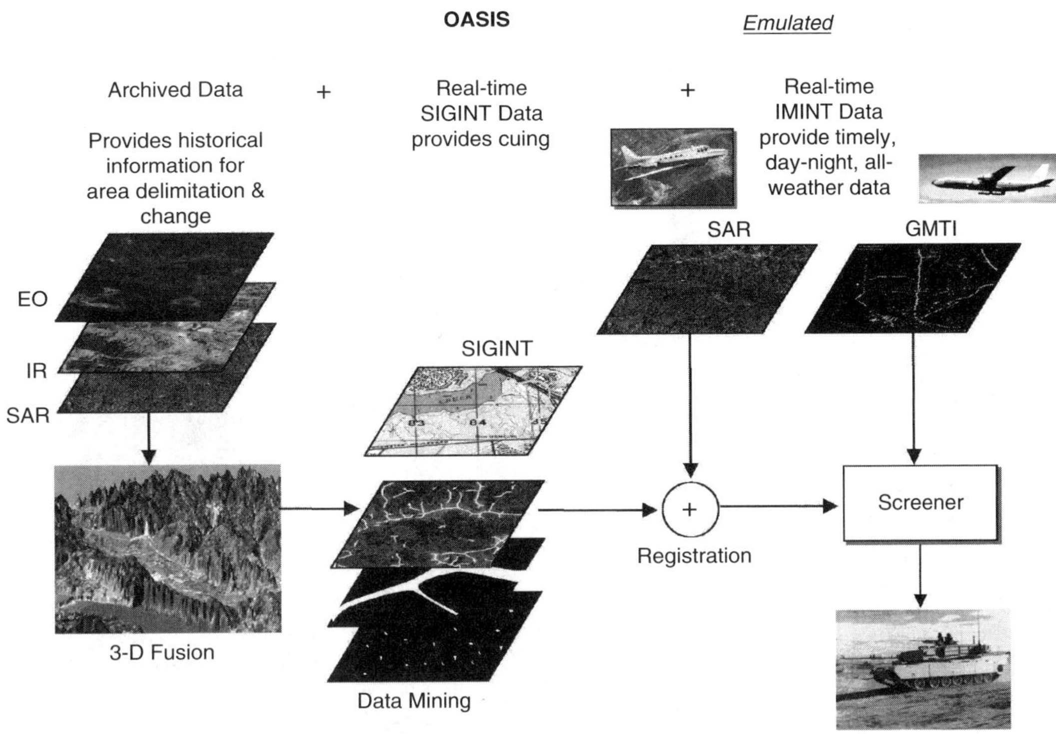

FIGURE 2.8 OASIS ATR and visualization.

TABLE 2.2 Synopsis of NRM Expected Performance

Experimental Configuration	Max Comm BW Requirement (MB/s)	Max Throughput Requirement (GFLOPS)	Processors Employed	Result Turn-Around Time
1 m data	26	0.7	1	1.6
1 m data with HDVI	26	2.2	2	2.6
1/4 m data	410	2.5	2	2.8
1/4 m data with HDVI	410	10	10	7

TABLE 2.3 Synopsis of NRM Performance

Experimental Configuration	Comm BW Measured (MB/s)	Throughput Measured (GFLOPS)	Processors Employed	Result Turn-Around Time
1 m data	26	0.7	1	1.4
1 m data with HDVI	26	2.2	2	2.5
1/4 m data	410	2.5	2	2.7
1/4 m data with HDVI	410	10	8	7.8

The sensor fusion application is OASIS (operator assisted integrated systems), which is an automatic target recognition and visualization suite (see Figure 2.8). OASIS processes real-time SAR data and archived data generated by sensors with different modalities like EO and IR [36]. A block diagram of the experimental test bed is shown in Figure 2.9. The experimentation resource network consisted of three

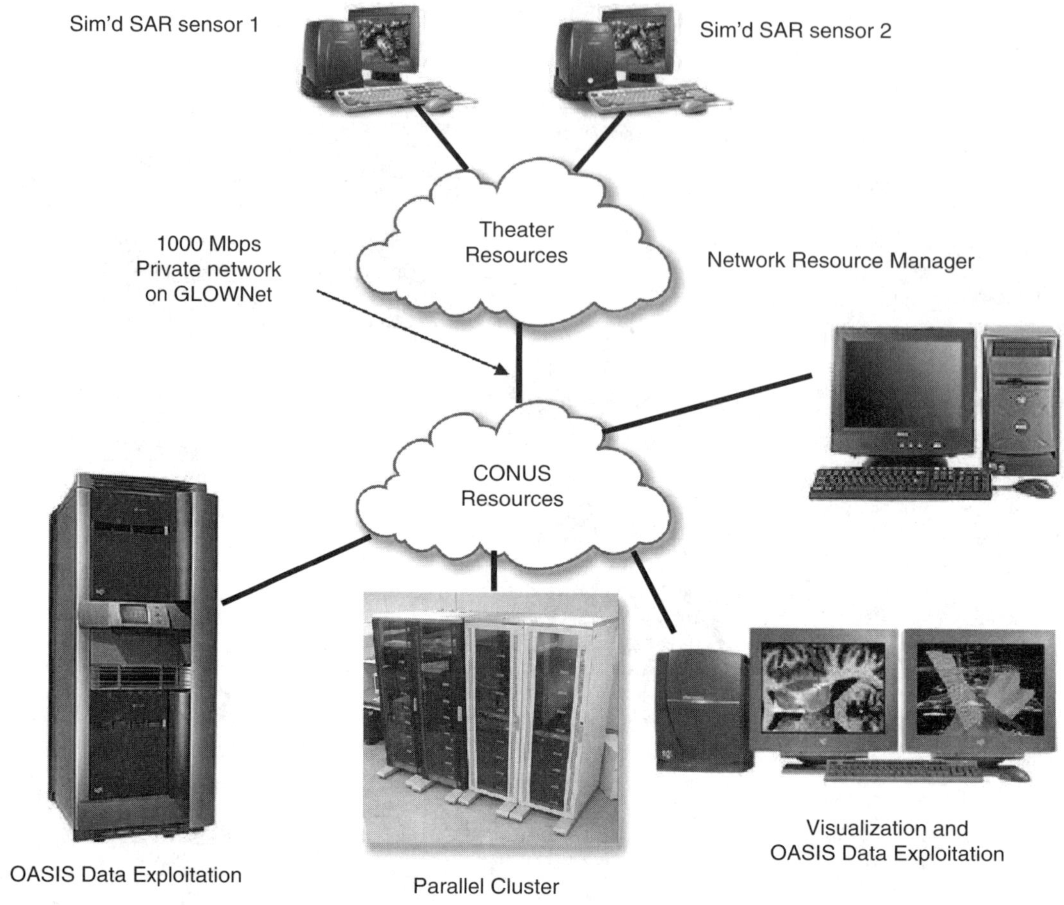

FIGURE 2.9 Experimentation resource network.

SGI O2 workstations, an eight-processor SGI Origin, an eight-node, dual Pentium3 class Beowulf cluster, and a PC workstation, which hosted the NRM.

For this experiment, two SGI O2s were used as sensor surrogates to transmit unprocessed complex SAR imagery generated with range and cross-range resolutions of 1 and 1/4 m, respectively. The sensor surrogates fed data into the OASIS processing chain. To keep the complexity of the system manageable, only the most computationally intensive stage was made remappable. This stage, the HDVI processing [3] (stage 3 in Figure 2.10), had six options for the NRM ranging from a single SGI processor to six Pentium3 class cluster processors. The HDVI processing was conducted on targets detected on the two images at both resolutions, and image formation was conducted on processors in the local area network. The performance metrics for the OASIS applications were determined with a combination of actual performance measurements and modeled performance analyses. Table 2.2 is a tabulated synopsis of the expected performance of the NRM and Table 2.3 shows the actual performance of the NRM. The expected and actual performance values compared very well.

Because this network was PE resource limited, the objective of the NRM was to use the smallest fraction of PE bandwidth available across the network while meeting network conduit, PE utilization, latency, throughput, and network-wide bandwidth usage constraints. It is clear from the results that the NRM was able to tailor the communication and computation solution it delivered based on the particular application needs and the constraints imposed. The successful completion of this experiment has initiated further research and development to give the NRM greater functionality, automation, and flexibility.

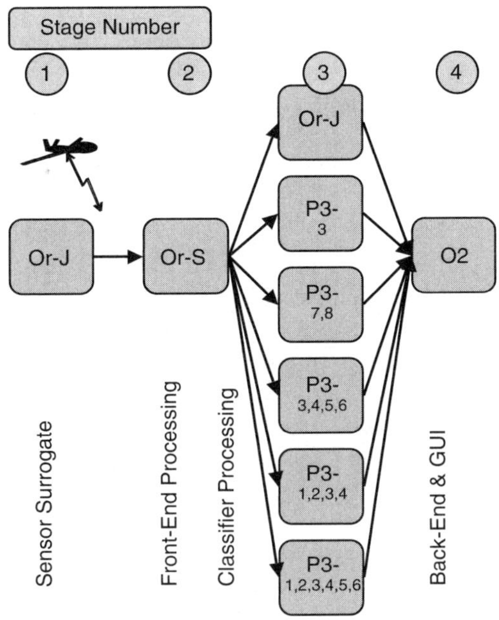

FIGURE 2.10 Graph of OASIS application onto the experimental resources.

Acknowledgments

The authors thank the members of the Precision Targeting via Collaborative Networking team at MIT Lincoln Laboratory for formulating many of the concepts discussed in this chapter. The authors also thank Dr. Mari Maeda, formerly of DARPA/ITO, and Dr. Gary Koob of DARPA/IPTO for their encouragement and support of this project.

References

1. Usoff, J., Beavers, W., and Cox, J., Wideband networked sensors processing, in *Proc. High Performance Embedded Computing Workshop*, November 2001.
2. Cuomo, K.M., Pion, J.E., and Mayhan, J.T., Ultrawide-band coherent processing, *IEEE Trans. Antenna Propagation*, 47, 1094, June 1999.
3. Benitz, G.R., High-definition vector imaging, *MIT Lincoln Lab. J., Special Issue Super-Resolution,* 10:2, 147, 1997.
4. Nguyen, D.H. et al., Super-resolution HRR ATR Performance with HDVI, *IEEE Trans. Aerospace Electron. Syst.*, 37:4, 1267, October 2001.
5. Chan, V.W.S., Optical space communications, *IEEE J. Selected Topics Quantum Electron.*, 6:6, 959, November/December, 2000.
6. Awduche, D. et al., RSVP-TE: extensions to RSVP for LSP tunnels, RFC 3209, http://www.faqs.org/rfcs/rfc3209.html, December 2001.
7. Ash, J. et al., Applicability statement for CR-LDP, RFC 3213, http://www.faqs.org/rfcs/rfc3213.html, January 2002.
8. Cormen, T.H., Leiserson, C.E., and Rivest, R.L., *Introduction to Algorithms.* McGraw-Hill, New York, 1993.
9. Strater, J. and Wollman, B., OSPF modeling and test results and recommendations, Mitre Technical Report 96W0000017, Mitre Corporation, 1996.

10. Perkins, C., *Ad Hoc Networking*, Addison-Wesley, Boston, 2001.
11. Floyd, S. and Jacobson, V., Random early detection gateways for congestion avoidance, *IEEE/ACM Trans. Networking*, 1:4, 397, August 1993.
12. Stevens, W., TCP slow start, congestion avoidance, fast retransmit and fast recovery algorithms, RFC 2001, http://www.faqs.org/rfcs/rfc2001.html, January 1997.
13. Stadler, J.S., Performance enhancements for TCP/IP on a satellite channel, in *Proc. IEEE Military Commun. Conf. 1998 (MILCOM98)*, 1, 270, October 1998.
14. Allman, M., Glover, D., and Sanchez, L., Enhancing TCP over satellite channels using standard mechanisms, RFC 2488, http://www.faqs.org/rfcs/rfc2488.html, January 1999.
15. Berrou, C., Glavieux, A., and Thitimajshima, P., Near Shannon limit error-correcting coding and decoding: turbo codes. 1, in *Conf. Rec. IEEE Int. Conf. Commun. 1993 (ICC 93)*, 2, 1064, May 1993.
16. Foschini, G.J., Layered space-time architecture for wireless communication in a fading environment when using multiple antennas, *Bell Labs Tech. J.*, 1:2, 41, Autumn 1996.
17. Raleigh, G.G. and Cioffi, J.M., Spatio-temporal coding for wireless communications, in *Proc. IEEE Global Telecommun. Conf. 1996 (GLOBECOM 96)*, 3, 1405, November 1996.
18. Sisterson, L.K. et al., An architecture for semi-automated radar image exploitation, *Lincoln Lab. J.*, 11:2, 175–204, 1998.
19. Cooley, J.A. et al., Software-based erasure codes for scalable distributed storage, in *Proc. 20th IEEE Symp. Mass Storage Syst.*, 157–164, April 2003.
20. Luby, M.G. et al., Practical loss-resilient codes, in *Proc. 29th ACM Symp. Theory Computing*, 150–159, 1997.
21. Byers, J.W., Luby, M.G., and Mitzenmacher, M., Accessing multiple mirror sites in parallel: using tornado codes to speed up downloads, in *Proc. IEEE INFOCOM 1999*, 275–283, March 1999.
22. Silberschatz, A. and Galvin, P., *Operating System Concepts*, 5th ed., Addison-Wesley, Reading, MA, 1998.
23. Wind River Systems, Inc. http://www.windriver.com/, accessed July 2003.
24. FSMLabs (Finite State Machine Labs), Inc. http://www.fsmlabs.com/, accessed July 2003.
25. Mercury Computer Systems, Inc. http://www.mc.com/, accessed July 2003.
26. Abbott, D., *Linux for Embedded and Real-Time Applications*, Newnes, Amsterdam, 2003.
27. Object Management Group. http://www.omg.org/, accessed July 2003.
28. Hoffmann, H., Kepner, J., and Bond, R., S3P: Automatic, optimized mapping of signal processing applications to parallel architectures, in *Proc. High Performance Embedded Computing Workshop 2001*, September 2001.
29. The vector, signal, and image processing library. http://www.vsipl.org/, accessed July 2002.
30. Reuther, A.I. and Goodman, J.I., Resource management for digital signal processing via distributed parallel computing, in *Proc. High Performance Embedded Computing Workshop 2002*, September 2002.
31. Goodman, J.I. et al., Discrete optimization using decision-directed learning for distributed networked computing, in *Proc. IEEE Asilomar Conf. Signal, Syst. Computers*, 1189–1196, November 2002.
32. Neuman, B.C. and Ts'o, T., Kerberos: an authentication service for computer networks, *IEEE Commun.*, 32:9, 33, September 1994.
33. Breitbart, Y. et al., Topology discover in heterogeneous IP networks, in *Proc. IEEE INFOCOM 2000*, 265–274, March 2000.
34. Astic, I. and Foster, O., A hierarchical topology discovery service for IPv6 networks, in *Proc. 2002 Network Operations Manage. Symp.*, 497–510, April 2002.

35. Reuther, A.I. and Goodman, J.I., dynamic resource management for a sensor-fusion application via distributed parallel grid computing, in *Proc. High Performance Embedded Computing Workshop 2003*, 2003.
36. Avent, R.K., A multi-sensor architecture for detecting high-value mobile targets, in *Proc. 2002 SIAM Conf. Imaging Sci. (IS02),* March 2002.

3

Sensor Network Management

Linnyer Beatrys Ruiz
Pontifical Catholic University of Paraná and Federal University of Minas Gerais

José Marcos Nogueira
Federal University of Minas Gerais

Antonio A. F. Loureiro
Federal University of Minas Gerais

3.1 Introduction 3-1
3.2 Management Challenges 3-2
3.3 Management Dimensions 3-3
Dimensions for WSN Management • Management Levels • WSN Functionalities • Management Functional Areas
3.4 MANNA as an Integrating Architecture 3-15
Management Services, Functions, and Models • Functional Architecture • Information Architecture • Physical Architecture
3.5 Putting It All Together 3-25
3.6 Conclusion 3-25

3.1 Introduction

A wireless sensor network (WSN) consists of a large number of sensor nodes deployed over an area and integrated to collaborate through a wireless network. WSNs encourage several novel and existing applications such as environmental monitoring; health care; infrastructure management; public safety; medical; home and office security; transportation; and military [1, 2, 9, 17, 18]. These have been enabled by the rapid convergence of three technologies: digital circuitry, wireless communications, and the microelectromechanical system (MEMS). These technologies have enabled very compact and autonomous sensor nodes, each containing one or more sensor devices, computation and communication capabilities, and limited power supply.

Some of the applications foreseen for WSNs will require a large number of devices in the order of tens of thousands of nodes. Traditional methods of sensor networking represent an impractical, complex, and expensive demand on cable installation. WSNs promise several advantages over traditional sensing methods in many ways: better coverage, higher resolution, fault tolerance, and robustness. The ad hoc nature and deploy-and-leave vision make it even more attractive in military applications and other risk-associated applications, such as catastrophe, toxic zones, and disasters [2, 9]. Performing the processing at the source can drastically reduce the computational burden on application, network, and management. On the other hand, any solution must take into account specific characteristics of this type of network.

WSN management must be autonomic, i.e., self-managed (self-organizing, self-healing, self-optimizing, self-protecting, self-sustaining, self-diagnostic) with a minimum of human interference, and robust to changes in network states while maintaining the quality of services. Until now, WSNs and their applications have been developed without considering an integrated management solution. The task of building and deploying management systems in environments that will contain tens of thousands of network elements with particular features and organization and that deal with the aforementioned attributes is not trivial. This task becomes more complex due to the physical restrictions of the unattended sensor nodes, in particular energy and bandwidth restrictions.

In this chapter, the focus is on WSN management, which comprises a large number of devices in the order of tens of thousands of nodes. Clearly, the mechanisms associated with traditional management paradigms must be rethought. In this sense, a new paradigm called autonomic management is explored. The rest of this chapter is organized as follows. Section 3.2 presents an overview of network management and discusses the management challenges for WSNs. In Section 3.3, management dimensions (management levels, WSN functionalities, and management functional areas) are presented and discussed. A management architecture for WSNs called MANNA is presented in Section 3.4, as well as how it works. In Section 3.5, a simple example shows the different aspects together. Finally, Section 3.6 presents conclusions.

3.2 Management Challenges

One of the major goals of network management is to promote productivity of network resources and maintain the quality of the service provided. However, the management of traditional networks and of WSNs has several significant differences. This section discusses important characteristics of WSNs that make their management different from that of other networks.

A WSN is a tool for distributed sensing of one or more phenomenon that reports the sensed data to one or more observers. A WSN provides services for observers as well as for itself. It produces and transports application data, so, in this sense, the network provides service to itself. The objective of a WSN is to monitor and, eventually, control a remote environment. Sensor nodes execute a common application in a cooperative way (i.e., a clear, common goal in the overall network), which may not be the case in a traditional network.

The traditional computer networks are designed to accommodate a diversity of applications. Network elements are installed, configured by technicians, and connected in a network in a way to provide different kinds of services. Technicians' maintenance of components or resources is a normal fact. The network tends to follow well-established planning of available resources and the location of each network element is well-known. In a WSN this is not often the case because the network is planned to have unattended operation. In fact, the initial configuration of a WSN can be quite different from what was supposed to be in cases such as throwing the nodes into an ocean, forest, or other remote regions. In unpredictable situations, a configuration error such as a planning error may cause the loss of the entire network even before it starts to operate.

Energy is a critical resource in WSNs. Thus, all operations performed in the network should be energy efficient. Topology is dynamic because sensor nodes can become out of service temporarily or permanently (nodes can be discarded, lost, destroyed, or even run out of energy). In this scenario, faults are a common fact, which is not expected in a traditional network.

Depending on the WSN application, it may be interesting to identify uniquely each node in the network. Furthermore, one may be interested in a value associated to a given region and not to a particular node — for instance, in the temperature at the top of a mountain. A WSN is typically data centric, which is not common in traditional networks.

A managed WSN is responsible for configuring and reconfiguring under varying (and, in the future, even unpredictable) conditions. System configuration ("node setup" and "network boot up") must occur automatically; dynamic adjustments need to be done to the current configuration to best handle changes in the environment and itself. A managed WSN always looks for ways to optimize its functioning; it will monitor its constituent parts and fine-tune workflow to achieve predetermined system goals. It must perform something akin to healing — it must be able to recover from routine and extraordinary events that might cause some of its parts to malfunction. The network must be able to discover problems or potential problems, such as uncovered area, and then find an alternate way of using resources or reconfiguring the system to keep it functioning smoothly. In addition, it must detect, identify, and protect itself against various types of attacks to maintain overall system security and integrity. A managed WSN must know its environment and the context surrounding its activity and act accordingly. The management entities must find and generate rules to perform the best management of the current state of the network [22].

A managed WSN with this has various characteristics can be called an autonomic system [1], which is an approach to self-managed computing systems with a minimum of human interference. This term derives from the autonomic nervous system of the human body, which controls key functions without conscious awareness or involvement. The processors in such systems use algorithms to determine the most efficient and cost-effective way to distribute tasks and store data. Along with software probes and configuration controls, computer systems will be able to monitor, tweak, and even repair themselves without requiring technology staff — at least, that is the goal [1].

WSN management must be autonomic, i.e., self-managed and robust to changes in network states while maintaining the quality of service; that is, it must be capable of self-configuration, self-organization, self-healing, and self-optimization. However, the computational cost of autonomic processes can be expensive to some WSN architectures.

Probably, the fundamental issue about the management of a WSN is concerned with how the management can promote plant and resource productivity, and how it integrates in an organized way functions of configuration, operation, administration, and maintenance of all elements and services.

The task of building and deploying autonomic management systems in environments in which tens of thousands of network elements with particular features and organization will be present is very complex. This task becomes even more involved due to the physical restrictions of the sensor nodes, in particular energy and bandwidth restrictions. The management application to be built also depends on the kind of application being monitored. A good strategy is to deal with complex management situations by using management dimensions.

3.3 Management Dimensions

In general, for traditional networks, management aspects are clearly separated from network common activities, i.e., from the services they provide to their users. It is also said that an overlap of management and network functionalities exists, although the implementation can be thought of independently. This separation can be promoted by using two traditional management dimensions: management functional areas [14] and management levels [15].

The requirements to be satisfied by systems management activities can be categorized into functional areas. These facilities have come to be known as the specific management functional areas (SMFAs): fault management; configuration management; performance management; accounting management; and security management. This has proved to be a helpful way of partitioning the network management problem from an application point of view [14].

To deal with the complexity of management, management functionality with its associated information can be decomposed into a number of logical layers: business management; service management; network management; and network element management. The architecture that describes this layering is called the logical layered architecture (LLA) [15]. Management activities can be clustered into layers and decoupled by introducing manager and agent roles. A logical layer reflects particular aspects of management and implies the clustering of management information supporting that aspect. Typically, an interaction takes place between adjacent layers, but due to operational and management considerations other interactions may also occur between nonadjacent layers.

The use of the management dimensions is a good strategy to deal with complex management situations by decomposing a problem into smaller subproblems, in successive refinements steps, and to provide a separation between application and management functionalities through a management architecture. This will make possible the integration of organizational, administrative, and maintenance activities for a given network.

WSN management must be simple, adherent to network idiosyncrasies, including its dynamic behavior, and efficient in its use of scarce resources. The adoption of a strategy based on the traditional framework of functional areas and management levels will permit management integration in the future. However, for WSN management it is necessary to go further. Using management functional areas and management levels is not enough because WSNs are application specific.

The following discussion concerns how the traditional management dimensions can be applied in WSN management. Also, new dimension for WSN management is proposed that considers the general aspects of the different types of the networks.

3.3.1 Dimensions for WSN Management

WSNs are embedded in applications to monitor the environment and act upon it. Thus, the management application should try to be "compatible" with the kind of application being monitored. In order to have better development of WSN management services and functions, it is necessary to characterize the WSN and establish a novel management dimension. Thus, looking at the characteristics of various WSN applications, five main WSN functionalities are identified: configuration; sensing; processing; communication; and maintenance. These functionalities define a novel dimension for the management, as presented in Figure 3.1 [22]. Configuration is the first functionality before a network starts sensing the environment, processing, and communicating data. Maintenance treats specific characteristics of WSN applications during the entire network lifetime.

In this way, WSN management will have an organization that comes from abstractions offered by management functional areas, management levels, and WSN functionalities (configuration, sensing, processing, communication, and maintenance). The novel dimension introduced can be observed in the upper part of Figure 3.1.

The coordination among the three planes can be based on policies. Policy-based network management (PBNM) [7] is a feasible alternative because it allows the manager to set actions to be carried out by the network without worrying too much about network details. Managers can define suitable actions in due time and still have a global or local view of the network. PBNM helps to manage complex networks such as WSNs. The managers will only inform concerning what is expected, but not how it should be obtained. The agents will be intelligent to decide what to do as well as how and when to do it. Automatic services and functions can be executed toward self-management if appropriate conditions, such as residual energy level, are present.

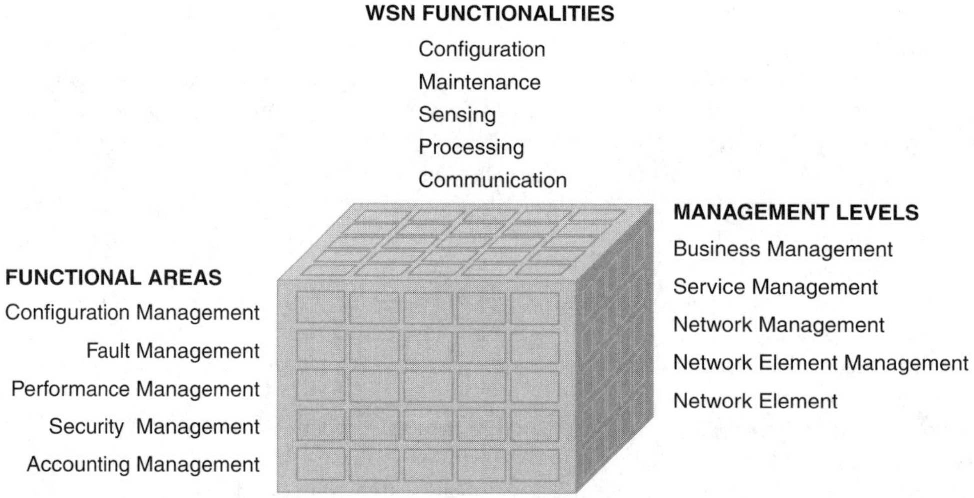

FIGURE 3.1 Management dimensions for WSNs. (From Ruiz, L.B., Nogueira, J.M., Louriero, A.A., *IEEE Commun. Mag.*, 41(2), 116–125, 2003. With permission.)

Three management dimensions must be considered in the definition of a management function, establishment of an information model, service composition, and development of a management application. The next subsections explain WSN management from the perspective of management level, WSN functionalities, and management functional areas.

3.3.2 Management Levels

Many traditional management systems use this model in a bottom-up approach; however, in WSN management, the LLA model is used in a top-down approach. After analyzing the business level issues, the necessities of the lower levels become clear. Similarly, it is only after defining the application, including the corresponding requirements on the service layer, that one can plan the network, network element management layers, and network elements. This is a key observation when reasoning about WSN management. A brief discussion concerning WSN management from the perspective of management level is now presented.

3.3.2.1 Business Management

Requirements that allow the characterization of a sensor network come from the objectives defined for the business management layer. Because WSNs depend on applications, business management deals with service development and determination of cost functions. It represents a sensor network as a cost function associated with network setup, sensing, processing, communication, and maintenance. WSN applications have enormous potential benefits for society as a whole and represent new business opportunities. Instrumentation of environments [2, 9] with numerous networked sensor nodes can enable long-term data collection at scales and resolutions that are difficult, if not impossible, to obtain otherwise. In the future, one can expect to have Internet end-points equipped with a variety of sensors to monitor the network and their own state, as well as fairly sophisticated computing capabilities to enable them to function as decision elements and not just as repeaters. As more aspects of society are connected to networks, their sensory components become more prominent.

3.3.2.2 Service Management

A WSN is used to monitor and, sometimes to control, an environment. WSN service management introduces new challenges due to scarce network resources, dynamic topology, traffic randomness, energy restriction, and a large amount of network elements. WSN services are concerned with functionalities (see Figure 3.1) associated with application objectives. Basic WSN services are sensing, processing, and data dissemination [21]. Two main issues are associated with WSN service management: quality of service (QoS) and denial of service (DoS).

Quality of service. QoS architectures can only be effective and provide guaranteed services if QoS elements can be adequately configured and monitored; mechanisms can be defined to help managers to deal with these elements. Also, such mechanisms must allow replacement of the current device-oriented management approach by a network-oriented or cluster-oriented approach. Thus, in addition to the management of elements (physical and logical resources), management applications must also manage QoS aspects. Components involved in QoS support to WSNs include QoS models, QoS sensing, processing, and QoS dissemination [22]. The larger the number of monitored QoS parameters is, the larger the energy consumption and the lower the network lifetime are.

QoS model. A QoS model specifies an architecture in which some of the services can be provided in WSNs. All other QoS components, such as QoS sensing, QoS processing, and QoS dissemination (e.g., signaling, QoS routing, and QoS MAC), must cooperate to achieve this goal. A management application can establish the QoS model and can control the QoS signaling that coordinates behavior of the other components. QoS-related tasks must be performed by using network management functions.

QoS sensing. QoS sensing considers the sensor device calibration, environment interference monitoring, and exposure (time, distance, and angle between sensor device and phenomenon). Meguerdichian [18] defines coverage area as a measure of QoS for a WSN. In the worst-case coverage, attempts are made to quantify the quality of service by finding areas of low observability to sensor nodes and detecting breach regions. In the best-case coverage, the management application must find areas of high observability to sensors and identify the highest accuracy. A denser network will lead to more effective sensing because of the higher accuracy of the network (e.g., areas of intersection and redundant information) and better fault tolerance.

On the other hand, this will lead to a large number of collisions and potentially to congestion situations, increasing latency and reducing energy efficiency. Congestion control must be based not only on the capacity of the network, but also on the accuracy level required at the observer. The traffic in a WSN is different from conventional networks: it is a collective communication operation with redundancy. Thus, the management application has the flexibility of meeting the performance demands by controlling the reporting rate of sensors, controlling the virtual topology of the network (by turning off some sensors), or optimizing the collective reduction communication operation (by data aggregation). The provision of QoS can rely on resource reservation. When an active node goes out of service due to operational problems, the management application activates a redundant node, defining a sort of resource reservation scheme. In case of a low density of sensors, the network coverage area can be committed, thus affecting the quality of the service. Resource reservation is being applied.

QoS dissemination. Reliable data delivery is still an open issue in the context of WSNs. QoS dissemination in WSNs is a challenging task because of constraints, mainly energy and dynamic topology of WSNs. The two components for QoS dissemination are QoS routing and QoS medium access control (MAC). QoS routing finds a path that satisfies a given QoS requirement, and QoS MAC solves the problem of medium contention that supports reliable unicast communication [29]. To support QoS, a link state information such as delay, bandwidth, cost, loss rate, and error rate in network should be available and manageable. One of the objectives of the management application is to obtain and to manage link state information in WSNs for monitoring QoS. This is very difficult because the quality of a wireless link is apt to change with the circumstances, such as residual energy, node distribution, density (all change along the network lifetime), and interference. Configuration characteristics such as coverage area, density, network organization, node deployment (distribution), latency, and communication range may degrade or deny the service.

QoS processing. Processing quality depends on the robustness and complexity of the algorithms used, as well as processor and memory capacities. The computing paradigm changes from one based on computational power to one driven by data. The way to measure processing performance changes from processor speed to the immediacy and accuracy of the response and energy consumption. Individual computers become less important than lower granularity and dispersed computing attributes.

The network quality of service can be measured by the energy consumption to execute a service with a determined quality level. In most WSNs, energy consumption is one of the main metrics. However, in some situations, during certain events the network must apply the maximum of energy possible in the delivery of information — for instance, in WSNs deployed over the havoc of a cave-in where as much information as possible is needed in the shortest time period. In this kind of application, to extend the network lifetime is not that important. However, without proper management mechanisms, the network can suffer the implosion problem (a large amount of data generating congestions, collisions, and data losses in the network).

Any situation that diminishes or eliminates the capacity of the network to perform its expected job is called DoS (denial of service). Some examples of incidental threats are hardware failures, software bugs, resource exhaustion, and unexpected environmental conditions. DoS aspects will be discussed in Subsection 3.3.4.4.

3.3.2.3 Network Management

This layer aims to manage a network, which is typically distributed over an extensive geographical area, as a whole. In the network management level, relationships among sensor nodes are to be considered. It is known that individual nodes are designed to sense, process data, and communicate, thus contributing to a common objective. In this way, nodes can be involved in collaboration, connectivity, and aggregation relationships. A WSN is composed of interconnected managed objects (physical or logical) capable of exchanging information. In these cases, the WSN is basically composed of two parts: physical resources and services. Service execution depends on the physical resource capabilities.

3.3.2.4 Network Element Management

Managed network elements represent the sensor and actuators nodes or other WSN entities, which execute management functions and provide sensing, processing, and dissemination services. The basic functions of a WSN management network element are

- Power management (how a sensor node uses its power)
- Mobility management (how the movement of sensor nodes is planned, run, and registered)
- State management (how a sensor node manages the three management states defined for a node: operational, administrative, and usage)
- Task management (how a sensor node balances and schedules the sensing, processing, and dissemination tasks given to a specific network state)

Each sensor node must be autonomous and capable of organizing itself in the overall community of sensor nodes to perform coordinated activities with global objectives.

Sensor nodes have strong hardware and software restrictions in terms of processing power, memory capacity, battery lifetime, and communication throughput. These are typical characteristics of mobile and wireless devices and not of wired network elements. Thus, software designed for a sensor node must consider these limitations, whereas an element for a wired network may have other restrictions such as performance and response time. The main physical restriction of a WSN is the available energy because batteries are often not recharged during the operation of a sensor node and all activities performed by the node must take energy consumption into account.

3.3.2.5 Network Element

The network element represents physical and logical components of a managed element. Physical resources include sensor or actuator nodes; power supply; processor; memory; sensor device; and transceiver. Logical resources include communication protocols; application programs; correlation procedures; and network services. Because applications may require networks with a large number of sensor nodes, a network element can deal with a single node component or a group of nodes. In such a case, a manageable element can be a cluster of nodes or a cluster-head node, rather than an individual node. The design of a sensor node is motivated by the need to create an inexpensive device with a small form factor and low power dissipation.

Understanding node capability allows function management to be structured and fine-tuned more efficiently. The physical aspects of a network element are described in the following.

- *Power supply.* Energy consumption patterns of individual nodes and of the entire network must be characterized and profiled. This process yields a better understanding of where to apply trade-offs in the design of the management. The most widely used power supply in a WSN is the battery, which is classified into the following types [23]:
  - Linear model — the battery is considered to be a bucket of energy that is linearly drawn from this bucket by the energy consumers
  - Dependent model — considers the rate at which energy is drawn from the battery to compute the remaining battery lifetime; at high discharge rates, the capacity of the battery is reduced
  - Relaxation model — takes into account a phenomenon seen in real-life batteries in which the battery's voltage recovers if the discharge rate is decreased
- *Computational module.* This module is composed of processor and memory. It is responsible for the collaborative processing between nodes to achieve the levels of service and reliability desired by the observer.

- *Sensor element.* Sensing devices can be classified into three groups: monitors (e.g., magnetometer, light sensor, temperature, pressure, humidity); motion detectors (e.g., accelerometer); and media processing (e.g., audio, video).
- *Transceiver.* The main types of a transceiver are radio frequency (RF), infrared, and optical. RF communication is based on electromagnetic waves with frequencies ranging from tens of kilohertz to hundreds of gigahertz. Of the most important factors in the design of RF communications is the size of the antenna. To optimize transmission and reception, an antenna should be at least $\lambda/4$, where λ is the wavelength of the carrier frequency. In optical laser communication, a transmitting device uses a laser beam to send information. An optical receiver, in the form of a photodiode or charge-coupled device (CCD) array, receives the signal and decodes the data. Optical communications can be classified into two types: passive (the laser signal is generated through a secondary source) and active (the transmitting device generates its own laser signal). A few points should be noted regarding the differences between optical and RF communication. Both forms of communication are based on sending electromagnetic waves through air. To compare RF to optical communication, one must conside the receiving end of the communication system. For both, a trade-off takes place between size and receiving performance [12].
- *Software.* This is used to represent a set of programs and procedures that becomes an autonomous system capable of executing the information processing, relaying, or routing.

3.3.3 WSN Functionalities

This section presents the novel proposed dimension for the WSN management, composed by the configuration, sensing, processing, communication, and maintenance functionalities. These WSN functionalities can be observed in the upper part of Figure 3.1. This novel dimension is obtained from the functional model defined in Reference 22, which presents a scheme to characterize WSNs considering that they are application dependent. Because a management solution depends on the features of the network, this solution must also be proposed considering the type of network. For this reason, WSN functionalities are serviceable in the development of the management application [22].

3.3.3.1 Configuration

This functionality involves procedures related to planning, placement, and self-organization of a WSN. The configuration functionality (predeployment) is related to the:

- Definition of WSN application requirements
- Determination of the monitoring area (shape and dimension)
- Characteristics of the environment
- Choice of nodes
- Definition of the WSN type
- Service provided

In the deployment phase, sensor nodes can be placed by dropping them from a plane, rocket, or missile, and placed one by one by a human or a robot. Any placement approach for sensor nodes must also take into account the expense and difficulty in redeploying nodes. This is chiefly due to the limited life span of nodes and to their generally nonreplaceable power sources [19]. Another problem is the optimal location of the access point (sink node or base station). An inefficient configuration management may adversely affect overall performance.

WSNs are application specific, which means that the configuration functionality changes from one WSN to another. Next, the configuration is discussed considering the possible types of WSN and the other two management dimensions.

Considering the network management level and management functional areas based on configuration functionality, WSNs can be classified in various ways. A WSN is said to be homogeneous when all nodes have the same hardware; otherwise, it is said to be heterogeneous. A WSN is hierarchical when nodes are grouped for the purpose of communication, and flat otherwise. When nodes are stationary, a WSN is static; otherwise it is dynamic. Note that the topology may be dynamic even when nodes are stationary because new nodes can be added to the network or existing nodes can become unavailable. A WSN is symmetric concerning signal transmission when each transceiver has the same transmission range, and asymmetric otherwise. A WSN is said to be regular concerning node placement when its nodes are placed in a grid; it is called irregular when its nodes are randomly distributed, presenting different densities on the monitored area, and it is balanced when its nodes are randomly distributed and present a uniform distribution. Depending on the number of nodes per area unit, a WSN can be sparse or dense.

Considering the network element management level and the management functional areas based on the configuration functionality, the sensor nodes in a WSN are spread over a region and communicate among themselves using point-to-point wireless communication, thus forming an ad hoc network. The nodes are autonomous when they are able to execute location discovery and self-configuration tasks without human intervention, for example, the location discovery. To relay information off the network, sensor nodes are equipped with a wireless communication device (transceiver). A wireless sensor node also comprises one or more sensor elements, and a battery, memory, and processor. The size of a node is an important consideration. Nodes need to have small form factors so that they may be located unobtrusively in the environment targeted for monitoring. The restriction in size is closely related to the amount of energy available to a node. A rugged and robust construction is required if nodes are dispersed in an inhospitable terrain such as a forest.

Software developed to execute in a wireless sensor node must take into account its hardware restrictions. Because of limited energy capacity, nodes are expected to be thrown away once their energy supply is exhausted. The system can have levels of redundancy built into it to allow failures or to increase accuracy. This can be achieved by using more sensor nodes than are strictly necessary to cover an area. Also, due to environmental nature, logistics, and deploying costs, the deployment of sensors can be a one-time operation; therefore, after nodes have been distributed in the field, human intervention is not an option. The three basic different types of sensor nodes are: common nodes responsible for collecting sensing data; sink nodes (monitoring nodes) responsible for receiving, storing, and processing data from common nodes; and gateway nodes that connect sink nodes to external entities called observers. WSNs can also include actuators that enable control of or actuation in a monitored area. In a hierarchical network, it is common to have a base station (BS) that works as a bridge to external entities.

Considering the service management level and the management functional areas, the WSN comprises three entities: observer, phenomenon, and environment. The observer is a network entity or a final user that wants to have information about data collected, processed, and disseminated by sensor nodes. Depending on the type of application, the observer may send a query to the WSN, and receive a response from it. These queries can be done with or without fidelity. The translation of the query could be performed by the application software or sensor nodes. The WSN may participate in synthesizing the query (e.g., filtering some sensor data or summarizing several measurements into one value), but these procedures are related to the processing functionality. The phenomenon is the entity of interest to the observer that is sensed and optionally analyzed or filtered by the WSN. The observer is interested in monitoring a phenomenon under some latency and accuracy restrictions. A sensor element generates data about a given phenomenon such as temperature, pressure, electromagnetic field, or chemical agents because it can be comprised of different sensor elements.

3.3.3.2 Sensing

The lowest level of the sensing application is provided by the autonomous sensor nodes. An important operation in a sensor network is data gathering. Sensing functionality depends on the type of the phenomenon. Thus, WSNs can be classified in terms of data gathering required by the application as continuous (when sensor nodes collect data continuously along the time), reactive (when they answer to an observer's query or gather data referring to specific events occurring in the environment), and periodic (when nodes collect data according to conditions defined by the application). Some approaches can coexist in the same network; this model is referred to as the hybrid collect model. An example of a continuous phenomenon is temperature and an example of an application in which the phenomenon is moving is a sensor deployed for animal detection. Other examples of phenomena are video; audio; pressure; mechanical stress; humidity; soil composition; luminosity; seismic; and chemical.

Whether gathering is continuous or not, WSNs are defined based on how the data will be transmitted to the observer. The sensing encloses the exposure (time, distance, and angle of phenomenon exhibition at the sensor), calibration, and sensing coverage. Depending on the density of the phenomenon, it will be inefficient if all sensor nodes are active all the time. A model that is well-suited to this case is the Frisbee model [5]. On the other hand, redundancy (overlapping in the sensor coverage) should be utilized in such a way that fault tolerance in the communication network is avoided and better accuracy can be found [26]. Nevertheless, the sensors can be mobile. In this case, the sensors are moving with respect to each other and to the observer as well, and they have direction, orientation, and acceleration.

3.3.3.3 Processing

Memory and processor of a sensor node form the computational module, which is a programmable unit that provides computation and storage for other nodes in the system. Depending on the communication constraints of the system, algorithms must be developed that will allow individual nodes or clusters of nodes to share and process data efficiently. The computational module performs basic signal processing (e.g., simple translations based on calibrating data or threshold filters) and dispatches the data according to the application. Processing can also involve correlation procedures such as data fusion, which combines one or more data packets received from different sensors to produce a single packet (data fusion). Data fusion helps to reduce the amount of data transmitted between the sensor nodes and the observer and allows design of a network that delivers required data while meeting energy requirements. Other possible tasks are security processing and data compression.

3.3.3.4 Communication

Individual nodes communicate and coordinate among themselves. Two types of communication are proposed: infrastructure and application. Infrastructure communication refers to the communication needed to configure, maintain, and optimize operation. The configuration and topology of the sensor network may be rapidly changing in the presence of a hostile environment, a large volume of assigned work, and nodes that fail routinely. Conventional protocols may be inadequate to manage such situations; thus, new protocols are required to promote WSN productivity. In a static sensor network, an initial phase of the infrastructure communication is needed to set up the network and an additional communication is needed to perform its reconfiguration. If the sensors are mobile, additional communication is needed for path discovery/reconfiguration.

Application communication (dissemination) relates to the transfer of sensed data (or information obtained from it). The amount of energy spent in transmitting a packet has a fixed cost related to the hardware and a variable cost that depends on the distance of transmission. Receiving a data packet also has a fixed energy cost. Therefore, to conserve energy, short distance transmissions are preferred. Because the access point (sink node or the BS) may be located far away, the cost to transmit data from a given node to the access point may be high. In a homogeneous and flat WSN, the sensor nodes can form a multihop network by forwarding each other's messages, which can provide different connectivity options. In a heterogeneous and hierarchical WSN, the cluster heads can form a single-hop network for reporting aggregated data to the BS. Within a cluster, measured data are sent to the cluster head by the sensor nodes under its control. All nodes in a cluster are identical except in the heterogeneous WSN, where the cluster head has a larger transmission capacity.

In terms of the data delivery required by the application interest, WSNs can be classified as continuous, when sensor nodes collect data and send them to an observer continuously along the time, and as on demand, when they answer an observer's query. A WSN is event driven when sensor nodes send data referring to events occurring in the environment and programmed when nodes collect data according to conditions defined by the application. Some approaches can coexist in the same network; such a model is referred to as the hybrid model. The cost of sending data continuously may lead to a more rapid consumption of the scarce network resources and, thus, shorten resource lifetime.

Multihop wireless capabilities will enable communication and coordination among autonomous nodes in unplanned environments and configurations. At the same time, wireless channels present challenges of dynamic operating conditions, power constraints for autonomously-powered nodes, and complicating interactions between high level behavior and lower level channel characteristics (e.g., increased synchronized communication will significantly degrade channel characteristics).

For any of the preceding models, the communication approach can be classified as:

- Flooding (sensors broadcasting their information to their neighbors, which in turn broadcast these data until they reach the observer)
- Gossiping (sending data to one randomly selected neighbor)
- Bargaining (sending data to sensor nodes only if they are interested)
- Unicast (sensor communicating to the sink node, cluster head, or BS directly)
- Multicast (sensors forming application-directed groups and using multicast to communicate among group members)

A major advantage of flooding or broadcast is the lack of a complex network layer protocol for routing and address and location management.

In a WSN, each sensor node puts its information onto a common medium. This requires careful attention to protocols in hardware and software. In master–slave protocols, one node gives the commands and another node or a collection of nodes executes them. The cluster head is usually the master and the common nodes (sensors and actuators) are slaves. This protocol allows tight traffic control because no node is allowed to transmit unless requested by the master, and no communication is allowed between slaves except through the master (e.g., medium control access protocol using a channel fixed allocation scheme). In a peer-to-peer network, all nodes are created equal. A node can be a master one moment and then be reconfigured at another time. Peer-to-peer configurations offer the greatest flexibility, but they are the most difficult to control. Any node can communicate directly to any other node.

3.3.3.5 Maintenance

Maintenance functionality is used in the WSNs that can configure, protect, optimize and heal themselves without a lot of input from the human operators who have, until now, been required to keep traditional networks up and running. Maintenance detects failures or performance degradations, initiates diagnostic procedures, and carries out corrective actions on the network. Its ability to discover changes in the network state enables the self-management to adapt and optimize the network behavior. Beyond corrective maintenance, the other types of maintenance are: adaptive (the system should adapt to meet the changes); preventive (the system should learn to anticipate the impact of those changes); and proactive (as it gets smarter, the system should learn to intervene so as to preempt negative events). An example of maintenance concerns the density of nodes in the WSN; in case of a high node density, the maintenance can turn off some nodes temporally.

The WSN state (e.g., topology, energy, coverage area) changes frequently. In the case of static networks, changes occur because nodes may become unavailable during operation. This dynamic behavior must be observed. The maintenance depends on the knowledge of the network state. Thus, maintenance functionality is needed to keep the network operational and functional to ensure robust operation in dynamic environments, as well as optimize overall performance. Maintenance provides dependability, the main attributes of which are reliability; availability; safety; security; testability; and performability.

WSNs have important characteristics depending on the application. Some of them are:

- Planning
- Deployment
- Coverage
- Accuracy

- Fidelity
- Density
- Self-organization
- Adaptation
- Location

The points described in this subsection will play an important role in the definition of the management services and functions.

3.3.4 Management Functional Areas

WSN management considers fault, security, performance, and accounting management functional areas extremely dependent on the configuration functional area. In WSNs, all operational, administrative, and maintenance characteristics of the network elements; the network, services; and business; and the adequate execution in the activities of configuration, sensing, processing, communication, and maintenance (as shown in Figure 3.1) are dependent on the configuration of the WSN. An error in the configuration or a forgotten requisite during the planning may compromise all the functionalities of the other areas. This idea is depicted in Figure 3.2, in which the configuration functional area plays a central role. As mentioned before, there are several significant differences in the management of traditional networks and WSNs. In this sense, management functional areas must revisit considering the WSNs features.

3.3.4.1 Configuration Management

Configuration management is a functional area of high relevance in WSN management. Because the objective of a sensor network is to monitor (acquisition, processing, and delivery of data) and, eventually, to control an environment, any problem or situation not anticipated in the configuration phase can affect the offered service. The configuration management must provide basic features such as self-organization, self-configuration, self-discovery, and self-optimization. Some management functions defined for network level configuration management are:

- Requirements specification of the network operational environment
- Monitoring of environmental variations
- Size and shape definition of the region to be monitored
- Node deployment — random or deterministic
- Operational network parameters determination
- Network state discovery
- Topology discovery
- Network connectivity discovery
- Control of node density
- Synchronization
- Network energy map evaluation
- Coverage area determination
- Integration with observer

Some management functions defined for network-element level configuration management are:

- Node programming
- Node self-test
- Node location
- Node operational state
- Node administrative state
- Node usage state
- Node energy level

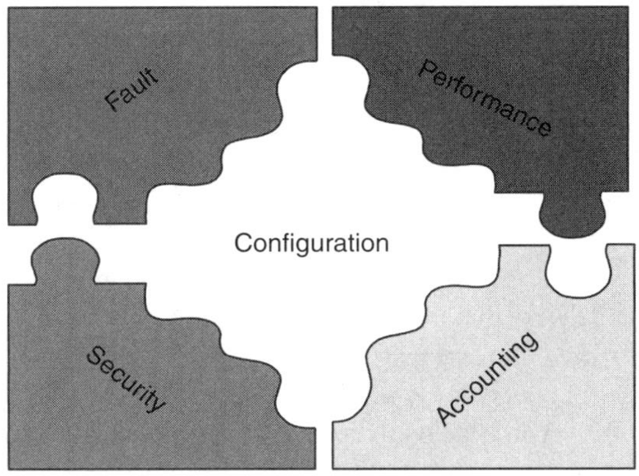

FIGURE 3.2 The role of configuration management. (From Ruiz, L.B., Nogueira, J.M., Louriero, A.A., *IEEE Commun. Mag.*, 41(2), 116–125, 2003. With permission.)

3.3.4.2 Fault Management

Faults in WSNs are not an exception and tend to occur frequently. This is one of the reasons why management of WSNs is different from the traditional network management. Faults happen all the time due to energy shortage, connectivity interruption, environmental variations, and so on. In general, sensor networks must be fault tolerant and robust and must survive despite occurrences of faults in individual nodes, in the network, or even in services provided. In addition to events caused by energy problems, other events can happen in a wireless sensor network related to communication; quality of service; data processing; physical equipment fault; environment; integrity violation; operational violation; security; and time-domain violation. Therefore, even if a node has an adequate energy level to execute its function, it may decide not to do that for other reasons. Fault management must provide basic characteristics such as self-maintenance, self-healing, and self-protection.

Failures will be frequent in a WSN, and fault management is a critical function. Several characteristics of sensor networks suggest that faults, common in traditional computer networks, will be even more common in this kind of network..

- Large-scale deployment of cheap individual nodes means that node failures from fabrication defects will not be uncommon.
- Attacks by adversaries will be likely because these networks will often be embedded in critical applications. Worse, attacks will be made easier because these networks will often be deployed in open spaces or enemy territories, where adversaries can manipulate the environment (so as to disrupt communication by jamming) and also have physical access to the nodes.
- Ad hoc wireless communication by radio frequencies means that adversaries can easily put themselves in the network and disrupt infrastructure functions (such as routing) taken by the individual nodes.

Fault management, an essential component of any network management system, will play an equally, if not more, crucial role in WSNs.

In the majority of applications, failure detection is vital not only for fault tolerance, but also for security. If, in addition to detecting a failure, one can also determine (or gather indications) that it has malicious origin, the observer can be alerted to an attack.

3.3.4.3 Performance Management

The challenge is to perform this task without adversely consuming network resources. In performance management, a trade-off must be considered: the higher the number of managed parameters, the higher the energy consumption and the lower the network lifetime are. On the other hand, if parameter values are not obtained, it may not be possible to manage the network appropriately.

The configuration (in terms of sensor capabilities, number of sensors, density, node distribution, self-organization, and data dissemination) plays a significant role in determining the performance of the network. Performance management must consider the self-service characteristic. As such, the performance of the network and provided service are best measured in terms of meeting the accuracy and delay requirements of the observer, as well as consumed energy.

The accuracy indicates the reliability or exactness of a result; it can also be defined as the fraction of valid results from all results obtained. The accuracy of a measurement at a network element (sensor) is specific to the physical transducer and the nature of the phenomenon. At the network level, accuracy depends on the delay in data delivery due to network congestion, route length, duty cycle of the sensors, or aggregation processing of data. Accuracy at the service level depends on the metric chosen by the application for establishing the coverage area and amount of energy to be spent in gathering and disseminating data. At the observer, it is likely that multiple samples will be received from different sensor nodes and with different data quality. Thus, additional performance metrics include:

- Coverage area
- Exposure
- Goodput (the ratio of the total number of packets received by the observer to the total number of packets sent by all sensors over a period of time [25])
- Sensor cost
- Scalability
- Produced data quality

In some applications, in addition to information about some features of the phenomenon, it might be necessary to know where (sensor location), when (data–time), and how (sensor calibration, exposure) to manage the WSN performance.

Regardless of the application, certain critical features can determine the efficiency and effectiveness of a sensor network [24]. These features can be categorized into quantitative features and qualitative features. Qualitative features include network settling time; network join time; network depart time; network recovery time; frequency of updates (overhead); memory requirement; and network scalability. Qualitative critical features include knowledge of nodal location; effect of topology changes; adaptation to radio communication environment; power consciousness; single- or multichannel; and preservation of network security.

3.3.4.4 Security Management

Security functionalities for WSNs are difficult to provide because of their ad hoc organization, intermittent connectivity, wireless communication, and resource limitations. A WSN is subject to different safety threats: internal, external, accidental, and malicious. Information or resources can be destroyed; information can be modified, stolen, removed, lost, or disclosed and service can be interrupted. Even if the WSN is secure, the environment can turn it insecure or vulnerable. Security management must provide self-protection, reliability, disposability, privacy, authenticity, and integrity.

Determining if a fault or collection of faults is the result of an intentional DoS attack presents a concern of its own — a point that becomes even more difficult in large-scale deployments, which may have higher nominal failure rates of individual nodes than small networks will. The robustness against physical challenges may prevent some classes of DoS attacks. Each layer of the protocol stack is vulnerable to different DoS attacks and has different options available for its defense.

3.3.4.5 Accounting Management

Accounting management includes functions related to the use of resources and corresponding reports. It establishes metrics and quotes and limits what can be used by functions of other functional areas. These functions can trace the behavior of the network and even make inferences about the behavior of a given node. Accounting management must be considered self-sustaining.

A WSN contains an energy producer (battery) and some energy consumers (transceiver, computation module, and sensor devices). Operations of the application or management can be measured or counted in terms of energy consumption. Given the node characteristics, the average sensor lifetime determines the cost of running a sensor network. One way to reduce total energy consumption is to cut down the number of high-energy operations at the cost of an increase in the number of low-energy operations. The measured cost can be amortized using prediction models [10]. Some functions related to accounting management include: discovery, counting, storing, and data reporting of a parameter; network inventory; determination of communication costs; energy consumption; and traffic checking.

3.4 MANNA as an Integrating Architecture

The MANNA architecture [22] was proposed to provide a management solution to different WSN applications. It provides a separation between both sets of functionalities, i.e., application and management, making integration of organizational, administrative, and maintenance activities possible for this kind of network.

The approach used in the MANNA architecture works with each functional area, as well as each management level, and proposes the new abstraction level of WSN functionalities (configuration, sensing, processing, communication, and maintenance) presented earlier (Figure 3.1). As a result, it provides a list of management services and functions that are independent of the technology adopted.

The MANNA architecture establishes some automatic services, which feature self-managing, self-organizing, self-healing, self-optimizing, self-protecting, self-sustaining, and self-diagnostic, with a minimum of human interference. It is robust to changes in the network state and establishes some services to maintain the quality of the provided services.

3.4.1 Management Services, Functions, and Models

The definition of management service* is a task that consists of finding which activities or functions must be executed, when, and with which data. Management services are executed by a set of functions, and they need to succeed to conclude a given service. Management functions represent the lowest granularity of functional portions of a management service, as perceived by users. The conditions for executing a service or function are obtained from the WSN models.

The WSN models, defined in the MANNA architecture, represent aspects of the network and serve as a reference for the management. These models provide an abstract vision of the system through which is possible to hide all nonrelevant aspects given a certain objective.

Figure 3.3 represents a scheme to construct the management, starting at the definition of management services and functions that use models to achieve their goals. A management service can use one or more management functions. Different services can use common functions that use models to retrieve a

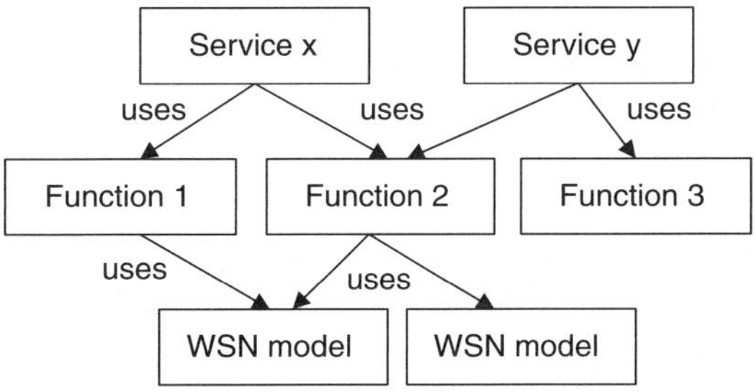

FIGURE 3.3 Services, functions, and WSN models. (From Ruiz, L.B., Nogueira, J.M., Louriero, A.A., *IEEE Commun. Mag.*, 41(2), 116–125, 2003. With permission.)

*Note that the term management service is different from the service management functional area.

network state concerning a given aspect. Therefore, the management functions use and generate management information.

MANNA architecture considers the three management dimensions in the definition of the management functions and in the development of the functional, physical, and information architectures (see Figure 3.1). A partial list of the management functions, in no particular order, follows. The complete list can be obtained from Reference 21.

- Environmental monitoring function
- Monitored area definition function
- Coverage area supervision function
- Node deployment definition function
- Node deployment function [4]
- Environmental requirements acquisition function
- Network operating parameters configuration function
- Topology map discovery function
- Network connectivity discovery function
- Aggregation function
- Data fusion function
- Node density control function
- Priority of action definition function
- Management operation schedule function
- Cooperation discovery function
- Synchronization function
- Energy map generation function
- Network coverage area definition function
- User interface function
- Self-test function
- Node localization discovery function
- Node operating-state control function
- Node administrative-state control function
- Node usage-state control function
- Node mobile function
- Navigation plan function
- Energy-level discovery function

Some functions allow one to obtain characteristics related to the efficiency and effectiveness of a WSN. Some of them are quantitative functions defined to obtain parameters presented by Subbarao [24], such as network settling time function; network join time function; network depart time function; network recovery time function; frequency of updates (overhead) function; memory requirement function; network scalability function; and energy consumption function.

The distributed management MANNA architecture is based on two paradigms: policy-based management and autonomic management. In most of the management applications, the MANNA architecture uses automatic services and functions executed by a management entity invoked as a result of information acquired from a WSN model. This is called self-management. Management functions can also be semiautomatic when executed by an observer assisted by a software system that provides a network model or invoked by a management system. They can be manual when executed outside the management system. Five possible states are defined for a function:

- Ready (when the necessary conditions to execute a function are satisfied)
- Not ready (when the necessary conditions to execute a function are not met)
- Executing (when the function is being executed)
- Done (when the function has a successful execution)
- Failed (when a failure occurs during execution of the function)

Locations for managers and agents, as well as functions that they can execute, are suggested by the functional architecture. The MANNA architecture also proposes two other architectures: physical and information.

The following discussion concerns how the MANNA architecture can cope with different kinds of network and presents the functional, information, and physical architectures.

3.4.2 Functional Architecture

The functional architecture describes the distribution of management functionalities in the network among manager, agent, and management information base (MIB). In the architecture, it is possible to have a diversity of manager and agent locations. The management choice depends on the functional areas involved, the management level considered, and the application running in the WSN, i.e., depends on the network functionalities (Figure 3.1). This architecture introduces the organizational concept of a management "domain," which is an administrative partition of a network for the purpose of network management. Domains may be useful for reasons of scale, security, or administrative autonomy. Each domain may have one or more managers monitoring and controlling agents in that domain. In addition, managers and agents may belong to more than one management domain. Domains allow the construction of strict hierarchical, fully cooperative, and distributed network management systems.

3.4.2.1 WSN Manager

WSN management can be centralized, distributed, or hierarchical. In a centralized management network, a single manager collects information from all agents and controls the entire network. A distributed management network has several managers, each responsible for a subnetwork and communicating with other managers. In a hierarchical management network, intermediate managers distribute the management tasks. The management alternative to be chosen depends on the application running on the WSN. In any solution, it may be important to have a manager entity located externally to the WSN. The external manager has a global vision of the network and can perform complex tasks (automatic services and functions) that would not be possible inside the network. However, this manager can be the only one (centralized management) or it can collaborate with another manager localized inside the network (decentralized management).

3.4.2.2 WSN Agents

The development of a functional architecture raises the question of the most adequate location for an agent, given a particular kind of WSN. A possible alternative to the agent location is to place it close to the manager, i.e., external to the network. However, this may cause isolation of the management and make it difficult to integrate it in the future and to access other management systems.

Next, some possible configurations are explored:

- *Agents in flat and homogeneous WSNs.* A flat WSN has at least one sink node to provide network access. All network nodes have the same hardware configuration. Some possible alternatives for flat and homogeneous networks considering agent location in the WSN are:
  - Agents inside the network and external manager (Figure 3.4a)
  - Agents in the sink node (Figure 3.4b)
  - Agents and manager in the network; the two possibilities for manager organization are hierarchical (Figure 3.4c) and distributed (Figure 3.4d)

  In any of these proposals, the main concern is the large amount of traffic that may be generated in response to operation requests and in sending notifications. Another alternative is to place managers inside the network and allowing them to communicate among themselves. This defines a distributed management. In case of having agents as part of common nodes, some questions remain, such as how to distribute the agents, how to define domains for the agents, and how to deal with nodes with more than one agent.

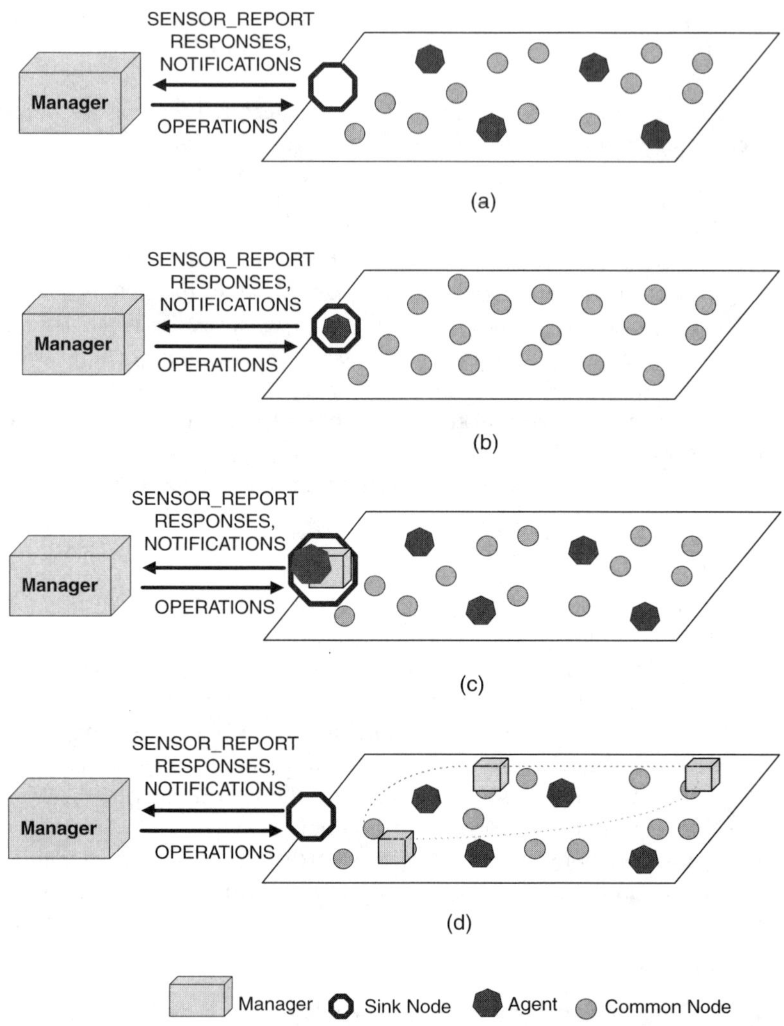

FIGURE 3.4 Manager and agent location in flat WSNs.

- *Agents in flat and heterogeneous WSNs.* In a heterogeneous WSN, nodes differ in their physical hardware capabilities. Agents can be placed in more powerful nodes as long as they present adequate location in the network. The sink node can host an intermediate manager or even present no management function. To establish a distributed management, agents can be placed in less powerful nodes and managers in more powerful ones.
- *Agents in hierarchical homogeneous or heterogeneous WSNs.* In this kind of network, there is no sink node. A cluster-head node is responsible for sending data to a base station. It also communicates with the observer. The cluster head may also execute correlation of management data. This computation may decrease the information flow and thus energy consumption. The correlation may also allow a multiresolution in which differences are filtered and a higher precision is obtained. Some possible alternatives for a hierarchical WSN considering the agent location include:
  - Agents in cluster heads and external manager (Figure 3.5a)
  - Agent in the base station (Figure 3.5b)
  - Agents in the network and intermediate manager (Figure 3.5c)
  - Agents and distributed managers in the network (Figure 3.5d)

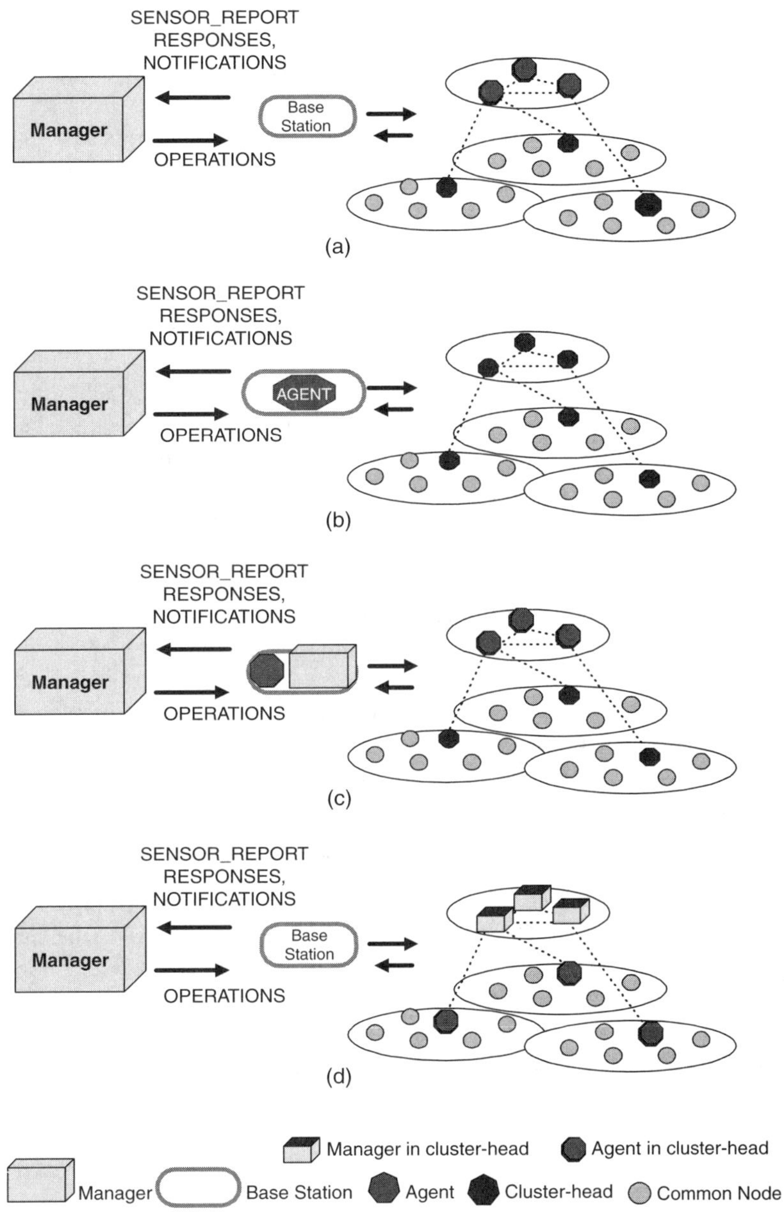

FIGURE 3.5 Agent location in hierarchical WSN.

3.4.2.3 Management Application

In the management architecture (functional, information, and physical), how the management entities receive and analyze information and react to it, which services and functions will be executed, and how the information is exchanged through the communication interface are defined. The type of management (centralized, hierarchical, or distributed) is also defined. Now, the "implosion problem" is explained and management aspects concerning WSN functionalities are addressed.

Centralized management for WSNs, as well as for traditional ad hoc networks, is not always appropriate. One main reason is the traffic concentration problem caused by a central manager that receives and originates management traffic. In addition, the response implosion problem may happen when a high volume of incoming replies is triggered by management operations or events. In case of WSNs, there will always be one access point (sometimes more than one), through which data go to the observer or to the management application. The access point represents a sink node or a base station that can make use of a gateway to communicate with the external environment.

To resolve the implosion problem for management and application, one possibility is to select only a subset of nodes sending data, known as fidelity. In the case of management, some agents are selected to send replies back. This approach may be suitable for densely populated sensor networks with a large number of sensor nodes, in which missing information from some nodes can be ignored with acceptable accuracy. The accuracy of the calculation might significantly degrade. In a sparse sensor network, or a network with a small number of nodes not collecting enough replies, however, the number of replies may not be small enough to be received without taking into account the response implosion problem. Another solution is to make a scheduled response approach [16].

A management solution depends on the features of the network. In some WSNs, only a few management functions can be implemented. In other cases, the management functions must be semiautomatic or manual because of restrictions in the computation. The MANNA architecture is built to provide a management solution to different WSN applications. Depending on the application, it may be interesting or not to use determinate management services, which also can be implemented as automatic, semiautomatic, or manual.

A management solution must also be proposed considering the type of the dissemination: continuous, on demand, programmed, or event driven (see Section 3.3.3.4). In a continuous monitoring scheme, agents are programmed to send monitoring data continuously to a manager. In an on-demand scheme, a manager sends a query to one or more agents, and it receives data back from those agent nodes. In an event-driven monitoring scheme, agents are programmed to send data to a manager only when an event happens and a local condition is satisfied.

Each one of these management solutions has pros and cons. In a continuous monitoring scheme, a management application that stops receiving data from a given node may be an indication of a problem, mainly if the previous sensor condition was normal. The cost of sending data continuously may lead to more rapid consumption of scarce network resources and thus shorten its lifetime. In an on-demand and programmed scheme, the monitoring node can become aware of a problem in the network after sending a query to the node. The cost of having this information is proportional to the number of queries sent or the number of programmed responses. Finally, the design of an event-driven monitoring scheme makes some assumptions about how events are generated. If they happen in an unpredictable way, then, again, there is the problem of consumption of network resources.

On the other extreme, if a node does not report an event, it may be an indication of a failure or of an event that did not occur. In both cases, the management application cannot differentiate them. The same is true for the on-demand network. In normal situations, an event-driven scheme only sends an event to the sink node when it happens. This is the minimum possible cost associated with an event when it must be sent to the management application.

In energy-constrained WSNs, event-driven networks represent an attractive option when compared to continuous networks because they typically send and receive far fewer messages. This translates to a significant energy saving because message transmissions are much more energy intensive when compared to sensing and (CPU) processing.

In terms of failure detection, event-driven networks present challenges not found in continuous and programmed networks. Under normal conditions, a management application of a continuous network receives sensing data at regular intervals. This stream of data not only delivers the content in which one is interested, but also works as an indication of how well the network is operating. If the management application receives data from every single node, then all is well (of course, assuming that the messages are authenticated and cannot be spoofed). If, however, the management application stops receiving data from certain nodes or entire regions of the network, a failure has occurred.

3.4.2.4 Issues Concerning Management Information Base Implementation and Usage

The description of objects present in the information model and the relationship among them are specified in the management information base. In the WSN, to update an MIB with the current network state may require measuring various parameters. In general, the collection of these parameters can have spatial and temporal errors. This is called the "uncertainty problem."

To have a higher precision in the network state, probabilistic measures should be performed with a higher granularity. As in any probing, this would take a finite amount of the system energy and could modify the network state. This is called the "probe effect"; in this way, better precision in management information requires modification of the state.

The MANNA architecture proposes the limitation in the scope as a method for reducing uncertainty and energy consumption while updating the MIB. Spatial limitation consists of defining a physical space inside which the data will be considered for management. Temporal limitation defines a time window (fixed or sliding) inside which the collected data are considered. Functional limitation selects the data of a certain functional network segment for management — for example, the data of a group of nodes or a group leader.

3.4.3 Information Architecture

To ensure common solutions for WSN management, the MANNA architecture defines an information model. WSN management has two kinds of management information: static and dynamic. Static management information describes the configuration of services, network, and network elements. Dynamic management information describes information that changes frequently.

In the MANNA architecture, static management information is based on object orientation and dynamic management information is described by WSN models (see Figure 3.3). From the management point of view, the MANNA functional architecture establishes the circumstances in which a manager will receive event notifications and how it can get its information (monitoring). It also becomes clear what kind of influence the management system has over the WSN resources and how to control them.

3.4.3.1 Static Information

Two types of object classes represent resources under the three different dimensions: managed object and support object. The managed object class directly relates with the network components and with the network. The support object classes play the role of supporting the management functions, i.e., making available to them the necessary information.

The specification of an object class is done through predefined syntactic structures called templates, based on the abstract syntax notation.1 (ASN.1) language, which is used to describe the objects and their characteristics. Object classes may be inherited or reused from standard objects; reuse allows future management integration. Some object classes and their new attributes, based on WSN characteristics, are listed next.

Support object classes. These classes can be programmed by the agent or can be present in the management application. They are mostly derived from the OSI reference model. Some support object classes include:

- Log
- State change record
- Attribute change value record
- Event record
- Event forwarding discriminator
- Management operation schedule
- Information log
- Management log
- Energy level severity assignment profile

- Current remaining energy level summary control
- Monitored object
- Current data object
- History data object
- Threshold data object
- Scanners

Managed object classes. The RFC3433 [3] describes managed objects for extending the entity MIB (RFC 2737) to provide generalized access to information related to physical sensors, which are often found in a networking equipment (such as chassis temperature, fan RPM, and power supply voltage). The RFC 3433 is used and other object classes defined. Some of the defined managed object classes follow:

- *Network* is composed by interconnected managed objects (physical or logical ones) capable of exchanging information. Examples of new attributes for this class include:
  - Network identifier
  - Composition type (homogeneous or heterogeneous)
  - Organization type (flat or hierarchical)
  - Organization period
  - Mobility (stationary, stationary nodes and mobile phenomenon, mobile node or mobile phenomenon)
  - Data delivery (continuous, event driven, on demand, programmed, or hybrid)
  - Type of access point (sink node or base station)
  - Localization type (relative or absolute)
  - Control (open or close)
  - Mission (critical or common)
  - Node distribution (regular, irregular, balanced, sparse or dense)
  - Node deployment (affected by many factors, some of which are the sensor node capabilities of individual nodes, radio propagation characteristics, and the topology of the region)

  Other constraints may include a degree of overlapping in the sensor coverage of two nodes so that they may collaborate.
- *Managed element* represents the sensor node and actuator nodes or other WSN entities that perform functions on managed elements and provide sensing, processing, and communicating services. Examples of new attributes of this class include:
  - Localization (relative or absolute)
  - Element type (common node, sink node, gateway, or cluster head)
  - Minimum energy limit
  - Mobility (direction, orientation, or acceleration)

  The problem is where to place the base station or sink node. Some approaches use a combination of computational geometry, computer-aided design, and numerical optimization methods.
- *Equipment* represents the physical components of a managed element. In this case, this class represents the physical aspects of the sensor node constitution, which is composed of memory, processor, sensor device, battery, and transceiver. The equipment class can be specialized in object classes. For instance,
  - Battery type (linear: the battery is considered to be a bucket of energy; energy is linearly drawn from this bucket by the energy consumers)
    - Discharge rate-dependent model (considers rate at which energy is drawn from the battery to compute the remaining battery life; at high discharge rates, battery capacity is reduced)
    - Relaxation model (takes into account a phenomenon seen in real-life batteries in which the battery's voltage recovers if the discharge rate is decreased)
    - Battery capacity
    - Remaining energy level
    - Energy density

  - Computational module composed by processor and memory (clock; state of use; available memory; endurance; AD channel; operating voltage; IO pins)
  - Sensor element (sensor type; current consumption; voltage range; min–max range; accuracy; temperature dependence; version; state current; exposure)
  - Transceiver (type; modulation type; carrier frequency; operating voltage; current consumption; throughput; receiver sensitivity; transmitter power)
- *System* is used to represent hardware and software, which constitute an autonomous system capable of executing the information processing and/or transference. Examples of new attributes include:
  - Operating system type
  - Version
  - Code length
  - Complexity
  - Total MIPS per available MIPS
  - Synchronization type (mutual exclusion, synchronization of processes)

  A notification of change in an attribute value must be reported upon the event occurrence, such as a software upgrade.
- *Environment* represents the environment in which the WSN is operating. Examples of new attributes include:
  - Environment type (internal, external, and unknown)
  - Noise ratio
  - Atmospheric pressure
  - Temperature
  - Radiation
  - Electromagnetic field
  - Humidity
  - Luminosity

  The environment can present static and dynamic features.
- *Connection* represents the actual connections and is expressed as an association between particular points. The direction of connectivity can be unidirectional (asymmetric) or bidirectional (symmetric). If an instance of this class is unidirectional, the point "a" will be the origin and the terminal point "z" will be the destination. The operational state will indicate the capacity to load a signal. An example of attribute for this class is the communication direction (simplex, half duplex, full duplex). The network topology describes the connections that may exist, and it is expressed as relationships between a set of points.
- *WSN observer* represents the entity that requires WSN services. It may be a human user applying for the use of services via some human–machine communication or it may be some computer-based organizational system.
- *WSN goals* are the benefits provided to users that are obtained by carrying out WSN activities and using WSN services. They can be defined as accuracy, latency, fidelity, etc.
- *WSN management context* defines the environment in which WSN management services are carried out. The definition includes the description of the entity responsible for managing the network, what is managed, and how it can be managed. The WSN management context is described by using three dimensions: management functional areas, management levels, and WSN functionalities.

3.4.3.2 Dynamic Information

In a WSN, network conditions can vary dramatically along the time. In this case, the use of models established by MANNA is of fundamental importance for the management, although its updating cycle can be extremely dynamic and complex. Based on the information obtained with these models, services and functions are executed according to management policies. Dynamic management information is described by WSN models and needs to be obtained frequently. Because acquisition of this information has a cost in terms of energy consumption, an important aspect is to determine the adequate moment, frequency, and fidelity for updating that information. Furthermore, the information collected may not be valid at the moment at which it is processed by the management entity due to delays, omissions, and uncertainty present in WSNs. Static information is needed in order to obtain the WSN models.

In the following, some network models are presented. They always represent dynamic aspects of the network. The dynamic information represented in the network models could or could not be stored in MIBs. Some of the WSN models (map) follow:

- *Network topology map* represents the topology map and the reachability of the network.
- *Residual energy* represents the remaining energy in a node or in a network.
- *Sensing coverage area map* describes the actual sensing coverage map of the sensor elements.
- *Communication coverage area map* describes the present communication coverage map from the range of transceivers.
- *Cost map* represents the cost of energy necessary for maintaining desired performance levels.
- *Production map* represents nodes that are producing.
- *Usage standard map* represents the activity of the network. It can be delimited for a period of time, quantity of data transmitted for each sensor unit, or the number of movements made by the target.
- *Dependence model* represents the functional dependency that exists among the nodes.
- *Structural model* represents aggregation and connectivity relations among network elements.
- *Cooperational model* represents relations of interaction among network entities.

3.4.4 Physical Architecture

The physical architecture defines how management information is exchanged between management entities. It can be seen as the implementation of the functional architecture. In doing so, physical aspects such as the management protocol, physical location of agents, agent functionalities, implemented management service, and supported interfaces for WSNs are defined. The interface among management entities should use a light-weight protocol stack. The MANNA architecture does not define a protocol stack for these interfaces, but provides protocol profiles that may be adequate for each application type.

Application layer. Although the simple network management protocol (SNMP) [28], common management information protocol (CMIP) [13], Web-based management protocol (WBM) [8], and the ad hoc network management protocol (ANMP) [6] allow management in a decentralized and event-oriented way, the structure of managed components is always rather rigid. In these paradigms, management intelligence always resides in the management instance, while the information is generated in the managed instances.

An alternative method would be to delegate management functionalities to the managed systems. A solution for supporting this feature in the implementation of the physical architecture is management by delegation (MbD) [11]. Other alternatives are intelligent agents and mobile agents. In the model of mobile agents, data stay at the local place while the processing task is moved to the data locations. The management functions are performed locally and only the resulting data are sent to the manager. By transmitting the code instead of data, the mobile agent model offers several important benefits:

- Network bandwidth requirements are reduced, which is especially important for real-time applications and when communication uses low-bandwidth wireless channels.
- Agents can migrate to another node when the hosting node is compromised.
- Network scalability is supported.
- Agents can migrate to regions of interest independently of the movement of nodes, if they are mobile.
- Extensibility is supported — that is, mobile agents can be programmed to carry out task-adaptive processes, which extend the capability of the system.

- More stability is achieved because mobile agents can be sent when the network connection is alive and return results when the connection is re-established along with the network data.
- The delay in management actions is reduced.
- Managers are not required to instruct agents all the time.
- The main management part does not reside in the manager.
- Agent cloning offers means for robustness and fault tolerance.

Transport layer. For all protocols described in the application layer, the correct reception of data messages is not assured [27]. Unlike traditional networks (e.g., IP networks), reliable data delivery is still an open research question in the context of WSNs.

Network layer. This should be designed considering power efficiency, and that WSNs are mostly data centric. Data aggregation is useful only when it does not hinder the collaborative effort of sensor nodes. Energy-efficient routes can be found based on the available power in the nodes and the energy required for transmitting data in the link along the route.

Data-link layer. This is responsible for the multiplexing of data streams, data frame transmission and reception, medium access, and error control. Medium access control has two goals: (1) to create the network infrastructure to establish communication links for data transfer and give the sensor network self-organizing ability; and (2) to share communication resources fairly and efficiently between sensor nodes. Simple error control codes with low complexity encoding and decoding might present the best solutions for sensor networks. Open research issues for MAC protocols in WSNs are: determination of low bounds on the energy required for sensor network self-organization; error control coding schemes; and power-saving modes of operation [20].

Physical layer. This is responsible for frequency selection, carrier frequency generation, signal detection, modulation, and data encryption. The 915-MHz ISM band has been widely suggested for sensor networks.

3.5 Putting It All Together

Consider that a management entity has just received the topology and energy messages. It calculates the sensing and communication range area maps and detects the existence of a high node density because there are lots of intersections among the sensing range of the nodes. The management entity faces a redundancy problem of the sensing data received. On one hand, redundancy provides a mechanism for fault tolerance and multiresolution (gives better accuracy), but on the other hand, it represents a waste of resources.

This redundancy problem can be detected by the MANNA architecture using the WSN models, in particular, the "topology map," "energy map," "communication coverage area map," and "sensing coverage area map." Based on these maps, maintenance services may be executed. These services are automatic and executed by a set of functions that use and generate the management information. In this case, one of the functions invoked is the "node administrative state control function."

This function represents the intersection of the three abstraction dimensions for the configuration functional area, network element management level and sensing functionality. The function allows locking the redundant nodes in the administrative state. For this, the agent assigns the value "locked" for the administrative state attribute of the objects (present in the MIB), which represents such nodes acting over the nodes and removing them from sensing, processing, and dissemination services. Figure 3.6 shows a diagram that represents this process.

3.6 Conclusion

Monitoring applications based on wireless sensor networks represent a new important class of applications that can provide data to different kinds of observers. Furthermore, WSNs must deliver the data of interest according to different parameters, such as power efficiency and latency.

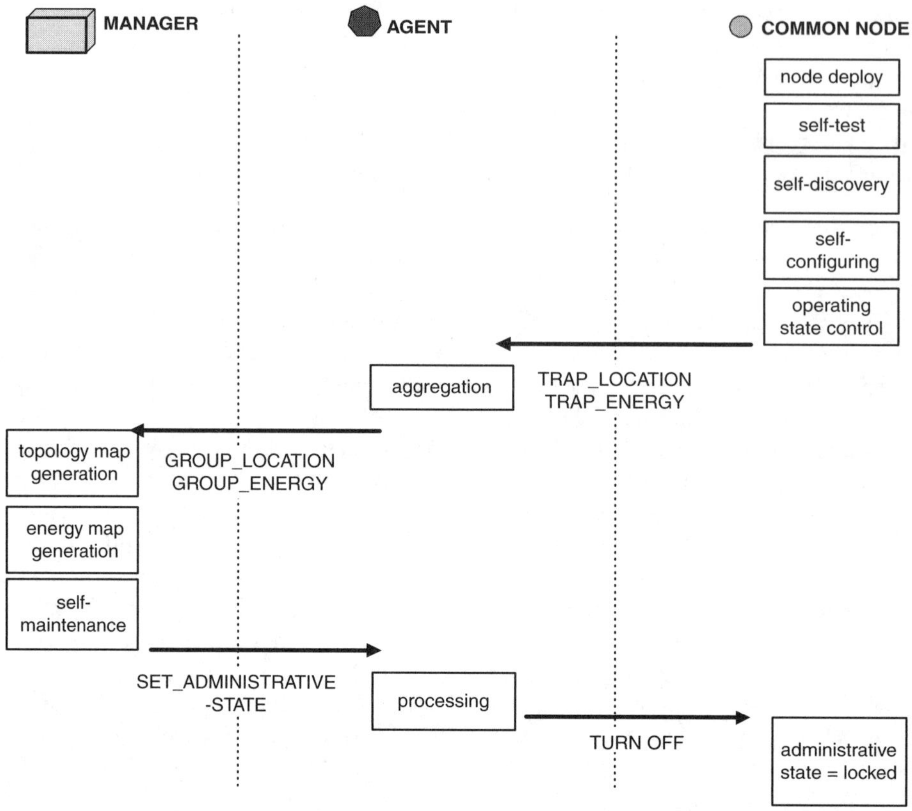

FIGURE 3.6 Applying the MANNA architecture: an example.

Management of WSNs is a new research area that only recently started to receive attention from the research community. This chapter discussed the issue of WSN management and presented autonomic management using the MANNA architecture, which is based on the traditional framework of functional areas and management levels. Adopting this strategy will permit management integration in the future. In the management architecture, the models were built that represent the network state (e.g., WSN topology map, WSN energy map, WSN coverage area map, and WSN production map). These models are important in different applications specified and designed for WSNs.

The fundamental issues about management of WSNs are concerned with how the management application promotes resource productivity and quality of services. Nevertheless, an important aspect is to verify the impact of the management services over the WSN lifetime, latency, goodput, and coverage area.

The important point to be stressed is that, although introduction of management has a cost, this must not affect the network behavior considerably. In fact, the goal is to have the benefits brought by the management solution outweighing the overhead introduced by the management application. Another interesting aspect is that the monitoring scheme to be chosen depends fundamentally on the kind of application monitored. Thus, the management requirements also change among sensor networks.

References

1. Autonomic computing. Available in http://www-3.ibm.com/autonomic/index.shtml.
2. B.R. Badrinath, M. Srivastava, K. Mills, J. Scholtz, and K. Sollins. Special issue on smart spaces and environments. *IEEE Personal Commun.*, 7(5), October 2000.

3. A. Bierman, D. Romascanu, and K.C. Norseth. RFC 3433 on entity sensor management information base (draft-ietf-entmib-sensor-mib-02.txt). Available in ftp://ftp.rfc-editor.org/in-notes/rfc3433.txt.
4. S.B.B. Deb and B. Nath. A topology discovery algorithm for sensor networks with applications to network management. Technical report DCS-TR-441, Department of Computer Science, Rutgers University, May 2002.
5. A. Cerpa, J. Elson, D. Estrin, L. Girod, M. Hamilton, and J. Zhao. Habitat monitoring: application driver for wireless communications technology. In *ACM SIGCOMM Computer Communication Review,* 31(2), 20–41, 2001.
6. W. Chen, N. Jain, and S. Singh. ANMP: ad hoc network management protocol. *IEEE J. Selected Areas Commun.*, 17(8), 1506–1531, August 1999.
7. J. Conover. Policy-based network management. *Network Computing*, November 1999.
8. Distributed Management Task Force (DMTF). Web-based management. Available in http://www.dmtg.org.
9. D. Estrin, R. Govindan, and J. Heidemann. Embedding the Internet. *Commun. ACM*, 43(5), 39–41, May 2000.
10. S. Goel and T. Imieli. Prediction-based monitoring in sensor networks: taking lessons from mpeg. Technical report, Rutgers University, 2001.
11. G. Goldzmidt and Y. Yemini. Distributed management by delegation. *Proc. 15th Int. Conf. Distributed Computing Syst.*, 333–340, June 1995.
12. S.E.-A. Hollar. Cots dust. Master's thesis, University of California, Berkeley, 2000.
13. International Organization for Standardization. ISO/IEC ITU-T X.711 Information Technology — Open System Interconnection — CMIP, specification 1991.
14. International Telecommunication Union (ITU). CCITT recommendation X.700, management framework for open systems interconnection (OSI) for CCITT applications, 1992.
15. International Telecommunication Union (ITU). ITU-T M.3010 — principles for a telecommunications management network, May 1996.
16. D.B. Johnson and D.A. Maltz. Dynamic source routing in ad hoc wireless networks. In Imielinski and Korth, Eds., *Mobile Computing*, Vol. 353. Kluwer Academic Publishers, 1996, 153–181.
17. S. Lindsey, C. Raghavendra, and K. Sivalingam. Data gathering in sensor networks using the energy delay metric. In *Int. Workshop Parallel Distributed Computing: Issues Wireless Networks Mobile Computing*, San Francisco, April 2001.
18. S. Meguerdichian, F. Koushanfar, M. Potkonjak, and M.B. Srivastava. Coverage problems in wireless ad hoc sensor networks. In *INFOCOM*, 1380–1387, 2001.
19. S. Mehrotra. Distributed algorithms for tasking large sensor network. Thesis submitted to the faculty of Virginia Polytechnic Institute and State University, July 2001.
20. National Chiao Tung University Department of Computer, Information Science. Mobile computing and broadband networking lab. Wireless sensor network. Available in http://pds.cs.nctu.edu.tw.
21. L.B. Ruiz, T.R.M. Braga, F. Silva, J.M.S. Nogueira, and A.A.F. Loureiro. Service management for wireless sensor networks. *IEEE LANOMS — Latin Am. Network Operation Manage. Symp.*, 55–62, September 2003.
22. L.B. Ruiz, J.M.S. Nogueira, and A.A.F. Loureiro. MANNA: a management architecture for wireless sensor networks. *IEEE Commun. Mag.*, 41(2), 116–125, Feb. 2003.
23. A. Savvides, S. Park, and M.B. Srivastava. On modeling networks of wireless microsensors. In *Joint Int. Conf. Measurement Modeling Computer Syst.*, 318–319, Cambridge, MA, June 2001.
24. M.W. Subbarao. Ad hoc networking critical features and performance metrics. Technical report, Wireless Communications Technology Group, NIST, September 1999.
25. S. Tilak, N. Abu-Ghazaleh, and W. Heinzelman. A taxonomy of wireless microsensor network models. *ACM Mobile Computing and Commun. Rev.* (MC2R), 6(2), April 2002.

26. M.A. Vieira, L.F. Vieira, L.B. Ruiz, A.A. Loureiro, and A.O. Fernandes. Scheduling nodes in wireless sensor network: a Voronoi approach. *IEEE LCN — Local Computer Network*, October 2003.
27. C.-Y. Wan, A. Campbell, and L. Krishnamurthy. PSFQ: a reliable transport protocol for wireless sensor networks. *WSNA'02*, 1–11, September 2002.
28. W. Stallings. *SNMP, SNMPv2, SNMPv3, ROMON, and ROMON2: Practical Network Management*. Addison-Wesley, Reading, MA, 3rd ed., 1998.
29. K. Wu and J. Harms. QoS support in mobile ad hoc networks. *Crossing Boundaries — GSA J. Univ. Alberta*, 1(1), 92–106, November 2001.

4

Models for Programmability in Sensor Networks

4.1 Introduction 4-1
4.2 Differences between Sensor Networks and Traditional Data Networks 4-2
4.3 Aspects of Efficient Sensor Network Applications 4-2
4.4 Need for Sensor Network Programmability 4-3
4.5 Major Models for System-Level Programmability.......... 4-4
Database Model • Active Sensor Model • Active Networks — Mobile Agents
4.6 Frameworks for System-Level Programmability............ 4-6
Directed Diffusion with In-Network Processing • Cougar • TinyDB • SQTL • Smart Messages — Spatial Programming • Maté • SensorWare • MagnetOS • DFuse
4.7 Conclusions 4-12

Athanassios Boulis
University of California at Los Angeles

4.1 Introduction

Several aspects of the form and operation of sensor networks have been encountered in the previous chapters, as well as strong indications of the great versatility that these systems exhibit and the multiple modes of operations supported in order to achieve their diverse goals. Reading the chapters on several different applications in this book only reinforces the observation that different applications require different distributed algorithms to be handled efficiently.

Having sensor networks with long lifetimes supporting multiple transient users with different needs implies that many different distributed algorithms will run in the network — algorithms that are not known *a priori*. This fact gives rise to the following question: How does one dynamically program the network to provide the users with the needed services efficiently? This chapter examines this problem and the different models proposed by researchers to address it. The discussion begins with some background on the differences of sensor networks with traditional data networks, immediately followed by a section on the general characteristics of efficient sensor network applications. These two sections allow one to motivate the need for dynamic programmability as well as the kind of programmability desired. A description of the different models to achieve such programmability and examples supporting frameworks then follow.

4.2 Differences between Sensor Networks and Traditional Data Networks

Although sensor networks are networks of computing devices, they are considerably different from traditional data networks. The first difference of sensor networks compared to traditional data networks is that they have severe energy, computation, storage, and bandwidth constraints. For example, the wireless sensor node designed by Rockwell Scientific [24] has a 133-MHz, 32-bit, Intel StrongARM 1100 CPU, 1 MB of FLASH memory, 1 MB of RAM, and a 100-Kbps radio, and must operate on two 9-V batteries. This is considered to be toward the high end of sensor network devices. A popular, low-end node design from UC Berkeley, the mica-II [12], uses a 7.37-MHz, 8-bit Atmel CPU with 128 KB of FLASH memory, only 4 KB of RAM, and a 35-Kbps Chipcon radio. The major resource problem in such networks is energy because these are static unattended networks and the nodes cannot have renewable energy sources. Energy is so important that algorithms designed for sensor networks often sacrifice response latency, accuracy, and other user-desired qualities to save energy and prolong the operational lifetime of the network.

The second difference of sensor networks compared to traditional data networks is their overall usage scenario and the implications that this brings to the traffic and interaction with the users. Typically, in traditional networks, users are connected to a node (or group of nodes) and require a service from another node. This two-entity communication model describes the overwhelming majority of traditional network traffic. The network acts as a medium bringing the two parties together. The interaction model is also straightforward; the user interacts directly with the user or service at the other end. Certain actions from the user will produce certain data transfers to and from the other end. The most popular exceptions to these rules are free roaming mobile agents providing data mining or broker services. However, this is a small portion of today's data networks.

Sensor networks, on the other hand, are less like networks (i.e., in the sense that they loosely connect independent entities) and more like distributed systems. As stated earlier, the nodes tightly collaborate to produce information-rich results. The user will rarely be interested in the readings of one or two specific nodes, but will be interested in some parameters of a dynamic physical process. To achieve this efficiently, the nodes must form an application-specific distributed system to provide the user with the answer. This is a departure from the two-entity model: there are no clear sources and destinations based on user desires — only the users and the *whole network*. The nodes involved in the process of providing the user with information are constantly changing as the physical phenomenon is changing. In conclusion, the sensor network is not there to connect different parties together, as in the traditional networking sense, but rather to provide information services to users.

4.3 Aspects of Efficient Sensor Network Applications

The preceding remark leads to the user-interaction topic. Apart from the user input, the physical phenomena now play a central role in the actions inside the network. The actions in each individual node are affected from external physical stimuli and information from other nodes, as well as direct input from the user. Actually, it is desirable to operate in a fashion in which a node's actions are affected largely by physical stimuli detected by the node or nearby nodes. Frequent long trips to the user are undesirable because they consume time and energy. Tennenhouse [27] calls this decentralized (i.e., not all traffic flows to/from user), autonomous (i.e., user is out of the loop most of the time) way of operating "proactive computing" (as opposed to interactive). The term "proactive" is also adopted to denote an autonomous and noninteractive nature. In order for sensor networks to realize their full potential and efficiently use their limited resources, they have to be viewed as distributed proactive systems.

Another efficient design principle is to keep communications localized. Apart from the apparent benefit of saving valuable communication energy, the algorithms can be made more robust by taking advantage of the broadcast nature of the channel combined with the ability to process inputs from all neighbors

— not just selected neighboring nodes. Finally, algorithms can benefit from acknowledging and exploiting the inherent energy–accuracy–latency trade-off present in sensor networks. That is, the more energy one is willing to give, the more accuracy and less latency is achieved, or by keeping the energy consumption constant, one can trade high accuracy for lower latency. Operating in the trade-off space, an algorithm becomes more flexible in accommodating user needs.

Successful applications for sensor networks employ one or more of the preceding design aspects to achieve their goal. Some examples include target tracking algorithms [8, 28]; edge detection algorithms [9, 22]; and periodic aggregation algorithms [4]. Sensor network algorithms' diversity is interesting to those who study them. Some of these algorithms might use common services such as a wake-up protocol [25] or a geographic routing protocol [17], but in essence they are deeply different. From the communication patterns (e.g., cluster based, tree structured, nonhierarchical) to the computation tasks (e.g., custom fusion of sensing data, keeping and processing state of neighbors), these algorithms are as diverse as the problems they tackle. Even in algorithms tackling the same general problem, one can find very different solutions (e.g., edge detection tackled by Chintalapudi and Govidan [9] and by Nowak and Mitra [22]).

Efficiently designed sensor networks are application-specific distributed systems that require a different distributed proactive algorithm as an efficient solution to each different application problem. Given the nature of sensor networks (i.e., diverse solutions for diverse problems), several generic questions come to mind:

- How does one deploy different algorithms into the network?
- What is the programming model that will implement these algorithms?
- What general support does one need from a programming framework?

4.4 Need for Sensor Network Programmability

Researchers who develop sensor network algorithms have shown little concern about how to program them. Most of the time, the proposed algorithms are assumed to be hard-coded into the memory of each node. In some platforms, the application developer can use a node-level OS (e.g., TinyOS [13]) to create the application, which has the advantages of modularity, multitasking, and a hardware abstraction layer. Nevertheless, the developer must still create a single executable image to be downloaded manually into each node. However, it is widely accepted that sensor networks will have long-deployment cycles and serve multiple transient users with dynamic needs. These two features clearly point in the direction of dynamic sensor network programming.

What kind of dynamic programmability is wanted for sensor networks? Hard-coding a few algorithms into each node that are tunable through the transmission of parameters is not flexible enough for the wide variety of possible sensor network applications. An ability to download executable images into the nodes is not feasible because most of the nodes will be physically unreachable or reachable at a very high cost. An ability to use the network in order to transfer the executable images to each and every node is energy inefficient (because of the high communication costs and limited node energy) and cannot allow multiple users to share the sensor network.

Ideally, it is desirable to be able to program the sensor network dynamically as a whole — an aggregate — and not as a mere collection of individual nodes. This means that a user connected to the network at any point will be able to inject instructions into the network to perform a given (probably distributed) task. The instructions will task individual nodes according to user needs, network state, and physical phenomena, *without any intervention from the user*, other than the initial injection. Furthermore, because multiple users should be able to use the sensor network concurrently, several resources/services of the sensor node should be abstracted and made sharable by many users/applications. This kind of programmability is called "system-level programmability." The next section presents the two main models adopted by researchers who try to provide system level programmability.

4.5 Major Models for System-Level Programmability

Before delving into individual research efforts by describing several frameworks and their properties, the two major models for system level programmability will be described: (1) the database model and (2) the active sensor model. Most research efforts fall into one of these models and some frameworks can exhibit characteristics from both.

4.5.1 Database Model

One approach of programming the sensor network as an aggregate is a distributed database system. Multiple users can inject database-like queries to be distributed autonomously into the network. The sensor network is viewed as a distributed database and the query's task is to retrieve the needed information by finding the right nodes and, possibly, to process the data in predefined ways (e.g., aggregate the data) as they are routed back to the user. The strong point of the database approach is that it offers an intuitive way to extract information from a sensor network hiding the complications of *embedded* and *distributed* programming. The user simply describes the information needed. The way in which data are retrieved in nodes and the distributed algorithm needed to retrieve and process the data are not specified. The user "magically" sees the requested information in the use node.

The model's limitation is that only predefined ways to process the data exist, thus implying that only certain types of applications (i.e., applications studied by the specific researchers that are mainly aggregation applications) are addressed in the most efficient way by the database model. If a new way to process and react to the data is needed by application N&U (new and unexplored), this can only be done at the user node (assuming that the human-controlled user node is easily upgradeable). Consequently, the algorithmic pattern to address application N&U under the database model will be an iteration of the generalized steps: (1) partially processed data arriving to the user node; (2) data undergoing custom processing; and (3) based on the result, a new database query issued. In most cases, this is not the structure of the most efficient algorithm to solve an application problem. Recently researchers have tried to augment the language model (e.g., by using event triggers) to accommodate a richer variety of distributed algorithms and provide more flexibility to the user. Nevertheless, the user has no ultimate control over the distributed algorithm executed in the network; this prevents maximum efficiency in certain applications.

The database model is a good solution in the following cases: (1) used in the full-scale network for applications that are well-studied under this model and (2) used in subnetworks with small diameter (e.g., 3 to 4 hops) as a flexible local data retrieval system. For the latter case, imagine a powerful cluster head node with a few less capable nodes around it. The less capable nodes can easily run the framework to interpret and reply to database queries while the cluster head runs a more heavyweight framework (e.g., of the active sensor variety). The cluster head can use the database model to retrieve aggregated data easily from the nodes around it. These data can be further processed by the cluster head and participate in a custom, user-defined distributed algorithm among other cluster heads.

4.5.2 Active Sensor Model

The term coined in Levis and Culler [19] denotes an adaptation of the active networking idea in traditional data networks to the sensor network realm. The difference is that although active networking tasks are reacting only to reception of data packets, active sensor tasks need to react to many types of events, such as network events, sensing events, and timeouts. Active sensor frameworks abstract the run-time environment of the sensor node by installing a virtual machine or a high-level script interpreter at each node. For example, single instructions of the scripts (or bytecodes) can send packets, or read data from the sensing device. Moreover, the scripts (or bytecodes) are made mobile through special instructions, so nodes can autonomously task their peers.

Active sensor frameworks seek to remedy the limited flexibility problem found in the database model at the expense of increased responsibility for the programmer. They provide a language model powerful enough to implement any distributed algorithm while at the same time hiding unnecessary low-level details from the application programmer. Many of the frameworks also provide a way to share the resources of a node among many applications and users that might concurrently use the sensor network. The control of the distributed algorithm (which implies efficiency in any application) comes at a cost compared to the database model. The programmer must explore, define, and test the distributed algorithm for each application.

The difficulty in designing an active sensor framework lies in determining how to define the abstraction of the run-time environment properly so that one achieves compactness of code, sharing of resources for multiuser support, and portability in many platforms, while at the same time keeping a low overhead in delays and energy. Two major choices determine the run-time abstraction:

- Choice of virtual machine (interpreting machine-level bytecodes usually based around a stack architecture) or script interpreter (interpreting high-level ASCII scripts)
- Choice for number and content of native services provided

These choices affect ease of programming, mobile code compactness, time it takes to execute a task, and the memory footprint required in the sensor nodes to accommodate the framework. For example, the more services provided, the more compact the mobile code becomes but the greater the memory footprint becomes. Also, by providing more native services, the execution time of a task is reduced because it is not necessary to rely on interpreted code to implement these parts of the task. Choosing a virtual machine usually requires less memory footprint, but creates less compact code when compared to a high-level scripting language. Given the conflicting nature of the preceding "performance" criteria, it is clear that no one optimal design point exists; rather, the optimality is determined by specific implementation goals. Some of the frameworks discussed in Section 4.6, for example, make some different choices because they target different hardware platforms.

The process of populating the sensor network with viral pieces of code as the active sensor model dictates resembles the operation of multiple collaborating mobile agents, replicating/migrating to the nodes at which the distributed algorithm should be executed. For this reason, the next subsection offers a general discussion on mobile agent (MA) frameworks.

4.5.3 Active Networks — Mobile Agents

Traditional distributed applications are designed as a set of processes (mostly network unaware) cooperating within assigned execution environments. MA technology, however, promotes the design of applications made up of network-aware entities that can change their execution environment by transferring while executing. In recent years, several research groups have created mobile systems based around the notion of an agent that consists of procedures and state data that can migrate from machine to machine. Some of these, such as Agent Tcl [10], have been built on top of interpreted scripting languages; others, such as Aglets, have relied on Java, which provides code mobility via applets and object serialization. The interest in this area is propelled by the advantages agents offer in Internet applications. The advantages fall into three different categories, as reported by Cabri et al. [7], among others:

- Bandwidth and delay savings because computation is moved to the data
- Flexibility because agents do not require the availability of specific code
- Suitability for mobile computing because agents do not require continuous network connections

Thus, when considering MAs, one overwhelmingly sees them in an Internet-application environment with the possibility of mobile endpoints. Consequently, mobile agents are viewed as free-roaming entities that are mostly autonomous with no point of control and should perform well under intermittent connections and mobility. The major design issue in such systems is how the agents communicate and collaborate. Basically, four coordination models classify mobile agents in their current Internet-motivated world:

- *Client/server model.* Direct connection is that of involved agents; the main advantage is the low overhead in delay and implementation. The main disadvantage is that agents are spatially and temporally coupled.
- *Meeting-oriented model.* Agents interact by opening and joining abstract meeting points. The model achieves spatial uncoupling but preserves temporal coupling.
- *Blackboard-based model.* The agents interact by leaving messages in predefined blackboards. Temporal uncoupling is achieved, but some weak spatial coupling still exists because the agents must know each other's names.
- *Linda-like model.* The blackboard is extended by introducing associative mechanisms into the shared data space, thus making the messages' content addressable. Spatial and temporal uncoupling is achieved.

Clearly, the advantages and disadvantages coupled with these models revolve around the notion of the agent's spatial and temporal coupling with its peers or lack thereof. This is understandable, if one remembers the previous discussion on mostly autonomous agents with intermittent network connections. Spatial and temporal uncoupling is desirable, even at the cost of more complex (thus less secure and less efficient) designs.

In the realm of sensor networks, however, these concerns and classifications are becoming irrelevant. The concern is mainly with building reconfigurable and distributed applications that can be reconfigured and relocated. The pieces of mobile code in active sensor frameworks (i.e., the equivalent of mobile agents) are envisioned to perform very tight collaboration with each other, thus departing from the autonomous agent model. In addition, this kind of collaboration will happen among locally clustered nodes, making the peer-to-peer direct communication easier. Furthermore, intermittent connections and mobility are not issues that the framework should hide, but instead should let the algorithm deal with them in an application-specific manner. Remember that efficiently designed applications in sensor networks do not rely on data from specific nodes; rather, they can handle inputs from a greatly varying set of nodes. If data are not available from certain nodes due to intermittent connections or mobility, the application simply keeps on working. For these reasons, the server/client model or the more general peer-to-peer direct communication model is an acceptable choice.

In conclusion, the MA paradigm is associated with the notion of a single agent migrating from node to node, performing part of a given task in each node while sparsely communicating with *specific* remote services or other MAs. The active sensor model, on the other hand, is associated with multiple simple lightweight agents that tightly collaborate to implement a distributed algorithm; their behavior and position is influenced by physical events as well as by user needs. Most of the time, the communication is not tied to specific nodes but rather to a statistically chosen set of nodes.

4.6 Frameworks for System-Level Programmability

This section looks into individual research efforts, beginning with database model frameworks. It continues with active sensor frameworks and concludes with a framework that mixes both notions.

4.6.1 Directed Diffusion with In-Network Processing

Early sensor network research has shown the benefits of attribute-based naming (e.g., geographical information) and routing in the operation of sensor network applications. Directed diffusion [15] was the first protocol to implement such ideas. Heidemann et al. [11] incorporate data-driven, low-level naming with directed diffusion, along with in-network processing ideas, to task the sensor network. The in-network processing is limited to aggregation filters that take n stream input data and produce m stream output data. The application programmer can use simple APIs to use the directed diffusion and custom filtering mechanisms. More specifically, the commands *subscribe*, *unsubscribe*, *publish*, *unpublish*, and *send* implement the publish/subscribe mechanism of directed diffusion, while the commands *addFilter*, *removeFilter, sendMessage,* and *sendMessageToNext* register and utilize custom filters for in-network processing. The initial implementation of the system does not contain a method to upload filters dynamically to the nodes. Although the authors do not explicitly categorize their work in the database model, one can see most of its main notions.

4.6.2 Cougar

Other systems, such as Cougar [2], focus more on transferring the sensor querying language (SQL) semantics of traditional databases to the distributed setting of sensor networks. In this case, the naming system developed in Heidemann et al. [11] is replaced by an SQL equivalent. Each node is equipped with a fixed database query resolver. As queries arrive at a node, the local resolver decides on the best distributed plan to execute the query and distributes the query to the appropriate nodes.

4.6.3 TinyDB

The more recent and probably more advanced system that follows the database model is the TinyDB [21] developed in Berkeley. The developers' main focus is aggregate queries (e.g., min, max, average); thus, they provide special optimizations for them (e.g., exploit the shared medium, perform what they call "hypothesis testing"). A query has the following general form:

```
SELECT expr1, expr2 ...
FROM sensors
WHERE pred1 [AND | OR] pred2 ...
GROUP BY groupexpr1, groupexpr2 ...
SAMPLE PERIOD t
```

The select clause lists the attributes or aggregates of attributes to retrieve from the sensors. Aggregates and nonaggregates cannot appear in the same select clause unless the nonaggregate fields appear in the "group by" clause. "Sensors" is the standard table containing one attribute for each type of sensor existing in the network. It is the common table on which queries are computed on the "where" clause, which filters out readings that do not satisfy the Boolean expression of predicates. The group clause is used in conjunction with aggregate expressions to specify a partitioning of readings before aggregation. For example, one might query:

```
SELECT buildingID, AVG(temp)
GROUP BY buildingID
```

to collect the average temperature from each building, instead of the average temperature over all sensor readings. Finally, the "sample period" clause specifies the time between reevaluation of the query with freshly sampled data.

TinyDB has recently added new language features to provide more flexibility to the programmers. To move beyond passive querying, clauses were added to spawn queries autonomously based on predefined events and also to create internal storage points in the network. Even with these additions, though, the declarative nature of TinyDB remains. The programmer has no ultimate control over the distributed algorithm executed in the network because its details are taken care of by the underlying TinyDB system.

4.6.4 SQTL

Jaikaeo et al. [16] developed the sensor querying and tasking language (SQTL). Starting from a database-like system, the researchers realized the limitations of a declarative language to the implementation of arbitrary distributed algorithms into the sensor network. Thus, they augmented their initial language with imperative style commands to help task the network.

SQTL fits in a more general architecture for sensor networks called sensor information networking architecture (SINA) [26], which uses SQL-like queries as well as SQTL programs. Some of its main features include: (1) hierarchical clustering; (2) attribute-based naming; and (3) a spreadsheet paradigm for organizing sensor data in the nodes. SQL-like queries use these three features to execute simple querying and monitoring tasks. When a more advanced operation is needed, SQTL plays the essential role by programming the sensor nodes and allowing proactive population of the program. In SINA, SQTL is used as an enhancement of simple SQL-like queries; thus, the framework still revolves around a database-like model.

4.6.5 Smart Messages — Spatial Programming

The Rutgers researchers have developed a mobile code platform for embedded systems called smart messages (SM) [3]. They used SM to develop their suggestion for a programmable sensor network framework, which they call spatial programming (SP) [14]. First, the characteristics of SMs will be presented and then the SP model will be discussed.

SMs are entities that carry code, data, and execution state (in order to resume execution from the same point upon migration of the SM). The code is written in Java language supporting a few extra commands relevant to the SM environment. The run-time environment consists of a KVM (Sun's Java virtual machine for embedded devices) modified to support the new commands. Apart from the mobile code entities (the smart messages), the SM environment also supports the abstraction of tags, which are essentially SM-persistent storage and are used as universal names. From naming underlying devices and OS services to naming nodes or application ports for specific data, tags do not have a specific structure. Tags can be used to access the sensor data, name the node, or leave next-hop information behind from a previously executed routing protocol.

The run-time environment also includes a manager for the tag space (essentially a name-based memory). The basic execution model of SMs is that one main agent for an application does the job by hopping from node to node, doing some portion of the work each time. Other agents (i.e., SM) perform supporting functions (e.g., routing). The new commands added to the basic Java language to create the extension of SMs are:

- Four commands to create, delete, read, and write tags
- One command to create a new SM or replicate yourself
- One command to block on a tag (used for synchronization)
- Two commands to migrate (to next hop or arbitrary)

The block command can block only on one tag thus allowing a program to wait only on a single event. Furthermore, only one smart message executes at each moment. If another is to be executed, the current active one must block or complete execution.

Based on the SM platform, researchers from Rutgers introduced a programming model for a network embedded system (a term that includes sensor networks) called spatial programming. SP is more a resource-based routing scheme than a programming model. The SM platform is augmented with a way to refer to nodes by spatial and arbitrary content properties of the node. The abstraction of spatial reference (SR) is introduced, which has the form "space:content_tag." Simple operations are defined on the space portion of an SR. For instance, one can take the difference of two spaces simply by writing space1-space2. Space can also be created with the use of the "rangeof" function, which receives a point and a radius as arguments. An SR can refer to multiple nodes (as it covers a certain space). One can reference individual nodes within an SR by using the "[i]" indexing convention. Another key point is the reference consistency; once an SR is created (and thus some nodes are referred with that name) SR_name[i] is always the same node.

Resources in nodes (e.g., sensor modules, software services) are accessed as variable names, which can be written and read. The names do not follow a particular structure so the applications must know in advance the custom way to access them. A weak point of the SP architecture concerns resource sharing, which is absent from the system; the applications must explicitly negotiate any sharing. Obviously, this method is error prone and at times impossible to follow because applications will not always have knowledge of each other. Finally, questions are posed concerning the actual programming model in SP. How is the code distributed in the network? How is collaborative operation between agents facilitated? The examples developed by the researchers to illustrate their framework present centralized applications (executing only at one node) that access resources remotely, much like RPC calls. This kind of execution is not the most desirable one, as was discussed in the first section of this chapter.

4.6.6 Maté

An active sensor framework for sensor networks called Maté is currently being developed in Berkeley [19]. Maté is a tiny virtual machine built on top of TinyOS [13]. TinyOS is an operating system, designed specifically for the Berkeley-designed family of sensor nodes, generically named "motes" [12]. Maté's goal is to make a sensor network composed of motes dynamically programmable in an efficient manner. This includes the capability to dynamically instruct a mote to execute any program, as well as expressing this program in a concise way. This is achieved by building a virtual machine (VM) for the motes. The VM supports a very simple, assembly-like language to be used for all needs of mote tasking. Programs (called capsules) written on the VM language can be injected to any node and perform a task. Furthermore, the capsules have the ability to self-transfer by using special language commands. This model seems extremely similar to the author's in SensorWare. Indeed, Maté shares the same goals as other active sensor frameworks, as well as the same basic principles to achieve these goals. However, as discussed in Section 4.5, design choices differentiate active sensor frameworks.

Maté, like its substrate TinyOS, was built with a specific platform in mind: the extremely resource-limited mote. The main restriction for the developer of mote-targeted frameworks (such as an OS or a VM) is memory. The newest version of a mote, called mica, offers 128 Kbytes of program memory and 4 Kbytes of RAM. An older version called rene2 has 16 Kbytes of program memory and 1 Kbyte of RAM. With an ingenious architecture, Maté supports both platforms. Because it is so constrained by memory, Maté must sacrifice some features that would make programming easier and more efficient.

First, a stack-based architecture with an ultracompact instruction set (all instructions are 1 byte) reminiscent of a low-level assembly language or the byte code of the Java VM is adopted. This kind of model makes programming of even medium-sized tasks difficult. Furthermore, due to the ultracompact instruction set, many 1-byte instructions are needed to express a medium complexity algorithm, leading in turn to large programs, compared to a higher-level, more abstracted scripting language. The size of programs is important because the code is transmitted/received using the radios of the nodes spending energy for every transmitted/received bit. Second, the behavior of a program when radio packets are received is rather rigid. A handler to process such events is essentially stateless in Maté. Thus, if a new pattern of packet processing is needed, a new handler must be transferred through the network. This imposes an overhead in energy consumption and execution time. Third, because there is only one context (i.e., handler) per event (e.g., clock tick, reception of packet), multiple applications cannot run concurrently in one mote.

Other active sensor frameworks that target richer platforms (e.g., Rockwell Scientific's node [24] includes a 1-Mbyte of program memory and 128 Kbytes of RAM) have the luxury of providing much richer native services to support easy programming with a high-level scripting language, as well as concurrent multitasking of a node so that multiple applications can concurrently execute in a sensor network. One such framework is present in the next subsection.

4.6.7 SensorWare

SensorWare [5, 6] is another active sensor framework developed at UCLA. This framework uses a high-level scripting abstraction based around Tcl [23] and a highly expandable run-time environment. The run-time environment provides multiple services that achieve the sharing of the sensor node's resources among multiple applications. The programming model is event based with event handlers to react to various high-level, application-specific events that occur during a period of interest. The expandability in SensorWare is achieved through the abstraction of virtual devices.

Almost everything in SensorWare is a device (e.g., sensor modules, localization procedure, routing protocols, neighborhood discovery). All devices have a unified interface to interact with them. More specifically, the programmer can act on the device, query the device, describe and name an event the device can produce, and dispose a previously defined event name. The programmer can use the wait command to wait on any of the previously described events. The scripts are made mobile through special commands and data can be carried with the scripts in the form of parameters passed by value. SensorWare has many features to enhance efficiency, flexibility, and ease of programming, the most important of which are:

- Custom script compression based on semantic information
- Script cashing and selective script population
- Addressing tied with routing
- Ability to register scripts as dynamic devices for seamless script coordination

A small code sample of SensorWare scripts follows.

file1:

#code_id 32 small code used as a parameter to other scripts

```
send neighbor $parent "here is your packet"
```

file2:

#code_id 33 this script is an example

```
parameter total_time small_code
set neighbors_num [llength [query neighbor]]
```

#spawn to all neighbors small_code

```
spawn neighbor 0 $small_code
interest timer t1 $total_time
set index 0
while {index<neighbors_num} {
  wait packet t1
  case {$event_name} {
    packet {
      debug "received packet: $event_body"
      }
    t1{
      debug "not all neighbors replied"
      exit.
      }
  }
  incr index
}
```

To invoke the example, do the following from a terminal (user node):

```
load small_code file1
load example_code file2
carry 5000 $small_code
spawn neighbor [id-n] $example_code
```

The preceding invocation commands simply load the code from the two files into Tcl variables, set the parameters passed to the code of the spawn command, and spawn the code in file2 in the current node. The code of file2 gets the parameters and assigns them to local names, finds out the number of neighbors by querying the neighbor device, and then spawns the small_code (which was passed as a parameter) into all its neighbors. The small code simply sends a message back to the current node. Back in the code of file2 one waits for a packet received or the timer named t1 to expire. According to which event is taking place, different messages are output. SensorWare has been used to implement complex applications such as the distributed estimation algorithm described in Boulis et al. [4] among others.

4.6.8 MagnetOS

MagnetOS [1] was developed at Cornell University and, although it is classified as an operating system for networked embedded systems, it can be seen as a method to program a sensor network dynamically. MagnetOS' key idea is a single system image. The entire network is seen as a unified Java virtual machine by the applications. The system consists of a static and a dynamic component. The static component is a partitioning service that partitions regular Java applications into objects that can be distributed into the network. The dynamic part in each node then provides services for application monitoring, object creation, and migration.

The programmer should write normal Java applications, oblivious of the distributed nature of the execution environment; MagnetOS will take care of partitioning and distribution of the application. The application is partitioned according to the objects that the programmer has defined. Thus, an object becomes a mobile application component. The objects are gradually distributed in the network following automatic object migration policies. In MagnetOS, two algorithms perform the automatic object migration: NetPull and NetCenter. NetPull watches communication at the one-hop neighborhood level and migrates components toward links with the greatest communication. NetCenter performs the same monitoring at the network level and can migrate a component several hops at a time.

Apart from the inefficiencies that the automatic code migration can create (e.g., slow convergence to a satisfactory distribution, oscillations of component placement), MagnetOS has the major drawback of completely hiding the distributed nature of the application. Despite the claim that the application can be defined with a single image in mind, the choice of object definition can greatly affect the efficiency of the distributed application because the number and type of object directly affects the partitioning of the application. The complete elimination of the distributed nature of an application from the mind of the programmer is an exciting goal, but very distant or even unattainable for a sufficiently diverse set of applications.

4.6.9 DFuse

Kumar et al. [18] aspire to generalize and facilitate the data fusion process (termed "aggregation" by other researchers) by providing a framework called DFuse. The framework consists of an API to define arbitrary fusion processing and an algorithm for automatic fusion point placement and relocation. The API allows the fusion application to be specified as a directed dataflow graph along with the definition of the fusion functions. The API hides many programming details common to fusion applications, such as buffer management, time stamping, and exception mechanism for error control. Furthermore, using the automatic placement algorithm considerably eases the deployment of such an application. The algorithm decides where the fusion points should be placed and periodically re-evaluates the placement. DFuse is evaluated in its current implementation of iPAQs + Linux + Stampede (a distributed programming system) by measuring the delay of the API's basic commands and by measuring the ability of the placement algorithm to optimize the fusion process.

DFuse seems successful because it restricts itself to a certain type of application without making overstatements on its general application. Certainly the restrictions on the dataflow graphs that the programmer can define (i.e., sources and sinks of the fusion computation are fixed) limit the type of applications that can benefit from DFuse; nevertheless, the framework presents an interesting combination of the database and active sensor models. The arbitrary definition of fusion algorithms brings an element of the imperative active sensor model, while the definition of dataflow graphs and the automatic placement of fusion points bring an element of the declarative database model.

4.7 Conclusions

Issues that concern sensor network programmability along with the major two models for dynamic system level programmability in sensor networks have been discussed. From the individual frameworks examined one can conclude that, when efficiency is the major concern in a large and diverse set of applications, the imperative active sensor model with explicit acknowledgment of the distributed nature of the applications is the solution. On the other hand, when ease of programming in a limited set of applications (e.g., aggregation) is the major concern, the declarative database model is the solution. The research community is currently moving toward a macroprogramming vision for dynamically programming the sensor network. This vision combines elements from both existing models. It will use an active sensor framework as an underlying mechanism to execute arbitrary complex distributed algorithms into the network and a declarative framework that will enable the automatic creation of these algorithms based on well-studied run-time primitives. The declarative part can include database-like queries or dataflow graphs to make the programming task easier. Such elements have already been seen in the DFuse framework for a restricted number of applications, but the generalized large-scale implementation of a macroprogramming framework is still far from realization.

References

1. Barr R. et al., On the need for system-level support for ad hoc and sensor networks, *Operating Syst. Rev., ACM*, 36(2):1–5, April 2002.
2. Bonnet P., Gehrke J., and Seshadri P., Querying the physical world, *IEEE Personal Commun.*, 7, 10–15, October 2000.
3. Borcea C. et al., Cooperative computing for distributed embedded systems, *Proc. 22nd Int. Conf. Distributed Computing Syst. (ICDCS)*, July 2002.
4. Boulis A., Ganeriwal S., and Srivastava M., Aggregation in sensor networks: an energy accuracy trade-off, *First IEEE Int. Workshop Sensor Network Protocols Applications* (SNPA 2003), Anchorage, AK, May 11, 2003.
5. Boulis A., Han C., and Srivastava M., Design and implementation of a framework for efficient and programmable sensor networks, in *Proc. MobiSys 2003*, San Francisco, CA, May 6–8 2003.
6. Boulis A. and Srivastava M.B., A framework for efficient and programmable sensor networks, in *Proc. OPENARCH 2002*, New York, June 2002.
7. Cabri G., Leonardi L., and Zamponelli F., MARS: a programmable coordination architecture for Mobile agents, *IEEE Internet Computing*, 4(4), 26–35, July–August 2000.
8. Chen J.C., Yao K., and Hudson R.E., Source localization and beamforming, *IEEE Signal Process. Mag.*, 19, 30–39, March 2002.
9. Chintalapudi K.K. and Govidan R., Localized edge detection in sensor fields, *First IEEE Int. Workshop Sensor Network Protocols Applications* (SNPA 2003), Anchorage, AK, May 11, 2003.
10. Gray R.S., Agent Tcl: a flexible and secure mobile-agent system, *Proc. 4th Annu. Tcl/Tk Workshop '96*, Monterey, CA, 10–13 July 1996.
11. Heidemann J. et al., Building efficient wireless sensor networks with low-level naming, *Proc. Symp. Operating Syst. Principles*, 146–159, October 2001.
12. Hill J. and Culler D., A wireless embedded sensor architecture for system-level optimization, Intel Research IRB-TR-02-00N, 2002.
13. Hill J. et al., System architecture directions for networked sensors, *Proc. ASPLOS-IX*, 93–104, Cambridge, MA, November 2000.
14. Iftode L. et al., Programming computers embedded in the physical world, *Proc. 9th IEEE Workshop Future Trends Distributed Computing Syst.* (FTDCS 2003), May 2003.
15. Intanagonwiwat C., Govindan R., and Estrin D., Directed diffusion: a scalable and robust communication paradigm for sensor networks, *Proc. 6th Ann. Intl. Conf. Mobile Computing and Networking, MobiCOM '00*, 56–67, Boston, MA, August 2000.

16. Jaikaeo C., Srisathapornphat C., and Shen C., Querying and tasking of sensor networks, *SPIE's 14th Annu. Int. Symp. Aerospace/Defense Sensing, Simulation, Control (Digitization of the Battlespace V)*, Orlando, Florida, April 26–27, 2000.
17. Karp B. and Kung H.T., GPSR: greedy perimeter stateless touting for wireless networks, in *Proc. 6th Ann. Intl. Conf.* Mobile Computing Networking, MobiCom, 243–254, Boston, MA, 2000.
18. Kumar R. et al., DFuse: a framework for distributed data fusion, in *Proc. ACM SenSys* 2003, Los Angeles, CA, Nov. 5–7, 2003.
19. Levis P. and Culler D., Maté: a tiny virtual machine for sensor networks, *Proc. 10th Int. Conf. Architectural Support Programming Languages Operating Syst.* (ASPLOS X), 85–95, October 5–9 2002.
20. Madden S.R. et al., TAG: a tiny aggregation service for ad-hoc sensor networks, *OSDI Conference*, 2002.
21. Madden S.R. et al., Supporting aggregate queries over ad-hoc wireless sensor networks, *Workshop on Mobile Computing and Systems Applications*, 2002.
22. Nowak R. and Mitra U., Boundary estimation in sensor networks: theory and methods, *2nd Int. Workshop Inform. Process. Sensor Networks*, Palo Alto, CA, April 22–23, 2003.
23. Ousterhout J.K., *TCL and the TK toolkit*, Addison-Wesley, Boston, 1994.
24. Rockwell WINS nodes, http://wins.rsc.rockwell.com/.
25. Schurgers C. et al., Optimizing sensor networks in the energy–latency–density design space, *IEEE Trans. Mobile Computing*, 1(1), 70–80, January–March 2002.
26. Srisathapornphat C., Jaikaeo C., and Shen C., Sensor information networking architecture, *Int. Workshop Pervasive Computing (IWPC'00)*, Toronto, Canada, August 21–24, 2000.
27. Tennenhouse D., Proactive computing, *Commun. ACM*, 43(5), 43–50, May 2000.
28. Zhao F., Shin J., and Reich J., Information-driven dynamic sensor collaboration for tracking applications, *IEEE Signal Process. Mag.*, March, 61–72, 2002.

5

Miniaturizing Sensor Networks with MEMS

5.1 Introduction 5-1
5.2 MEMS Basics 5-2
Micromachine Fabrication Techniques • Highly Integrated Processes
5.3 Sensors 5-4
Selection Criteria • Integrated Circuit Sensors • Nanosensors
5.4 Communication 5-5
RF Communication • Optical Communication
5.5 Micropower Sources 5-10
Energy Storage • Energy Harvesting
5.6 Packaging 5-12
5.7 Systems 5-13
5.8 Conclusion 5-15

Brett Warneke
Dust Networks

5.1 Introduction

As sensor network nodes decrease in size, denser networks can be deployed and entirely new sensor network applications will be enabled. Furthermore, smaller, lighter nodes will facilitate more network deployment methods, such as microaerial vehicles (MAV) and even air-borne dispersal. An additional side effect of miniaturization techniques based on semiconductor batch fabrication is that the manufacturing cost of the sensor nodes can be reduced for large quantities, which will allow for denser and more extensive sensor networks. These factors of discrete size and large, dense networks will enable new methods of interacting with the environment and provide more information from more places in a less intrusive way than before. Application areas enabled by miniaturized sensor nodes are numerous and include defense and intelligence networks; tracking the movements of birds, small animals, and even insects; fingertip accelerometer virtual keyboards; and interfaces for the disabled.

Sensor nodes can be divided into four major components: sensors; communication; power source; and circuits for computation, data storage, and sensor signal processing. The volume of the sensor node circuits is being reduced through dramatic process scaling and greater integration of mixed signal functions into a single chip. Microelectromechanical systems (MEMS) are similarly reducing the size and cost of sensors, some communications components, and power supplies while also reducing the power consumption of the former two. Furthermore, MEMS techniques can reduce packaging size and facilitate tighter integration.

5.2 MEMS Basics

MEMS is based on microfabrication techniques developed for microelectronics. By extending these processes, micromachining techniques have been developed to fabricate micron-scale mechanical features that are often controlled or sensed electrically, forming microelectromechanical systems. Through highly integrated processes, these electromechanical components can be fabricated alongside microelectronics, yielding complex systems.

In order to provide some background on MEMS, several of the fundamental micromachining processes will first be described, followed by the highly integrated processes that are more advantageous for miniaturizing sensor network nodes. This chapter will deal only with micromachining processes based on semiconductor microfabrication techniques because they have more promise of inexpensive batch fabrication and are more easily integrated with microelectronics for a small system size. For more information, Pierret [1] provides a good introduction to microfabrication technologies; Petersen [2] has produced the seminal paper on micromachining; Muller et al. [3] and Trimmer [4] provide collections of classic papers in the field; and references 5 through 11 are reference textbooks on micromachining and MEMS.

5.2.1 Micromachine Fabrication Techniques

Most micromachining processes begin with a substrate 100 to 600 μm thick, usually composed of silicon, other crystalline semiconductors, or quartz. Upon this substrate a number of process steps are performed, such as thin film deposition; photolithography; etching; oxidation; electroplating; machining; and wafer bonding. One of the key concepts of planar micromachining is that of sacrificial and structural layers — the former refers to thin films that are etched away to allow structures patterned in the structural layers to move. Common elemental structures include cantilevers, membranes, and plates suspended on thin or narrow flexural beams.

Bulk micromachining involves removing relatively large portions of the substrate, including the entire thickness, typically with a silicon etchant such as ethylene diamine pyrochatechol (EDP); tetramethylammonium hydroxide (TMAH); sublimated XeF_2, HNA (HF + HNO_3 + acetic acid); SF_6 plasma; or deep reactive ion etch (DRIE). A simple bulk process would involve first depositing a masking material such as SiO_2, photolithographically patterning it, and then placing the wafer in the silicon etchant for a specific period of time. If the etchant etches laterally as well as vertically (such as an isotropic etchant), the mask material will be undercut, potentially releasing structures such as cantilevers. Bulk micromachining often results in structures that move vertically. Figure 5.1 illustrates this process, except with masking layers resulting from a CMOS process.

Surface micromachining consists of depositing and patterning a series of sacrificial and structural layers on top of the wafer, followed by a final release step that etches away the sacrificial layers. A basic process would start with a silicon wafer, deposit 1 μm of SiO_2, and pattern it to form places for the structural layer to be attached to the substrate. Next, 2 μm of low-stress polysilicon would be deposited and patterned to form the microstructures. In the final step, an HF etch would remove the SiO_2 and release the structures. Surface micromachining usually produces structures that move laterally, but vertical motion is also possible.

A third style of micromachining, which combines the deep etches of bulk micromachining yet yields structures more similar to surface micromachining, begins with silicon-on-insulator (SOI) wafers. These wafers contain a several-micron thick "buried oxide" that isolates a relatively thin silicon device layer from the bulk substrate. The device layer, which is where the transistors or microstructures are formed, is usually only a couple of microns thick for CMOS wafers, but for MEMS processes it can be much thicker, such as 50 μm. After patterning a photoresist mask, the device layer is etched in a DRIE that can achieve high aspect ratios — up to 100:1. This allows the formation of deep, narrow trenches. Finally, a timed oxide etch removes the buried oxide from beneath the structures to be released. Because the structural material is single crystal silicon, very flat beams and plates can be made with no residual stress,

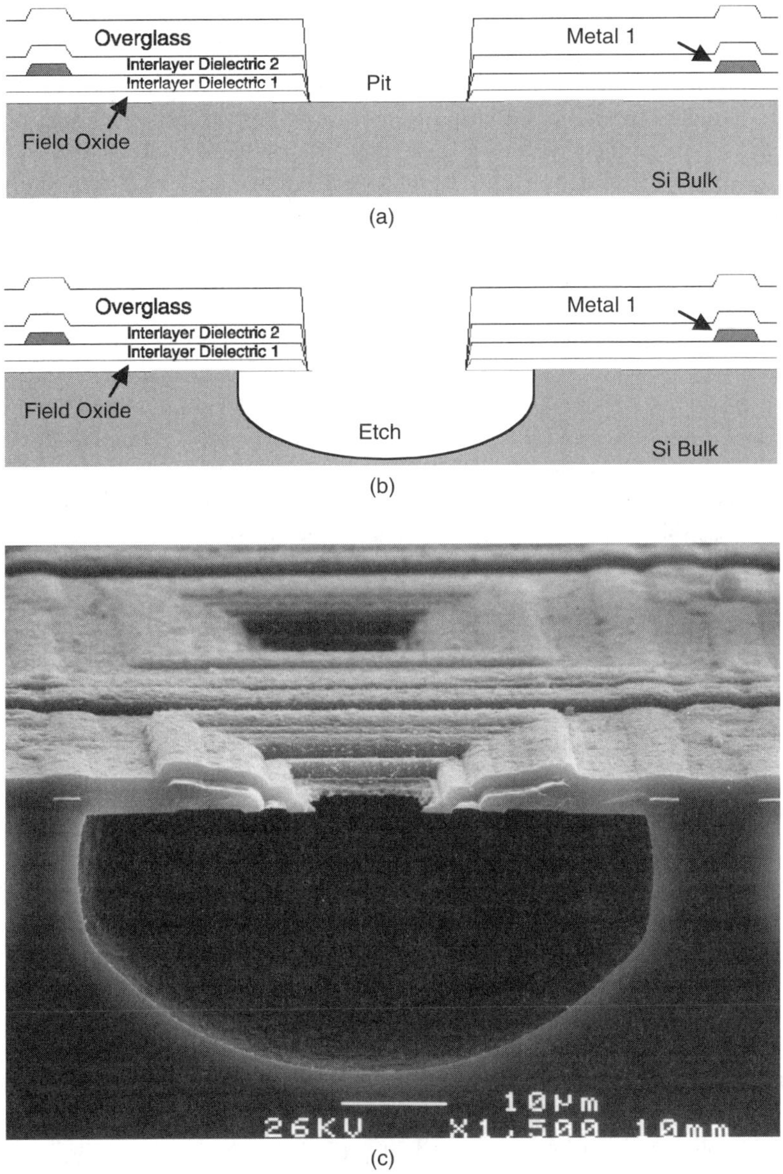

FIGURE 5.1 Cross sections of bulk micromachining in standard CMOS. (a) The wafer as it appears when it returns from the CMOS foundry with the various dielectric layers patterned so that the silicon substrate is exposed. When the wafer is then placed in an isotropic silicon etchant, such as XeF2, the silicon is dissolved and the dielectric layers become undercut, as shown in (b) and (c). (From Warneke, B. and Pister, K.S.J., *Sensors Actuators A*, 89(1–2), 142–151, 2001. With permission.)

while the thick device layer drastically reduces the compliance of beams in the vertical axis, which is advantageous for lateral structures.

5.2.2 Highly Integrated Processes

By integrating disparate components together into a single process, significant reductions in the size of the sensor node may be possible. Of particular interest are processes that combine CMOS transistors with micromachining capabilities. Analog Devices has successfully commercialized a process based on a standard BiCMOS process with a 4-μm low-stress polysilicon structural layer inserted into the flow before the interconnect metallization is deposited. An additional mask is used at the end to protect the oxide over the circuits during the sacrificial oxide release etch [12].

A number of techniques have been demonstrated to perform postprocess micromachining on foundry CMOS. One approach utilizes poly-SiGe microstructures and poly-Ge sacrificial layers on top of a CMOS wafer. These films can be deposited at temperatures low enough that the CMOS aluminum interconnects are not damaged and poly-Ge can be etched with hydrogen peroxide, which does not attack the CMOS layers [13].

One of the simplest techniques of adding micromachining to CMOS requires only a single maskless postprocess etch [14]. By stacking the contact, via, and overglass cut layers, a region of silicon will be exposed when the chip returns from the foundry. The silicon can then be sacrificially etched by bulk Si etchants such as XeF_2 [15] with the oxides and metals acting as the mask and structural layer (Figure 5.1). However, this method does not work in submicron processes that use tungsten plugs in the vias.

CMOS high aspect ratio micromachining facilitates maskless postprocessing in submicron processes by using the top metal layer as a mask for a high aspect ratio reactive ion etch that removes any oxide not protected by metal. In this way, narrow trenches down to the silicon substrate can be made. An isotropic plasma Si etch then releases the microstructures formed by the CMOS thin films.

5.3 Sensors

5.3.1 Selection Criteria

A large amount of MEMS research and product development has been in the area of sensors, so a wide variety of measurands using numerous detection techniques are available with micromachined sensors [7–9]. Examples include thermal sensors [16]; accelerometers [17]; gyroscopes [12]; pressure sensors [18]; microphones [19]; radiation detectors; magnetic sensors; flow sensors; and chemical and biological sensors. However, when selecting or designing a sensor for use in a miniature sensor network node, several criteria should be considered:

- *Volume of the complete sensor.* Although the active sensing element may be small, the complete system necessary to operate the sensor or interface it to the environment may be much larger. For example, a chemical sensor may require a sample gathering and preparation system much larger than the active region.
- *Energy consumption.* Because, for a given lifetime, energy needs directly affect the size of the power system and thus the sensor node, the energy required to make a measurement with the sensor should be minimized. The energy consumed is determined by the power consumption integrated by the time that the sensor has power applied to make a particular measurement.
- *Power consumption.* The first approach to reducing the energy consumption of a sensor is to reduce the power consumed by the device during operation, primarily by placing a high priority on minimizing the power consumption throughout the design of the sensor to guide trade-off decisions. For example, power considerations can affect the choice of detection technique — a piezoresistive sensor can have a large DC current, whereas a capacitive sensor will have no such component; however, the detection circuits are likely to have a high frequency excitation signal that consumes dynamic power. Nevertheless, power consumption cannot be considered in isolation because it is possible for the lowest power sensor to consume more energy per sample if smaller currents increase the time necessary to reach a stable measurement and thus add to the sample time and energy.
- *Suitability for power cycling.* One of the most straightforward methods of reducing the power consumed by a device is to turn it on only when necessary for as long as needed. It is therefore important that a sensor can be turned on and off relatively quickly. The gains become greater for sensors that are not sampled as frequently, such as a temperature sensor likely to be accessed once

a minute or less due to slow thermal time constants. Certain sensors, such as chemical or light, may need to integrate the measurand over a significant period of time, so their usage needs to be evaluated carefully. Additionally, low-frequency sensors such as seismometers can have long system time constants that prevent rapid power cycling. Some systems may benefit from a threshold or course sensor with a reduced energy consumption, which then triggers the sensor node to activate a higher energy consuming device with greater resolution.

- *Fabrication and assembly compatibility with the rest of the system.* A sensor that can be fabricated in the same substrate as other system components, such as the integrated circuits or communication devices, can greatly assist in building a compact node. If monolithic fabrication is not possible, assembly compatibility is also beneficial. For example, flip-chip bonding of heterogeneous substrates can yield small, integrated systems. These are all areas in which MEMS-based sensors can aid in miniaturizing a system.
- *Packaging requirements.* Some sensors may need contact with the environment, such as humidity and chemical sensors, which can limit the miniaturization potential.

5.3.2 Integrated Circuit Sensors

A number of measurands can be sensed by standard integrated circuits, which makes these sensors extremely easy to integrate with minimal additional volume. Temperature can be determined through the temperature dependence of subthreshold MOSFETs or the p–n junction of a diode or bipolar transistor. The proportional to absolute temperature (PTAT) circuit [20, 21] is most commonly used to extract the temperature signal, but other approaches that provide a digital output have been implemented, including using a counter to measure the frequency of a temperature-dependent ring oscillator [67].

Similarly, p–n junctions can also be used as photodiodes and phototransistors to measure light intensity, although a translucent window is necessary in the package. In addition, metal shields should be placed over sensitive circuits to prevent photogenerated carriers from interfering with their operation. Hall-effect sensors that detect magnetic fields can also be built from integrated circuits [22]. Sometimes ferromagnetic materials are deposited on top of the standard transistor process and patterned to form field concentrators that improve the responsivity of the sensor.

5.3.3 Nanosensors

Nanosensors can potentially provide further reductions in volume of the sensing element. The molecular scale and high relative surface area of nanowires allow precise control and sensitive detection of charged biological and chemical species [23]. In addition, nanowires can improve the responsivity of optical detectors by dramatically increasing the surface area of the detector; thermocouple-style temperature sensors are being developed with silicon nanowires. Meanwhile, carbon nanotubes have been demonstrated as chemical [24] and infrared [25] sensors.

5.4 Communication

MEMS does not impact the communication of wired sensor networks, but it can help miniaturize wireless communication. The most common form of wireless communication in use today is radio frequency (RF) radiation, including microwave and millimeter wave. However, because the relatively long wavelengths inherently limit the size of a sensor node utilizing these frequencies, free-space optical communication can be advantageous for building tiny sensor nodes.

5.4.1 RF Communication

The primary reasons to use RF communication are that it does not require line of sight and readily allows omnidirectional links. In some applications, such as asset tracking or supply chain monitoring, in which the node may be enclosed, these benefits are imperative. Nevertheless, RF does have limitations that make it less efficient for tiny, energy-constrained devices:

- Efficient antennas need to be a significant fraction of a wavelength, resulting in antennas that are many centimeters long at RF and microwave frequencies. Millimeter wave frequencies can yield more reasonably sized antennas, but the circuit efficiencies are lower and the transmission attenuation is greater.
- A small RF antenna will have very low antenna gain because beam divergence is fundamentally limited by diffraction, which is dependent on wavelength. To achieve the same milliradian collimation of an inexpensive laser pointer would require a 100-m diameter parabolic antenna at 1 GHz.
- RF transmitters have poor efficiency; a GMSK power amplifier has 50% slope efficiency (not including bias overhead), while the linear amplifiers used in CDMA systems have 10% slope efficiency. In addition, usually 1 to 100 mW of overhead is due to mixers, biasing, etc., although researchers are working to improve these efficiencies and build 100-μW radios that consume 5 nJ/(correct)b [26].
- The received power varies as the inverse of the distance raised to the second to seventh power due to multipath fading; for communication along the ground, such as cellular telephones, the average is four.

Together, these reasons make RF unattractive for tiny wireless nodes due to poor energy efficiencies and large radiators.

To illustrate these inefficiencies, the Bluetooth radio standard, which was designed for relatively low-power handheld devices, consumes about 100 nJ/b to transmit just tens of meters. Similarly, an IEEE 802.15.4 (draft) radio [27], which was actually designed for low-power wireless sensor networks, has a 100-m range; 0 dBm transmitted power (25 nJ/b); receiver sensitivity of –92 dBm; and 40 kbps data rate; it operates in the 902 to 928 MHz band. When actively communicating, it consumes 1 μJ/b on the transmit side and 2 μJ/b on the receive side, not including the power-up overhead time and idle periods.

Nevertheless, for those applications that do require RF nodes, MEMS can reduce the size of the transceiver [28, 29]. Figure 5.2 shows a block diagram of a typical wireless transceiver front end with a superheterodyne architecture. A relatively large number of high-Q, passive components are shown, including ceramic and SAW filters, discrete inductors, and discrete tunable capacitors (varactors) that cannot be fabricated with conventional integrated circuit processes. These components thus must be implemented with off-chip devices that end up dominating the size of the transceiver. Fortunately, micromachined components have been developed that may be able to replace each of these off-chip components; this will reduce the overall size of the transceiver through physically smaller components and the potential for integration with the integrated circuit chips.

Voltage-tunable high-Q capacitors can be fabricated by suspending a top aluminum plate on soft flexures over a bottom plate [30]. A DC bias on the resulting capacitor causes an electrostatic force to pull the top plate down, thus varying the capacitance. Such a structure has been demonstrated with a Q of 62 at 1 GHz.

There are a number of approaches for fabricating on-chip high-Q inductors. Two techniques improve the Q of normal planar inductors (which is 1 to 3 at 1 GHz): the first utilizes a NiFe thin film under the spiral to act as a core to increase the magnetic flux and thus the Q (6.6 at 4 MHz [31]); the second method uses a front- [32] or back-side silicon etch to remove the lossy substrate from underneath the spiral and achieve Qs of 5 at 1 GHz and 60 to 80 at 40 GHz. The latter approach is more readily integrated with circuits because it can be implemented with a postprocess etch, while the former requires adding nonstandard metal depositions into the process. More exotic fabrication techniques can be used to build three-dimensional inductors. Figure 5.3 shows a four-turn inductor fabricated on a silicon substrate with 5 μm-thick copper traces electroplated around an alumina insulating core with a 650 × 500 μm cross section. Direct-write laser lithography is used to pattern the top and sidewall photoresist. This device achieves 14 nH of inductance and a Q of 16 at 1 GHz, while a similar one-turn coil obtains an inductance of 4.8 nH and a Q of 30 at 1 GHz [33].

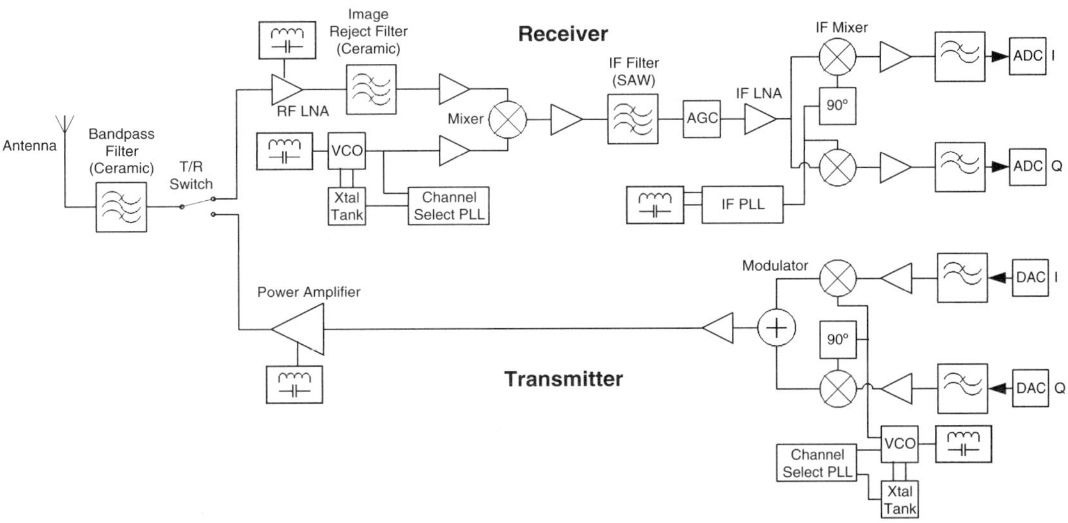

FIGURE 5.2 Diagram of a typical wireless transceiver front end showing the many off-chip, high-Q, passive components, such as filters, inductors, and capacitors, that could be replaced by micromechanical versions. Besides the component-size reduction, such components could potentially be integrated with the circuits for dramatic volume reductions.

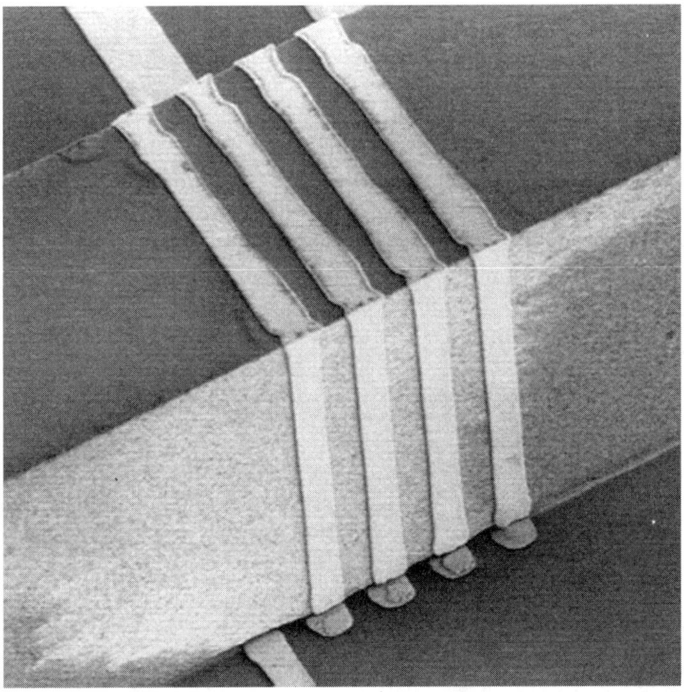

FIGURE 5.3 Four-turn inductor fabricated on a silicon substrate with electroplated copper around an insulating core. It has an inductance of 14 nH and a Q of 16 at 1 GHz. (From Young, D.J. et al., *Tech. Dig., Int. Electron Devices Meeting*, Washington, D.C., December 1997, 67–70. With permission.)

Some applications, such as replacing quartz crystals and ceramic and SAW filters, require even higher Qs than these devices can provide. Just as the macroscopic domain utilizes vibrating mechanical structures, the microscopic domain can achieve high Qs through vibrating resonators with a second-order response. Thin film bulk acoustic resonators (FBAR) are composed of a metal–piezoelectric–metal film stack, similar to a quartz crystal, and suspended on a thin membrane to provide acoustic isolation from the substrate. Such resonators can achieve a Q of 1200 at 1.9 GHz with an area of only $100 \times 100\ \mu m^2$ [34] and are currently in production.

An alternative method of building micromechanical resonators uses surface-micromachined polysilicon to suspend a flexural-mode beam over an electrode. The beam is electrostatically excited, resulting in the second-order resonance. Qs of 7450 have been achieved at 92 MHz [35]. These structures can also be mechanically coupled to form high-Q filters and filter-mixer structures [36] that allow multiple components of the system illustrated in Figure 5.2 to be replaced with a single passive micromechanical component. This could reduce size and power consumption of the transceiver. A similar fabrication process can also produce a ring or disk with a central anchor driven laterally through a submicron gap to obtain a Q of 9400 at 156 MHz. One of the major problems with micromechanical resonators is that they have relatively low power handling capabilities that limit their use in applications such as cellular telephones; however, these limits are high enough to be applicable to distributed wireless sensor networks that utilize short-range, multihop communication links.

Finally, the transmit/receive diplexer switch can be replaced with micromechanical relays that feature lower insertion loss ("on" impedance) and larger isolation ("off" impedance). The two major styles of switches are (1) cantilever beams with electrostatic pull-down electrodes and metal–metal contacts for DC operation; and (2) suspended membranes that are electrostatically deformed to increase the capacitive coupling through the structure dramatically [37]. Cantilever-style switches have been demonstrated with an actuation voltage of 30 V, >50 dB of isolation below 2 GHz, and <0.2 dB of insertion loss from DC to 40 GHz [38]. Besides their use as diplexers, the nearly ideal behavior of RF switches can be used to build small tunable filters, multiband antennas, true-time delay phased-array antennas, and even reconfigurable transceiver architectures [39].

It should be noted that although Figure 5.2 provides a good discussion point because of the large number of high-Q components that could be replaced by MEMS components, it is not the only transceiver architecture possible. For example, direct-conversion (zero-IF) [40] and subsampling [41] transceivers eliminate many of the filters. In addition, if the channel selectivity and other parameters of the radio band are relaxed, high-Q components may not be necessary, although the use of higher Q components can often lead to lower power consumption because of the reduced losses.

5.4.2 Optical Communication

Free-space optical communication has many advantages for miniature sensor nodes:

- Optical radiators such as mirrors and laser diodes can be made extremely tiny — 0.03-μm^3 lasers have been demonstrated [42].
- As mentioned earlier, optical transmission provides extremely high antenna gain, which yields higher transmission efficiencies.
- Although laser output slope efficiencies are only about 25%, the diode turn-on current overhead can be as low as 1 μW for vertical cavity surface emitting lasers (VCSELs), so the effective output efficiency can be much higher than RF power amplifiers.
- The received power only decays as the inverse of the distance squared, assuming line of sight.
- The high directivity of optical communication enables the use of spatial division multiple access (SDMA) [43], which is a simple network media access technique in which an imaging receiver can separately process simultaneous transmissions from different angles. SDMA thus requires no communication overhead and has the potential to be more energy efficient than RF media access methods such as frequency, time, and code division multiple access (FDMA, TDMA, CDMA).

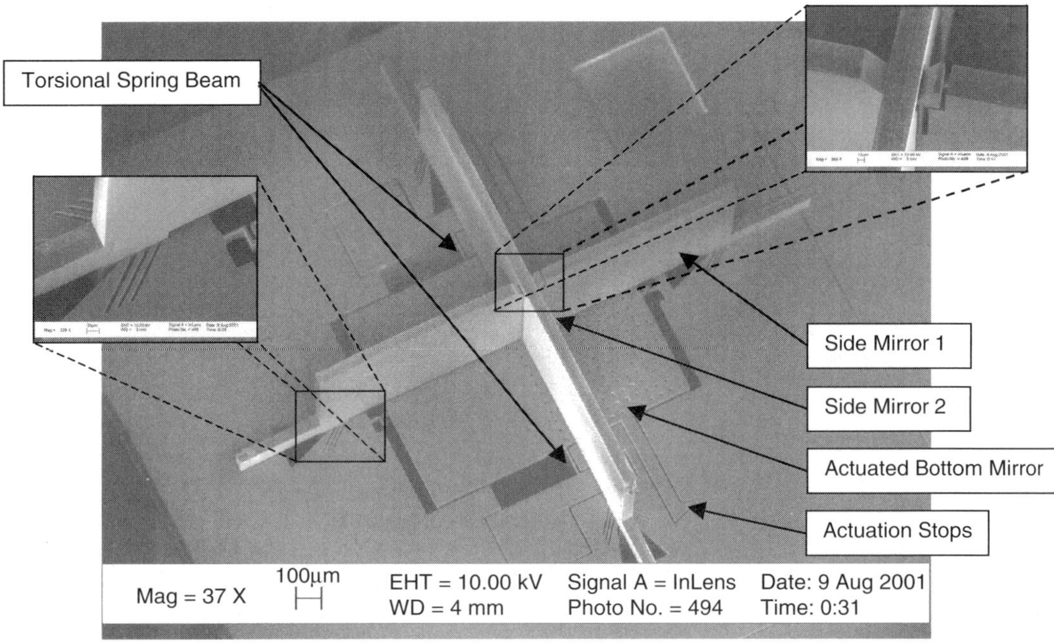

FIGURE 5.4 A quad-corner cube retroreflector (CCR) used for passive optical transmission. The electrostatically actuated bottom mirror rotates torsionally to disturb the orthogonality of the corner and switch the light reflected from the CCR from the "1" to "0" states. The insets show the spring locks that aid in assembly and maintain alignment. The device is fabricated on an SOI wafer with a 50-μm thick device layer using deep reactive ion etching. (From Zhou, L. et al., *IEEE J. Microelectromech. Syst.*, 12(3), 233–242, 2003. With permission.)

- It is extremely difficult to eavesdrop on collimated optical communication (low probability of detection and low probability of intercept), which is a significant security advantage.

The primary drawbacks of optical communication are that line of sight is necessary for all but the shortest distances and the narrow beams imply the need for accurate pointing. Fortunately, MEMS technology and clever algorithms can provide accurate pointing [44] and multihop, self-healing networking can allow messages to travel around certain obstacles.

The two primary methods of free-space optical transmissions are passive reflective systems and active-steered laser systems. The passive reflective system consists of three mutually orthogonal mirrors that form the corner of a cube (Figure 5.4) [45] — thus the name corner cube retroreflector (CCR). Light entering the CCR bounces off each of the mirrors and is reflected back to the sender parallel to the incoming beam. By electrostatically actuating the bottom mirror, the orthogonality can be disturbed, causing the reflection to no longer return to the sender. This behavior allows the CCR to communicate with an interrogator by simply modulating the reflected light and resembles the operation of a heliograph in which the operator bounces sunlight off a mirror to transmit Morse code messages to other ships. This is a concept that can be traced back to Greece in the fifth century B.C. Because the only energy consumed is that required to charge 3 pF of capacitance in the actuator, this is much more efficient than an approach that requires the generation of radiation, such as RF or lasers.

The device shown in Figure 5.4 is fabricated using deep reactive ion etching (DRIE) in an SOI wafer with a 50-μm device layer for flat, smooth mirror surfaces. It consumes 16 pJ/b transmitted, has a demonstrated range of 180 m, transmission data rates in excess of 4 kbps, and a size of $2 \times 2 \times 0.5$ mm, although it can be made smaller if less reflection is acceptable. One restriction with CCR-based communication is that it does not facilitate peer-to-peer communication, so a one-to-many network topology is required; however, distributed algorithms are under development to take advantage of such a network

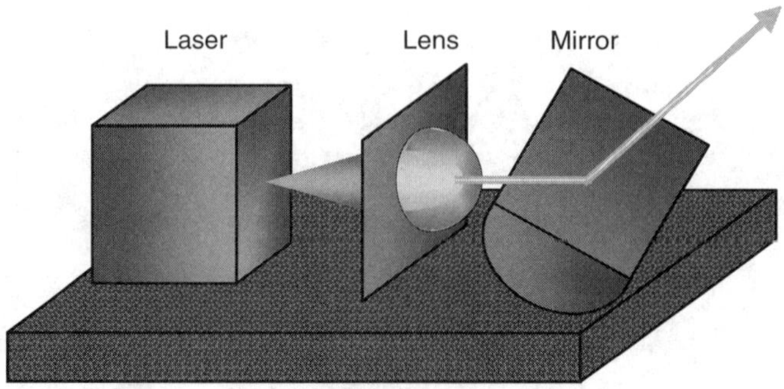

FIGURE 5.5 Conceptual diagram of a steered agile laser transmitter. A laser diode emits a beam that is collimated by a lens (may be micromachined) before bouncing off the MEMS beam steering mirror, which aims the beam toward the intended receiver.

for things such as sensor data compression. Furthermore, the communication range of a sub-mm CCR is theoretically limited to about 1 km in a practical implementation.

Active-steered laser communication utilizes a small laser diode, such as a VCSEL, a collimating lens, and MEMS beam-steering optics to transmit a tightly collimated light beam to a particular receiver (Figure 5.5). This facilitates peer-to-peer communication over a wide area, while maintaining many of the features of optical communication including high directivity and long-distance communication using little power. Because efficient lasers cannot be fabricated in silicon, monolithic integration is unlikely; however, micromachined structures can be used to aid in the alignment of a bare laser diode onto a chip [46]. On the other hand, three-dimensional micromachined collimating lenses have been demonstrated using reflowed photoresist [47]. The beam-steering optics are the most challenging part of the system because they should have close to hemispherical range, low actuation range, low cross-axis sensitivity, and be robust against shock. Current approaches use multilevel SOI MEMS for very flat mirrors, low cross-axis sensitivity, and robustness [48], but have only achieved up to 40° of optical deflection angle with a rather high actuation voltage of 90 V [49].

Finally, to illustrate the dramatic differences between the various communication schemes discussed, Figure 5.6 compares the communication range vs. energy/bit consumption of CCR, green laser, and GSM RF communication.

5.5 Micropower Sources

Miniature sensor nodes can be powered from energy storage or energy scavenging devices or a combination thereof. In addition, to allow larger peak currents or integration of charge from energy harvesters to compensate for lulls such as nighttime for a solar cell, capacitors may be used in these systems to lower the effective impedance of a battery or energy harvester. High-density capacitors, such as the Ultracapacitor [50], can store up to 10 mJ/mm^3, which is less than 1% the energy density of lithium cells.

5.5.1 Energy Storage

From the system's perspective, a good microbattery should have the following features:

- High energy density
- Large active volume to packaging volume ratio (i.e., a thin film on top of a 500-μm silicon wafer would not be desirable)
- Small cell potential (0.5 to 1.0 V) so digital circuits can take advantage of the quadratic reduction in power consumption with supply voltage
- Ability to configure efficiently into a series of cells to provide a variety of potentials for the various components of the system without requiring the overhead of voltage converters
- Rechargeable in case the system has an energy harvester

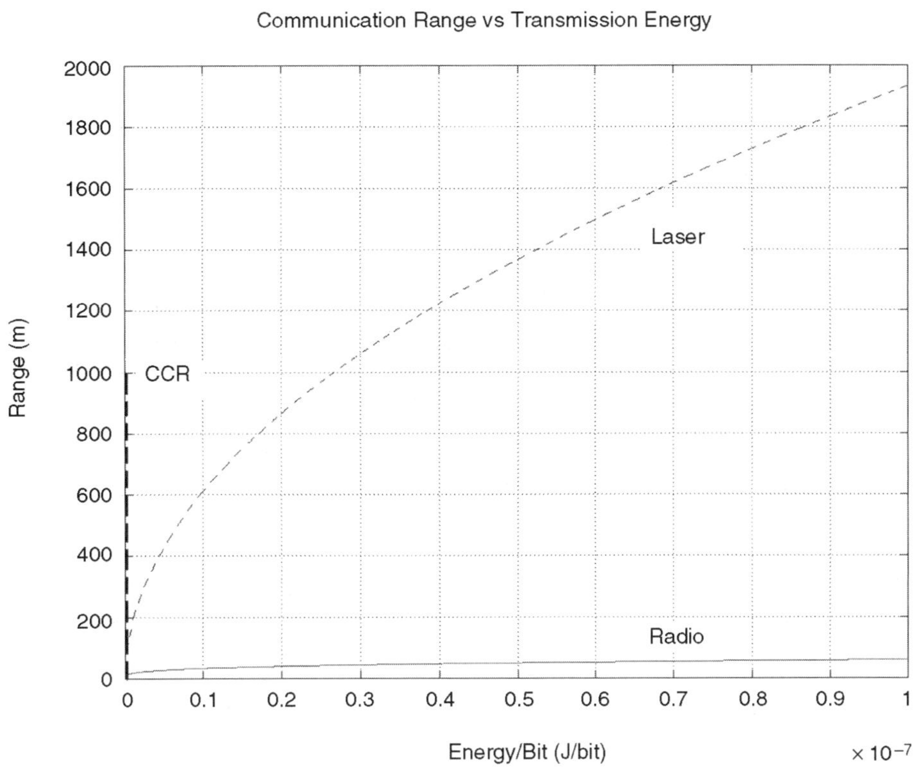

FIGURE 5.6 Communication range vs. transmission energy for RF (GSM, 1 GHz, isotropic, path loss $n = 4$); laser (532 nm green, 1 mW, 1 Mbps, 200 × 200 μm receiver aperture); and CCR (400 μm passive, 16 pJ/b independent of distance up to 1 km).

A variety of tiny batteries are being developed, including thin-film vanadium oxide and molybdenum oxide [51] that are fabricated using spin-casting sol-gel techniques and micromachined cavities containing an electrolyte, although the latter devices do not have high energy densities [52]. Nickel-zinc batteries have been developed with a footprint of 2 mm^2, <100 μm thick, and a capacity of 20 mJ/mm^2 with a discharge rate of >1 mA/mm^2 [53]. Another potential candidate chemistry is rechargeable thin-film lithium energy cells. Researchers at Oak Ridge National Laboratory have built 1 cm^2 × <15 μm Li-$LiCoO_2$ batteries with a 40,000 charge/discharge cycle life and a capacity up to 24 mJ/mm^2 [54–56]. A derivative process at the Jet Propulsion Laboratory uses microfabrication techniques to generate batteries as small as 50 × 50 μm with a 0.25-μm cathode film and capable of energy densities of 1.4 mJ/mm^2 [57].

One of the highest energy density battery chemistries available is the Zn-air cell. It is also available in the smallest button cell package: the Energizer IEC-PR63 weighs 0.2 g (including packaging); is 0.051 cm^3; $V_{oc} = 1.4$ V; and contains 33 mAh (160 J). TPL Inc. is using micromachining techniques to develop Zn-air volumetric batteries 2 mm in diameter and 0.5 mm thick with a capacity approaching 1 mAh (3 J/mm^3) [58]. With an areal capacity of 1.6 J/mm^2, the advantage of the volumetric approach is evident over the thin-film lithium batteries if maintaining a small footprint is a priority. To meet the demand for higher discharge current, TPL proposes to combine supercapacitors that store 30 mJ in a similar size in parallel with the batteries.

The biggest problem with current Zn-air cells is that the self-discharge is so high that, after the air terminal is opened, they have a shelf life of only a couple of weeks, although a micro Zn-air cell could potentially incorporate a micromachined air valve to control this self-discharge. The sensor node would then operate primarily off a capacitor that would be charged periodically by opening the air valve. An additional problem with Zn-air cells is that they are not rechargeable. This chemistry is thus only a candidate in short-term deployments. Even though process compatibility with the other components of the system may seem desirable, it may actually not be important due to the possibility of stacking the various components because the batteries do not need exposure to the environment.

In addition to chemical energy storage, radioactive isotopes provide another method of storing energy on a small sensor node; such techniques are already used extensively in deep space probes and satellites where long life and reliable operation are essential, just as in wireless sensor networks. Blanchard and coworkers [59] have demonstrated a micromachined radioactive battery based on a thin-film beta emitter coating a beam that performs a charge to mechanical conversion (as the beta particles leave, the beam acquires a positive charge, causing it to be attracted to the substrate). This is followed by a mechanical to electrical conversion using a piezoelectric material (the strain of the bending beam is converted to charge), also on the beam. Two companies [60, 61] have also proposed building millimeter-scale radioactive power sources based on beta emitters, the first of which is using betavoltaics — the direct conversion of beta particles to electricity by bombarding a p–n junction.

5.5.2 Energy Harvesting

Scavenging energy from the environment will allow the wireless sensor nodes to operate nearly indefinitely, without their batteries dying. Solar radiation is the most abundant energy source and yields around 1 mW/mm^2 (1 $J/day/mm^2$) in full sunlight or 1 $\mu W/mm^2$ under bright indoor illumination. Solar cells have conversion efficiencies up to 30%.

Vibration has been proposed as an energy source [62, 63] that can be scavenged. Vibration spectra of office windows, copy machines, microwave ovens, industrial motors, freeway traffic, and the human gait reveal that usable energy is there — typically on the order of 10 $\mu W/g$ of mass of the converter. Because the mass of a cubic millimeter of silicon is about 2 mg, this energy source is only feasible at the centimeter scale and above. The basic device used to extract energy from vibrations is a mass on a spring connected to a variable capacitor. In actual implementation, a lateral or gap-closing comb resonator is typically used. A precharged reservoir, such as a capacitor or rechargeable battery, a storage capacitor, and two switches form the basic charge-constrained conversion circuit.

More exotic energy sources that have been proposed include utilizing the excess heat from microrocket engine combustion [64]; using copper and zinc electrodes to generate power from seawater; and harvesting ATP for *in vivo* applications. For applications in which duty cycling is acceptable, solar cells or other power scavenging sources can be used to trickle charge a capacitor or battery, after that the stored energy can be used at much higher power rates than the charging pace.

5.6 Packaging

As the size of the sensor node decreases, the packaging considerations become more critical to prevent the package from dominating the volume and since nonstandard packaging is necessary. Some of the requirements of the packaging include:

- The microstructure, such as a CCR or accelerometer, must be protected while still being able to move.
- Electrical connections between various chips, such as bond wires or vertical interconnects from a battery, need to be facilitated and protected.
- Solar cells require clear packaging and possibly a lens to improve the collection efficiency.
- An optical receiver photodiode may require an optical filter.
- A CCR requires an antireflective (AR)-coated cover that allows illumination along its primary axis of [111].
- The packaging must add a minimum of extra volume.

- The deployment method used in the application will place certain requirements on the packaging. For example, micro air vehicle deployment would require the packaging to protect the sensor node from being dropped 100 ft.
- Toxic battery chemistries need sufficient shielding in case a human or animal swallows the node.
- Vibration harvesting devices need a solid mechanical connection to the environment.
- Sensors may require special access to the environment, so packages may require tailoring to the application. Examples include humidity, pressure, acoustics, strain, gaseous chemical and biological sensors, and fluidic sensors.

The use of a common substrate is also a consideration because it can ease assembly, but adds volume. The die substrates can be thinned to help reduce the impact of a common substrate.

Micromachining techniques can help meet some of the packaging requirements. For instance, microstructures such as accelerometers and resonators can be fabricated in sealed vacuum cavities by defining the cavity with a sacrificial layer; depositing a structural layer; removing the sacrificial layer through a small access hole; and then sealing the cavity by depositing a CVD, sputtered, or evaporated film or by growing an oxide on a polysilicon layer until the hole is sealed. Wafer bonding can also be used to protect microstructures within a hole or cavity in the wafer. A variety of microassembly technologies [65], such as pick-and-place methods for the microdomain, batch transfer, fluidic microassembly, and flip-chip bonding, facilitate the compact assembly of heterogeneous dies.

The CCR poses some of the most difficult packaging constraints because the device must be mechanically protected, allowed to move, and have good optical properties. Three options were proposed in Hsu [66]:

- A hemispherical cover can cause lensing effects if the diameter is too small, which affects the performance of the CCR.
- A flat plate elevated on short walls eliminates the lensing effects, but the plate must be large to avoid the edge blocking the light. Because the optimum axis of the CCR is at a 45° angle to the plate and the reflectivity of the plate increases as the angle of incidence increases, this approach is not optically efficient.
- A pyramid that has surfaces normal to the body diagonals of the CCRs can be used. Because the optimum incident angles for the CCRs are closer to normal to the package, reflections will be reduced.

Steered agile laser communication also requires a package that mechanically protects the micro-optical system, allows it to move, and has good optical properties. However, because an input optical beam is not necessary, a simple hemispherical cover is the best option.

For cubic millimeter sensor nodes, such as that shown in Figure 5.9, the best proposed solution at this time involves potting the node in an optical-quality polymer with some special molds as shown in Figure 5.7. This package provides many of the necessary features detailed previously, including providing access to the environment by molding holes in the polymer. An antireflective coating can probably be placed on the polymer at the end of the process.

5.7 Systems

A number of wireless sensor nodes have been developed that take advantage of MEMS to achieve a small size. Mason and colleagues [67] at the University of Michigan created a multisensor microcluster that measures temperature, pressure, humidity, and vibration/position. It includes a microcomputer, has a 50-m RF link, is less than 10 cm^3 (Figure 5.8), operates off a single battery, and consumes 530 μW average power and 10 mW while transmitting.

The microsystem contains a variety of chips: a commercial microcontroller (Motorola 68HC11); a power management chip; a commercial transmitter (RFM HX1005); a capacitive interface chip with an

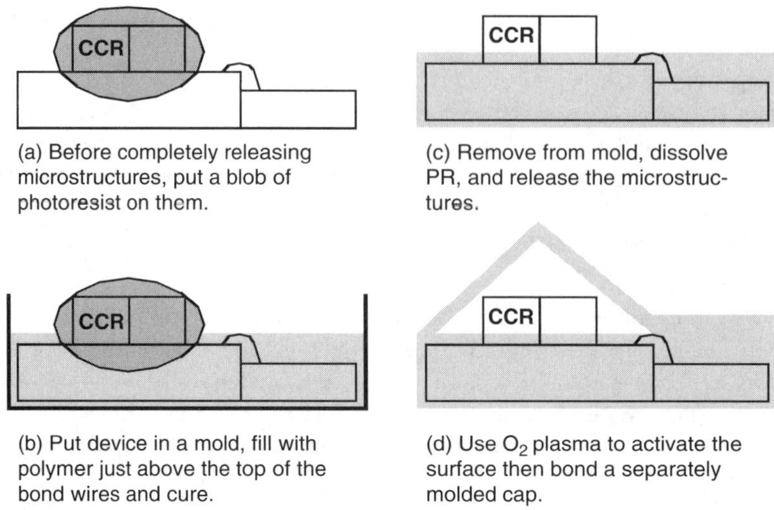

FIGURE 5.7 Polymer encapsulation process for cubic millimeter sensor node packaging.

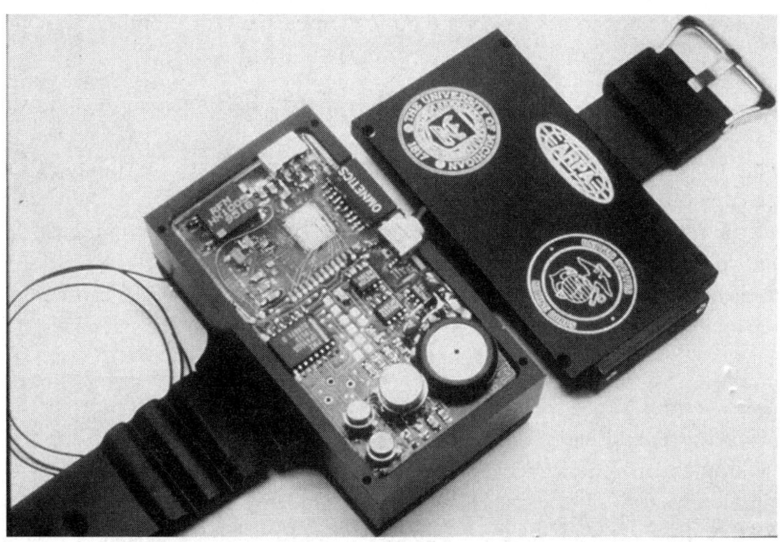

FIGURE 5.8 Multisensor microcluster containing MEMS pressure, humidity, and acceleration sensors and an RF transmitter with a 50-m range. The device is less than 10 cm^3. (Personal communication from K. Wise.)

integrated temperature sensor; a capacitive barometric pressure sensor; a capacitive relative humidity sensor; two accelerometers; a threshold accelerometer interface chip; and a lithium coin cell. The pressure sensor is fabricated using bulk micromachining and a silicon-glass dissolved-wafer process to create multiple diaphragms that segment the pressure range. The humidity sensor is fabricated with high-aspect-ratio micromolding and electroplating to form a series of interdigitated electrodes. A thin polymer film, whose dielectric constant varies as a function of moisture, fills the gaps between the electrodes and causes the capacitance to vary with humidity. A *z*-axis accelerometer is fabricated in a three-mask dissolved-wafer process and contains a proof mass suspended by torsional beams. At the end of the proof mass, a set of comb fingers is interdigitated with a set of fixed comb fingers that provide capacitive sensing of the movement of the proof mass. Finally, an array of threshold accelerometers, which are simply cantilever switches with varying proof masses and spring constants, is fabricated using the dissolved wafer process;

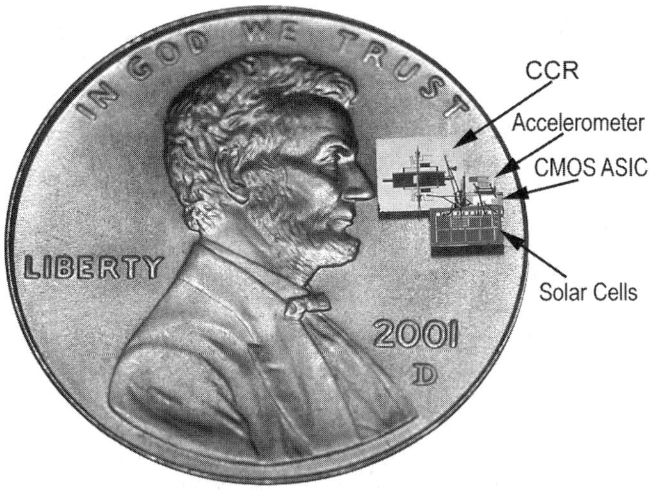

FIGURE 5.9 16-mm^3 Smart Dust mote, showing a 0.25-μm CMOS ASIC with optical receiver, ambient light sensor, and controller; solar power array; accelerometer; and CCR, each on separate die. (From Warneke, B.W. et al., *Proc. IEEE Int. Conf. Sensors 2002*, Orlando. With permission.)

p++ etch stop proof masses; oxide suspension beams; and gold contacts. A second-generation micro-cluster system reduced the volume to less the 5 cm^3, while forthcoming versions will be around 1 cm^3 and even down to 0.2 cm^3.

The Wireless Integrated Network Sensor (WINS) project at UCLA [68] developed a sensor node that included an infrared imager; seismometer; spectrum analyzer; RF transceiver; and lithium coin cells in a volume on the order of tens of cubic inches. The sensor integration relied on flip-chip bonding structures to a low temperature, cofired ceramic (LTCC) substrate that provided a platform for support of interface, signal processing, and communication circuits. In addition, the LTCC substrate provides small, embedded low-loss capacitors and high-Q inductors that are used by the transceiver. The infrared imager and seismometer were fabricated with bulk micromachining and flip-chip bonding. WINS also explored building a loop antenna on a CMOS die by removing the silicon substrate with a XeF_2 etch.

The PicoRadio project [26] at UC Berkeley is developing an ultralow energy transceiver for ubiquitous wireless data acquisition. The goal is to consume less than 5 nJ/(correct)b and less than 100 μW. The transceiver uses FBARs for low-phase noise oscillators [69] and filters, while vibration harvesting is being investigated for the power source [63].

The most extensive use of MEMS for miniaturizing wireless sensor nodes is the Smart Dust project [70] at UC Berkeley that seeks to push the volume of wireless sensor nodes aggressively down to a cubic millimeter. Figure 5.9 shows a 16-mm^3 autonomous solar-powered sensor node [71] with bidirectional optical communication. The system consists of four die: a 0.25-μm CMOS ASIC; a trench-isolation SOI solar cell array; a micromachined four-quadrant CCR; and a capacitive accelerometer. The ASIC contains an optical receiver that consumes 69 pJ/b; an ADC that uses 180 pJ/8-b sample; a photosensor for measuring ambient light; a finite state machine to control the system; and a 1-μW, 3.9-MHz integrated oscillator. A new DRIE SOI/CMOS process has been developed to allow integration of solar cells, CCR, and accelerometer along with high-voltage FETs. Figure 5.10 shows the resulting die combined with the same ASIC as in Figure 5.9 for a total device size of 6.6 mm^3.

5.8 Conclusion

Many aspects of wireless sensor network nodes can be miniaturized with MEMS technology. From the sensors to the wireless communication components and power supply, MEMS is reducing volume, improving performance, and reducing cost through batch fabrication techniques. In addition, MEMS

FIGURE 5.10 Mock-up of a 6.6-mm^3 autonomous Smart Dust mote. This mote has the same functionality as the one in Figure 5.9 — the CMOS ASIC is identical but process integration allowed the devices on the other die to be fabricated on a single die, thus reducing the size.

packaging and assembly techniques can help build miniature systems out of these small components. By miniaturizing sensor networks, not only will new applications be enabled, but they can also be deployed in more places, with higher densities and less interference to the monitored area, thus allowing improved data gathering. In this way the physical world can truly be instrumented.

References

1. Pierret, R.F., *Introduction to Microelectronic Fabrication*, Addison-Wesley, Menlo Park, CA, 1990.
2. Pierret, K., Silicon as a mechanical material, *Proc. IEEE*, 70(5), 420–457, 1982.
3. Muller, R.S., Howe, R.T., Senturia, S.D., Smith, R.L., and White, R.M. (Eds.), *Microsensors*, IEEE Press, New York, 1991.
4. Trimmer, W.S., *Micromechanics and MEMS: Classic and Seminal Papers to 1990*, IEEE Press, New York, 1997.
5. Madou, M., *Fundamentals of Microfabrication*, CRC Press, Inc., Boca Raton, FL, 2002.
6. Elwenspoek, M. and Jansen, H.V., *Silicon Micromachining*, Cambridge University Press, 1999.
7. Sze, S.M., *Semiconductor Sensors*, John Wiley & Sons, Sommerset, NJ, 1994.
8. Ristic, L.J. (Ed.), *Sensor Technology and Devices*, Artech House, London, 1994.
9. Kovacs, G.T.A., *Micromachined Transducers Sourcebook*, WCB McGraw-Hill, San Francisco, 1998.
10. Senturia, S.D., *Microsystem Design*, Kluwer Academic Publishers, Norwell, MA, 2001.
11. Gad–El-Hak, M. (Ed.), *The MEMS Handbook*, CRC Press, Inc., Boca Raton, FL, 2001.
12. Geen, J.A. et al., Single-chip surface-micromachined integrated gyroscope with 50°/hour root Allan variance, *2002 IEEE Int. Solid-State Circuits Conf. Dig. Tech. Papers*, 45, 426–427, 2002.
13. Franke, A.E., King, T.-J., and Howe, R.T., Integrated MEMS technologies, *MRS Bull.*, 26(4), 291–295, *Mater. Res. Soc.*, 2001.
14. Parameswaran, M. et al., A new approach for the fabrication of micromechanical structures, *Sensors Actuators*, 19, 289–307, 1989.
15. Warneke, B. and Pister, K.S.J., *In situ* characterization of CMOS postprocess micromachining, *Sensors Actuators A (Physical)*, 89(1–2), 142–151, 2001.
16. Baltes, H. et al., Micromachined thermally based CMOS microsensors, *Proc. IEEE*, 86, 1660–1678, 1998.

17. Yazdi, N., Ayazi, F., and Najafi, K., Micromachined inertial sensors, *Proc. IEEE*, 86(8), 1640–1659, 1998.
18. Eaton, W.P. and Smith, J.H., Micromachined pressure sensors: review and recent developments, *Smart Mat. Struct.*, 6(5), 530–539, 1997.
19. Rombach, P. et al., The first low voltage, low noise differential silicon microphone, technology development and measurement results, *Sensors Actuators A (Physical)*, A95(2–3), 196–201, 2002.
20. Timko, M.P., A two-terminal IC temperature transducer, *IEEE J. Solid State Circuits*, SC-11(6), 784–788, 1976.
21. Hosticka, B.J., Fichtel, J., and Zimmer, G., Integrated monolithic temperature sensors for acquisition and regulation, *Sensors Actuators*, 6(3), 191–200, 1984.
22. Nakamura, T. and Maenaka, K., Integrated magnetic sensors, *Sensors Actuators,* A22(1–3), 762–769, 1990.
23. Cui, Y. et al., Nanowire nanosensors for highly sensitive and selective detection of biological and chemical species, *Science*, 293, 1292–1298, August 2001.
24. Kong, J. et al., Nanotube molecular wires as chemical sensors, *Science*, 287(5453), 28 622–625, 2000.
25. Xu, J.M., Highly ordered carbon nanotube arrays and IR detection, *Infrared Phys. Technol.*, 42, 485, 2001.
26. Rabaey, J. et al., PicoRadios for wireless sensor networks: the next challenge in ultra-low-power design, *2002 IEEE Int. Solid-State Circuits Conf. Dig. Tech. Papers*, San Francisco, 45, 200–201, 2002.
27. http://www.amis.com/wireless/ASTRX1.html.
28. Nguyen, C.T.-C., Katehi, L.P.B., and Rebeiz, G.M., Micromachined devices for wireless communications, *Proc. IEEE*, 86(8), 1756–1768, 1998.
29. Rebeiz, G.M., *RF MEMS: Theory, Design, and Technology,* John Wiley & Sons, Sommerset, NJ, 2002.
30. Young, D.J. and Boser, B.E., A micromachined variable capacitor for monolithic low-noise VCOs, *Tech. Dig., 1996 Solid-State Sensor Actuator Workshop*, Hilton Head Island, SC, 86–89, 1996.
31. Von Arx, J.A. and Najafi, K. On-chip coils with integrated cores for remote inductive powering of integrated microsystems, *Dig. Tech. Papers, 1997 Int. Conf. Solid-State Sensors Actuators (Transducers'97)*, Chicago, IL, June 16–19, 1997, 999–1002.
32. Rofougaran, A. et al., A 1 GHz CMOS RF front-end IC for a direct-conversion wireless receiver, *J. Solid State Circuits*, 31, 880–889, 1996.
33. Young, D.J. et al., Monolithic high-performance three-dimensional coil inductors for wireless communication applications, *Tech. Digest, Int. Electron Devices Meeting*, Washington, D.C. December 1997, 67–70.
34. Ruby, R. Micromachined cellular filters, *IEEE MTT-S Int. Microwave Symp. Dig.*, 2, 1149–1152, June 1996.
35. Nguyen, C.T.-C., Vibrating RF MEMS for low power wireless communications (invited keynote), *Proc., 2000 Int. MEMS Workshop (iMEMS'01)*, Singapore, July 4–6, 2001, 21–34.
36. Wong, A.-C., Ding, H., and Nguyen, C. T.-C., Micromechanical mixer + filters, *Tech. Dig., IEEE Int. Electron Devices Meeting (IEDM),* San Francisco, CA, 1998, 471–474.
37. Goldsmith, C. et al., Characteristics of micromachined switches at microwave frequencies, *IEEE MTT-S Dig.*, 1141–1144, June 1996.
38. Hyman, D. et al., Surface-micromachined RF MEMs switches on GaAs substrates, *Int. J. RF Microwave Computer Aided Eng.*, 9(4), 348–361, 1999.
39. Izadpanah, H. et al., Reconfigurable low power, light weight wireless system based on the RF MEM switches, *1999 MTT-S Int. Topical Symp. Technol. Wireless Appl. Dig.*, Feb. 21–24, 1999, Vancouver, BC, 175–180.
40. Abidi, A.A., Direct-conversion radio transceivers for digital communications, *IEEE J. Solid-State Circuits*, 30(12), 1399–1410, 1995.
41. Sheng, S. et al., A low-power CMOS chipset for spread spectrum communications, *1996 Int. Solid State Circuits Conf. Dig. Tech. Papers*, 346–347, Feb. 1996.

42. Painter, O. et al., Two-dimensional photonic band-gap defect mode laser, *Science*, 284(5421), 1819–1821, 1999.
43. Kahn, J.M. et al., Imaging diversity receivers for high-speed infrared wireless communication, *IEEE Commun.*, 88–94, Dec. 1998.
44. Last, M. et al., Toward a wireless optical communication link between two small unmanned aerial vehicles, *Int. Symp. Circuits Syst. 2003*, Bangkok, Thailand, 3, III-930-3, May 2003.
45. Zhou, L., Kahn, J.M., and Pister, K.S.J., Corner-cube retroreflectors based on structure-assisted assembly for free-space optical communication, *IEEE J. Microelectromech. Syst.*, 12(3), 233–242, 2003.
46. Lin, L.Y. et al., Micromachined integrated optics for free-space interconnections, *Proc. IEEE Microelectromech. Syst. Conf.*, Amsterdam, Netherlands, Jan. 29–Feb. 2, 1995, 77–82.
47. Toshiyoshi, H. et al., A surface micromachined optical scanner array using photoresist lenses fabricated by a thermal reflow process, *J. Lightwave Technol.*, 21(7), 1700–1708, 2003.
48. Zhou, L. et al., Two-axis scanning mirror for free-space optical communication between UAVs, *IEEE Conf. Optical MEMS*, Waikoloa, HI, August 18–21, 2003.
49. Milanovic, V., Last, M., and Pister, K.S.J., Laterally actuated torsional micromirrors for large static deflection, *Photonics Technol. Lett.*, 15(2), 245–247, 2003.
50. http://www.powercache.com.
51. Harreld, J. H., Dong, W., Dunn, B. Ambient pressure synthesis of aerogel-like vanadium oxide and molybdenum oxide, *Mat. Res. Bull.*, 33(4), 561–567, 1998.
52. Lee, K.B. and Lin, L., Electrolyte-based on-demand and disposable microbattery, *Proc. 15th Annu. Int. Conf. Microelectromech. Syst. (MEMS 2002)*, Las Vegas, Nevada, 20–24 Jan. 2002, 236–239.
53. Humble, P.H., Harb, J.N., and LaFollette, R.M., Microscopic nickel-zinc batteries for use in autonomous microsystems, *J. Electrochem. Soc.*, 148(12), A1357, 2001.
54. Neudecker, B.J., Dudney, N.J., and Bates, J.B., "Lithium-free" thin-film battery with *in-situ* plated anode, *J. Electrochem. Soc.*, 147, 517–523, 2000.
55. Oak Ridge Micro-Energy, Inc., http://www.oakridgemicro.com/.
56. Front Edge Technology, Inc., http://www.frontedgetechnology.com/.
57. West, W.C. et al., Fabrication and testing of all solid-state microscale lithium batteries for microspacecraft applications, *J. Micromechan. Microeng.*, 12, 58–62, 2002.
58. TPL, Inc., http://www.tplinc.com/.
59. Blanchard, J.P. et al., Radioisotope power for MEMS devices. *ANS Trans. Am. Nucl. Soc.*, 86, 186–187, 2002.
60. Qynergy, Corp., http://www.qynergy.com/.
61. TRACE Photonics, Inc., Charleston, IL.
62. Roundy, S., Wright, P., and Pister, K.S.J. Micro-electrostatic vibration-to-electricity converters, *Intl. Mech. Eng. Conf. Exp. 2002*, IMECE2002-39309, Nov. 17–22, 2002, New Orleans, LA.
63. Meninger, S. et al., Vibration-to-electric energy conversion, *Proc. 1999 Int. Symp. Low Power Electron. Design*, San Diego, CA, 16–17 Aug. 1999, 48–53.
64. Teasdale, D. et al., Thrust and electrical power from solid propellant microrockets, *Proc. 14th Annu. Int. Conf. Microelectromech. Syst. (MEMS 2001)*, Interlaken, Switzerland, Jan. 2001.
65. Cohn, M.B. et al., Microassembly technologies for MEMS, *Proc. SPIE*, 3511, *Micromachining Microfabrication Process Technol. IV*, Santa Clara, CA, 3511, 2–16; 21–22, 1998.
66. Hsu, V., *M.S. Report*, University of California, Berkeley.
67. Mason, A. et al., A generic multielement microsystem for portable wireless applications, *Proc. IEEE*, 86(8), 1733–1746, 1998.
68. Asada, G. et al., Wireless integrated network sensors: low power systems on a chip, *Proc. 1998 Eur. Solid State Circuits Conf.*, The Hague, 22–24 Sept. 1998, 9–16.
69. Otis, B. and Rabaey, J., A 300μW 1.9GHz CMOS oscillator utilizing micromachined resonators, *Proc. Eur. Solid-State Circuits Conf. (ESSIRC)*, Florence, Italy, Sept. 24–26, 2002.

70. Warneke, B. et al., Smart dust: communicating with a cubic-millimeter computer, *Computer Mag.*, IEEE, Piscataway, NJ, 44–51, Jan. 2001.
71. Warneke, B.A. et al., An autonomous 16-mm^3 solar-powered node for distributed wireless sensor networks, *Proc. IEEE Int. Conf. Sensors 2002*, Orlando, FL, June 12–14, 2002.

6

Sensor Network Architecture and Applications*

Chien-Chung Shen
University of Delaware

Chaiporn Jaikaeo
University of Delaware

Chavalit Srisathapornphat
University of Delaware

6.1 Introduction 6-1
6.2 Sensor Network Applications 6-1
Querying Applications • Tasking Applications
6.3 Functional Architecture for Sensor Networks 6-3
6.4 Sample Implementation Architectures 6-4
SINA (Sensor Information Networking Architecture) • TopDisc (Topology Discovery for Sensor Networks)
6.5 Summary 6-12

6.1 Introduction

The sheer number of sensor nodes and the dynamics of their operating environments (for instance, limited battery power and hostile physical environment) pose unique challenges in the design of sensor networks and their applications. Issues concerning how information collected by and stored within a sensor network can be queried and accessed are of particular importance. In this chapter, sensor network applications are categorized into two classes — querying and tasking — and a generic functional architecture, termed *sensor network architecture* (SNA), to facilitate these applications is introduced. In this architecture, functional components and their interrelationship, which should be available in sensor networks, are identified. Two existing implementation architectures, SINA [1] and TopDisc [2], are examined as a case study by describing how SNA's functional components are exploited, as well as application characteristics supported by them.

The following section describes the two categories of applications for sensor networks. Section 6.3 describes the functional architecture of SNA. Two sample implementation architectures, SINA and TopDisc, are described in Section 6.4. Section 6.5 concludes the chapter.

6.2 Sensor Network Applications

Based on the characteristics of their operations, applications of sensor networks can be divided into two classes: querying and tasking. The following subsections present sample applications for each class.

*Portions reprinted with permission from *IEEE Personal Communications Magazine*, 8, 4, 2001. © 2001, IEEE.

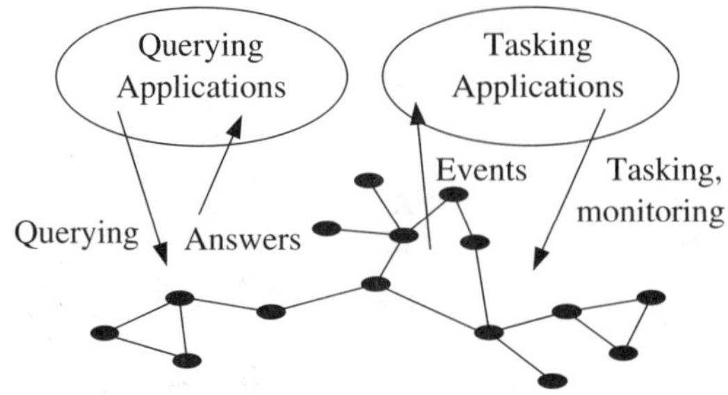

FIGURE 6.1 Querying and tasking applications in sensor networks. (From Shen, et al., *IEEE Personal Commun. Mag.*, 8(4), 52–59, 2001. With permission.)

6.2.1 Querying Applications

Querying applications concern how information collected by a sensor network can be retrieved based on specified criteria. For instance, environment sensing to extract information from the physical environments is one major application of sensor networks. Depending on its hardware capability, a sensor node can be programmed to collect temperature, humidity, light, pressure, chemical substances, or vibration information [3], and report it to the application. Applications may employ simple queries to obtain raw sensor data reported directly from each sensor node.

However, in some situations, complicated queries involving distributed data collection or aggregation become necessary. For example, to find out which region of the sensed area has the highest temperature, intelligent data collection, filtering, and aggregation could be carried out within the sensor network so that the observer will not need to obtain all raw data, thus conserving scarce system resources, such as battery energy and network bandwidth. In addition, the state of the sensor node, such as remaining energy level, operational status, or a list of neighboring sensors, can also be retrieved for management purposes [2]. The collected information could also be used to diagnose the health of sensors [4].

6.2.2 Tasking Applications

Tasking applications involve programming sensor nodes to perform specific actions upon certain events. Events can be physical environment changes, messages from nearby sensor nodes, or triggers from hardware/software modules inside a sensor node. A task can be as simple as asking individual sensor nodes to report information independently when they sense something unusual about their surrounding environments. More complex tasks may require distributed coordination, or even collaboration, among sensor nodes to achieve higher accuracy and/or efficiency. For instance, tracking a moving object in an area by simply having every single sensor node periodically and blindly monitor its surroundings can be very energy inefficient. If nodes surrounding the tracked object collaborate, more complete and accurate information can be collected with higher efficiency [5–7].

A similar idea of coordination can also be applied to reduce the number of nodes participating in data forwarding [2]. Modern equipment may have sensor modules operate in conjunction with actuator modules so that the behavior of sensor nodes can be controlled. In this case, tasking applications can utilize information obtained from sensor nodes to adapt nodes' behavior or movement pattern so as to achieve better sensing and networking performance. For environmental control applications, actuators can be controlled to affect the physical environments. An office building, for example, may have a sensor node installed in each room. These nodes then coordinate and send control signals to the air-conditioning unit, which, in turn, adjusts accordingly to achieve optimal comfort in all the rooms [8].

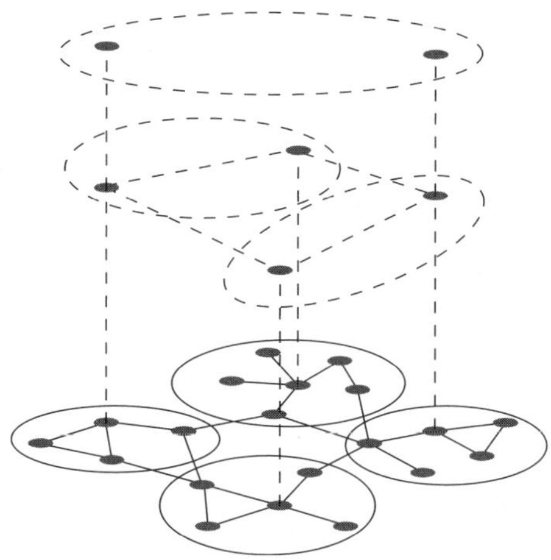

FIGURE 6.2 Clustering and a cluster hierarchy. (From Shen, et al., *IEEE Personal Commun. Mag.*, 8(4), 52–59, 2001. With permission.)

6.3 Functional Architecture for Sensor Networks

Compared to conventional distributed databases in which information is distributed across several sites, the number of sites in a sensor network equals the number of sensor nodes, and the information collected by each node (e.g., sensor readings) becomes an inherent attribute of that node [9]. To support energy-efficient and scalable operations, sensor nodes could be autonomously clustered. Furthermore, the data-centric nature of sensor information makes it more effectively accessible via an attribute-based naming approach instead of explicit addresses [10]. In addition, as these sensors are integrated into and extract information from physical environments, many applications also require the location information to be passed along with their sensor data. As a result, a generic functional architecture for sensor networks consists of the following components.

Hierarchical clustering. To facilitate scalable operations within sensor networks, sensor nodes could be aggregated to form clusters based on their energy levels and proximity. The aggregation process could also be recursively applied to form a hierarchy of clusters (Figure 6.2). Within a cluster, a cluster head will be elected to perform information filtering, fusion, and aggregation, such as periodic calculation of the average temperature of the cluster coverage area. In addition, the clustering process should be reinitiated in case the cluster head fails or runs low in battery energy. In situations in which a hierarchy of clusters is not applicable, the system of sensor nodes is perceived by applications as a one-level clustering structure in which each node is a cluster head by itself. The clustering algorithm introduced by Estrin and colleagues [10] allows sensor nodes automatically to form clusters, elect and re-elect cluster heads, and reorganize the clustering structure if necessary.

Location awareness. Because sensor nodes are operating in physical environments, knowledge about their physical locations becomes mandatory. Location information can be obtained via several methods. Global positioning system (GPS) is one of the mechanisms that provide absolute location information. For economical reasons, however, only a subset of sensor nodes may be equipped with GPS receivers and function as location references by periodically transmitting a beacon signal telling their own location information so that other sensor nodes without GPS receivers can roughly determine their position in the terrain. Other techniques for obtaining location information are also available. For example, optical trackers [11] give high-precision and -resolution location information but are only effective in a small region.

Attribute-based naming. With the large population of sensor nodes, it may be impractical to pay attention to each individual node. Users would be more interested in querying which area has temperature higher than 100°F or what the average temperature in a specific area is, rather than the temperature at sensor ID#101. To facilitate the data-centric characteristics of sensor queries, attribute-based naming is the preferred scheme [10]. For instance, the name [type=temperature, location=N-E, temperature=103] addresses all the temperature sensors located at the northeast quadrant with a temperature reading of 103°F. These sensors will reply to the query, "which area has temperature higher than 100°F?" Note that not only can physical or location attributes be part of a name, but so can logical attributes such as unique IDs, temporary variables, and clustering roles (e.g., cluster head or cluster member). Therefore, the traditional addressing scheme using node IDs becomes a special case of attribute based naming.

With the integration of these three components, the following two sample queries may be effectively and efficiently carried out.

- *Which area has temperature higher than 100°F?* In theory, the query is broadcast to and evaluated by every node in the network. Despite possibly the best returned result, the query would suffer from long response time. In practice, each cluster head may periodically update the temperature readings of its members, and the query can now be multicast to and evaluated by cluster heads only. This results in better response time at the expense of less accurate answers. Queries under stringent timing constraints can be evaluated by cluster heads of a higher tier.
- *What is the average temperature in the southeast quadrant?* Similarly, the average temperature of each cluster can be periodically updated and cached by cluster heads. Furthermore, the query should be delivered to nodes located (named) in the southeast quadrant only.

6.4 Sample Implementation Architectures

Given the SNA functional architecture, two implementation architectures are described: SINA, which implements SNA to facilitate querying and tasking applications, and TopDisc, which is specifically designed to perform topology management of sensor networks.

6.4.1 SINA (Sensor Information Networking Architecture)

SINA [1] adopts a middleware-based approach to implementing SNA functional architecture. By modeling a sensor network as a collection of massively distributed objects, SINA modules, running on each sensor node, serve as a middleware working across all sensor nodes; provide adaptive organization of sensor information; and facilitate query, event monitoring, and tasking (Figure 6.3). SINA allows sensor applications to issue queries and command tasks into, collect replies and results from, and monitor changes within the networks. SINA provides the following mechanisms to facilitate querying and tasking of sensor networks: information abstraction; information gathering methods; sensor query and tasking language; and sensor execution environment. These mechanisms are explained in detail in the following subsections.

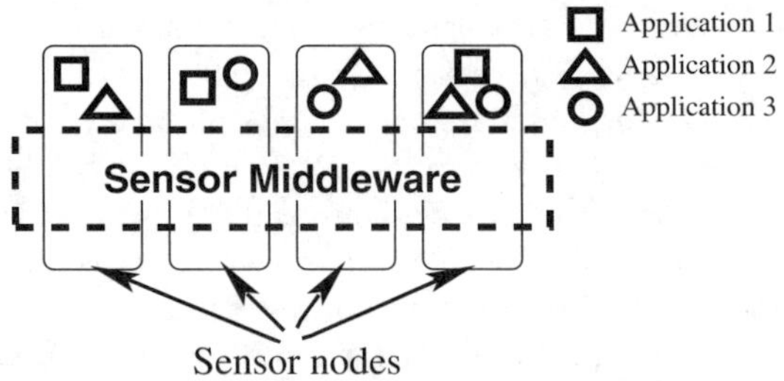

FIGURE 6.3 A model of sensor networks and SINA middleware. (From Shen, et al., *IEEE Personal Commun. Mag.*, 8(4), 52–59, 2001. With permission.)

6.4.1.1 Information Abstraction

In SINA, a sensor network is conceptually viewed as a collection of datasheets, each of which contains a collection of attributes of each sensor node. Each attribute is referred to as a cell, and the collection of datasheets of the network present the abstraction of an *associative spreadsheet.* In contrast to a conventional spreadsheet paradigm in which a data item is stored in a cell that is assigned an address according to its logical *x–y* coordinates, SINA refers cells via attribute-based names. Initially, a datasheet of each sensor node contains a few predefined attributes. Once these sensor nodes are deployed and form a sensor network, they can be requested by other nodes — for instance, from their cluster heads — to:

- Create new cells by evaluating valid cell construction expressions that may obtain information from other cells
- Invoke system-defined functions
- Aggregate information from other datasheets

Each newly created cell must be uniquely named and becomes a node's attribute, which can be a single value (e.g., remaining battery energy) or multiple values (e.g., history of temperature changes in the past 30 min). By incorporating a hierarchical clustering mechanism and an attribute-based naming scheme, SINA provides a set of operations to deal with data access and aggregation among sensor nodes. The mechanism of *associative broadcast* [12] has been employed to facilitate process interaction via attribute-based naming.

6.4.1.2 Information Gathering Methods

SINA provides a communication mechanism among sensor nodes to facilitate distributed applications. By providing efficient data dissemination and information-gathering supports suitable for specific application requirements, SINA abstracts low-level communications from high-level sensor applications. When users submit queries, it is not required to define how the information will be collected inside the network explicitly. SINA selects the most appropriate data distribution and collection method based on the nature of queries and current network status. Upon receiving users' queries, the *frontend* node — a special node directly connected to the user — has the responsibility to interpret and evaluate the queries by requesting information from other nodes.

With the sheer number of sensor nodes, collisions resulting from a large number of responses propagated back to the front-end node during a short period of time create the *response implosion problem* [9] depicted in Figure 6.4(a). The objective of the information-gathering mechanisms is to maximize the quality of responses in terms of their number and responsiveness while minimizing network resource consumption in conducting the query operations. Three primitive methods are provided to accomplish the information gathering task:

- *Sampling operation.* For certain types of applications (for instance, finding the average temperature over the entire network area), responses from every sensor node may cause a response implosion. To reduce the degree of the problem, some sensor nodes may not need to respond if their neighbors will. Nodes make autonomous decisions whether they should participate in this application based on a given response probability, as shown in Figure 6.4(b). This operation is also known as *Samplecast* [9]. An enhancement can be made to this approach if sensor nodes are not evenly distributed over the area. To prevent receiving more responses from dense areas, the response probability will be computed at each cluster head node based on the number of replies required from each cluster. This operation is called *adaptive probabilistic response (APR).*
- *Self-orchestrated operation.* In a network with a small number of nodes, responses from all nodes are necessary for the accuracy of the final result. Another approach to avoiding the response implosion problem is to let each node defer sending responses for some period of time. Despite some extra delay, this method aims to improve the overall performance by reducing the chances of collision. This operation is modified from the scheduled response approach described in Johnson and Maltz [13]. Assuming that nodes are distributed uniformly within the network

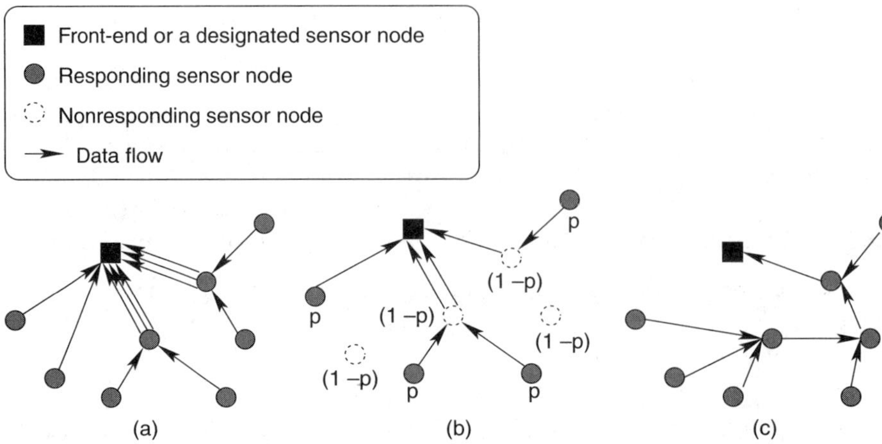

FIGURE 6.4 (a) The response implosion problem; (b) number of responses reduced by assigning sensor nodes a probability p to answer the request; (c) diffused computation operation allowing data aggregation at intermediate nodes. (From Shen, et al., *IEEE Personal Commun. Mag.*, 8(4), 52–59, 2001. With permission.)

terrain and that the number of nodes within h hops away from the front-end node proportional to h^2, the delay period at every node can be defined as

$$Delay = KH(h^2 - (2h-1)r)$$

where h is the length in number of hops away from the front end; r is a random number such that $0 < r \leq 1$; and H is a constant reflecting estimated delay per hop. To incorporate potential effects from queuing and processing delays, K is used as a compensation constant. Normally, K and H are combined and used as a single adjustable parameter.

- *Diffused computation operation.* For this operation, each sensor node is assumed to have knowledge about its immediate communicating neighbors only. Algorithms used for gathering information are constrained by the capability that each node can only communicate to other nodes in its surrounding area. Information aggregation logic is programmed as a script and disseminated among sensor nodes so that they know how to aggregate information en route to the front end. The conceptual data flow is depicted in Figure 6.4(c). Because data are aggregated at intermediate nodes on the way back to the front-end node, the consumption of valuable network bandwidth is reduced and the response implosion problem alleviated considerably. However, for large sensor networks, this diffusion approach might take a longer time to deliver results back to the front end.

The hierarchical structure enabled by SINA allows different information-gathering methods to be deployed in different levels within one application in order to optimize overall performance. The effects of the integration are discussed in Shen et al. [14].

6.4.1.3 Sensor Network Programming Languages

As part of SINA, sensor querying and tasking language (SQTL) [15] plays the role of a programming interface between sensor applications and SINA middleware. This is a procedural scripting language designed to be flexible and compact, with a capability of interpreting declarative query statements. In addition to sensor hardware access (e.g., `getTemperature`, `turnOn`), location-aware (e.g., `isNeighbor`, `getPosition`), and communication primitives (e.g., `tell`, `execute`), it also provides an event-handling construct, which is suitable for many sensor network applications in which sensor nodes are often programmed to process asynchronous events such as receiving a message or an event triggered by a timer. By using the "upon" construct, a programmer can create an event-handling block accordingly.

TABLE 6.1 Arguments Used by Actions in SQTL Wrapper

Argument	Meaning
`sender`	The sender of an SQTL message wrapper
`receiver`	Potential receivers specify by two following subarguments
`group`	Subargument of receiver to specify group of receivers; its possible value can be one of ALL_NODES, or NEIGHBORS
`criteria`	Subargument of receiver to specify selection criteria of receivers
`application-id`	Unique ID for each application in the same sensor network
`num-hop`	Number of hops away from a gateway node
`language`	Specify a language used in content
`content`	A payload containing a program, message, or return values
`with (optional)`	Tuples of parameters used in the program passed from sender to receiver
`parameter`	Repeatable subargument of "with"
`type`	Data type of the parameter
`name`	Name of the parameter
`value`	Value of the parameter

Source: From Shen, et al., *IEEE Personal Commun. Mag.*, 8(4), 52–59, 2001. With permission.

Currently, three types of events are supported by SQTL: (1) events generated when a message is received by a sensor node; (2) events triggered periodically by a timer; and (3) events caused by the expiration of a timer. These types of events are defined by the SQTL keywords "receive," "every," and "expire," respectively.

An SQTL message, containing a script, is meant to be interpreted and executed by any node in the network. In order to target a script to a specific receiver, or a group of receivers, the message must be encapsulated in an *SQTL wrapper* that acts as a message header for indicating the sender, the receivers, and a particular application running on the receivers, as well as parameters for the application.

The syntax of the extensible markup language (XML) is adopted for the SQTL wrapper, which defines an application layer header capable of specifying a complicated addressing scheme for attribute-based names. Table 6.1 summarizes common SQTL wrapper fields.

For applications that collect sensor information, a user may choose to invoke the built-in query interpreter instead of explicitly writing a procedural SQTL script. The query language has been adapted from structured query language (SQL) to serve as the primary mechanism for querying sensor networks. The following sample query statement, as delivered to all cluster heads in the network (encapsulated in the SQTL wrapper), would ask every cluster head to create a new cell called *avgTemperature* that maintains the average temperature among all of its cluster members:

SELECT avg(*getTemperature*())
  AS *avgTemperature*
  FROM CLUSTER-MEMBERS

As soon as an SQTL message containing such a query statement is received by target nodes, their execution environments (explained later) will pick the most appropriate information-gathering method available to evaluate the query.

Database techniques, such as view composition, materialization, and maintenance, could be adapted to maintain consistency among associated cells. A related work on querying a sensor network modeled as a device database may be found in Bonnet et al. [16].

6.4.1.4 SEE (Sensor Execution Environment)

Running on each sensor node, a sensor execution environment (SEE) is responsible for dispatching incoming messages, examining all arrival SQTL messages, and performing the appropriate operation for each type of action specified in the messages. SEE looks inside the `receiver` argument of a message and, based on its value, decides whether to forward the message to the next hop. Messages with "ALL_NODES" in their group subarguments will be rebroadcast to every sensor node in the network and those with "NEIGHBORS" will only be forwarded to the nodes' immediate neighbors.

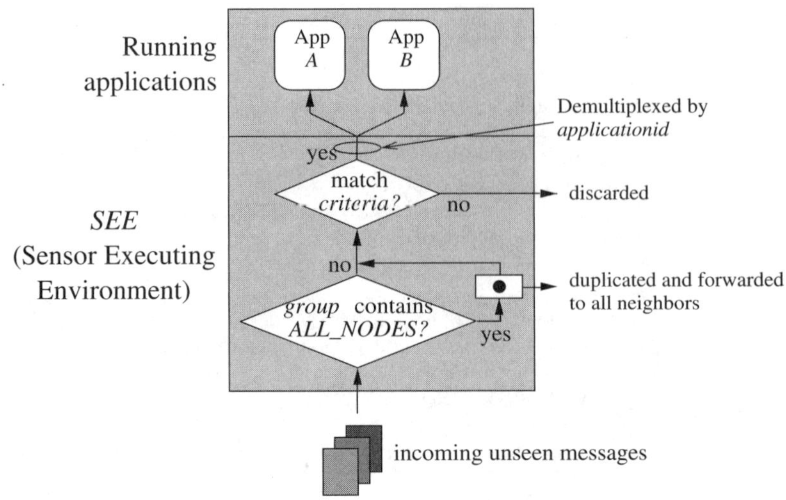

FIGURE 6.5 Dispatching of messages received by a sensor node. (From Shen, et al., *IEEE Personal Commun. Mag.*, 8(4), 52–59, 2001. With permission.)

SEE also prevents message looping by using a globally unique message ID, which is a combination of a unique node ID and message sequence number. An attribute-based name in the form of a list of attribute–value pairs indicated by the criteria field will be compared against the receiver's attributes stored in its datasheet. SEE only accepts the message if the node's attributes satisfy the criteria. This process of matching a message with its potential receivers when the message arrives at the receivers is termed *late binding* and is described by Bayerdorffer [12].

Once an SQTL script is injected from the front-end node to one or more sensor nodes, the script may push itself to other sensors in order to complete the assigned task. A `tell` message is then generated after a result is produced at each individual sensor node and is delivered back to the requesting node, which is normally the upstream node from which the script came. Figure 6.5 depicts the dispatching of incoming messages performed by SEE.

In addition to demultiplexing incoming SQTL messages, SEE also takes care of outgoing SQTL messages from all running applications. Outgoing messages will be distributed to target nodes specified in the `receiver` argument through the underlying communication mechanism. SEE may perform a translation of an attribute-based name into a unique, numeric link-layer address where applicable. Otherwise, broadcast will be used at the link layer.

6.4.1.5 Architectural View of SINA

Now the ways in which the three functional components defined in SNA are utilized and provided in SINA are examined. SINA provides an attribute-based naming mechanism by means of an associative spreadsheet in which nodes' attributes are defined in uniquely named cells. Destination groups are then determined by `criteria` fields that are part of SQTL. A mechanism for hierarchical clustering is not strictly tied to a particular algorithm and is intentionally left undefined for flexibility. A clustering algorithm such as the one described by Intanagonwiwat and colleagues [17] could be used. Once cluster heads have been elected, each node's cluster head role (i.e., whether it is a cluster member or a cluster head) will become one of its attributes. The clustering feature also allows different information-gathering methods to be used at different levels in the hierarchy in order to optimize overall performance. Similarly, mechanisms allowing nodes to obtain their location information are assumed, but not defined or used directly in SINA. It is left to the applications to target and query nodes' locations in the form of their attributes.

6.4.1.6 Sample Applications

SINA has been designed to support a wide range of sensor network applications. However, to illustrate its applicability to querying and tasking of sensor networks under this architecture, experiments were conducted on two sample applications: sensor network diagnosis and vehicle tracking; their behaviors and performance were studied using GloMoSim simulator. Results and more discussion of the two applications can be found in Shen et al. [14].

Diagnosis of sensor networks. Sensor network diagnosis is the process of querying the status of a sensor network and figuring out the problematic (group of) sensor nodes [4]. In order to monitor the status of a sensor network, one approach is to query as much information from as many sensor nodes as possible and then deliver the raw information to the manager for further processing, e.g., when a manager wants to know the remaining energy level within the network. In addition, to examine the correctness of results obtained from one sensing device, one possible method is to use the average of results obtained from other neighboring sensor nodes as a standard base to compare and diagnose the devices in doubt, given that the average has its deviation within an acceptable range. An example of using this method is to figure out which sensor node contains a faulty temperature-sensing device.

Coordinated vehicle tracking. The vehicle tracking application is to locate a specific vehicle or moving object and monitor its movement. To detect and identify an object, integrated results from more than one type of sensor, for instance, images from a camera, vibration from a seismic sensor, noise from an audio sensor, and so on, may be required. These results are to be processed and compared with the signature of the object of interest. However, the main interest is to program a coordination algorithm in the form of an SQTL script, which can be disseminated to all sensor nodes. The script controls the sensor nodes to detect the appearance of the interested object collaboratively in an effective and efficient manner. Thus, it is assumed that sensor nodes can obtain final processed results of detecting and identifying the tracked vehicle from the processing of combined sensing information.

A novice approach to tracking a moving object is to ask every sensor node to sense and detect the object's signature at the same time — an operation called the *ordinary vehicle tracking method.* However, this approach may waste sensor nodes' processing cycles, and thus inefficiently utilize a network's limited energy and shorten the overall network lifetime. In contrast, the coordinated vehicle tracking algorithm presented in Figure 6.6 is based on a suppression and reinitiation mechanism in order to achieve a better result of tracking, yet consume less network resources than the ordinary one. The main principle of the coordinated algorithm is to let the first sensor node detecting the vehicle suppress sensing activities of all other sensor nodes so that the others may stand by, which results in energy conservation. Furthermore, the node will need to reinitiate sensing activities of its neighbors in order to keep track of the moving vehicle. As long as the vehicle does not move faster than the propagation of this reinitiation message, the network can still monitor its trail. The tracking process is depicted in Figure 6.7 as well.

6.4.2 TopDisc (Topology Discovery for Sensor Networks)

TopDisc [2] provides a mechanism for data dissemination/aggregation and topology discovery in sensor networks. From an architectural point of view, TopDisc provides the same set of components specified by SNA. The following subsections describe the mechanism of TopDisc and present how its functional components are mapped to the SNA architecture. Finally, some sample applications supported by TopDisc are offered.

6.4.2.1 TopDisc Mechanism

TopDisc constructs an approximate topology of the network by collecting local topology information from *distinguished nodes* (or cluster heads) via a tree of clusters (TreC) rooted at the monitoring node. The mechanism is briefly described as follows. When TopDisc starts, all nodes are colored *white*, which means that they are undiscovered. The monitoring node initiates the topology discovery process by broadcasting a "topology discovery request." It then turns to *black*, which means that it is a distinguished

```
<execute>
   <sender> FRONTEND </sender>
   <receiver>      <group> NODE[0] </group>
                   <criteria> TRUE </criteria>
   </receiver>
   <application-id> 118 </application-id>
   <num-hop>        0    </num-hop>
   <language> SQTL </language>
   <with>
      <parameter type="clocktype" name="trackingTime"       value="600" />
      <parameter type="clocktype" name="reTrackingTime"     value="40" />
      <parameter type="clocktype" name="trackingFrequency" value="8" />
      <parameter type="object"    name="target"             value="Vehicle1" />
   </with>
   <content> <![CDATA[
      lastSensingResul  = false;
      timerApplication = createTimer(trackingTime);    // instantiate a timer
      timerApplication.start();                        // turn it on
      timerReTracking   = createTimer(reTrackingTime);
      execute (ALL_NODES, "TRUE", MESSAGE["content"]);   // re-broadcast
      if ((sensor1 = getMotionSensor()).turnOn()) {    // instantiate a sensor object
         upon {                  /                      /    and turn it on
            receive (msg) where msg["action"] == "tell" && msg["content"] == "suppress": {
               sensor1.standby();   break;
            }
            every (trackingFrequency): {
               if (sensor1.detect(target)) {
                  tell (ALL_NODES,          "TRUE", "suppress");
                  tell (NEIGHBORS,          "TRUE", "retrack");
                  tell (MESSAGE["sender"], "TRUE", "found");
                  lastSensingResult = true;
                  timerReTracking.start();
                  break;
               }
               else  lastSensingResult = false;
            }
            expire (timerApplication): sensor1.turnOff(); exit(0);
         }
         upon {    // After one sensor node sees the vehicle
            receive (msg) where msg["action"] == "tell" && msg["content"] == "retrack": {
               if (timerReTracking.expired()) {
                  sensor1.turnOn();
                  timerReTracking.start();
               }
            }
            receive (msg) where msg["action"] == "tell" && msg["content"] == "found":
               tell (MESSAGE["sender"], "TRUE", "found");
            every (trackingFrequency): {
               if (sensor1.detect(target)) {
                  tell (MESSAGE["sender"], "TRUE", "found");
                  if (!lastSensingResult)
                     tell (NEIGHBORS,       "TRUE", "retrack");
                  lastSensingResult = true;
                  timerReTracking.start();
               }
               else {
                  if (lastSensingResult)
                     timerReTracking.restart();
                  lastSensingResult = false;
               }
            }
            expire (timerReTracking)  : sensor1.standby();
            expire (timerApplication): sensor1.turnOff(); exit(0);
         }
      }
      else  exit(1);
   ]]> </content>
</execute>
```

FIGURE 6.6 Complete SQTL script for the coordinated vehicle tracking algorithm. (From Shen, et al., *IEEE Personal Commun. Mag.*, 8(4), 52–59, 2001. With permission.)

node. White nodes receiving a request from a black node become *gray* and rebroadcast the request with a random delay inversely proportional to the distance between the black node and themselves.

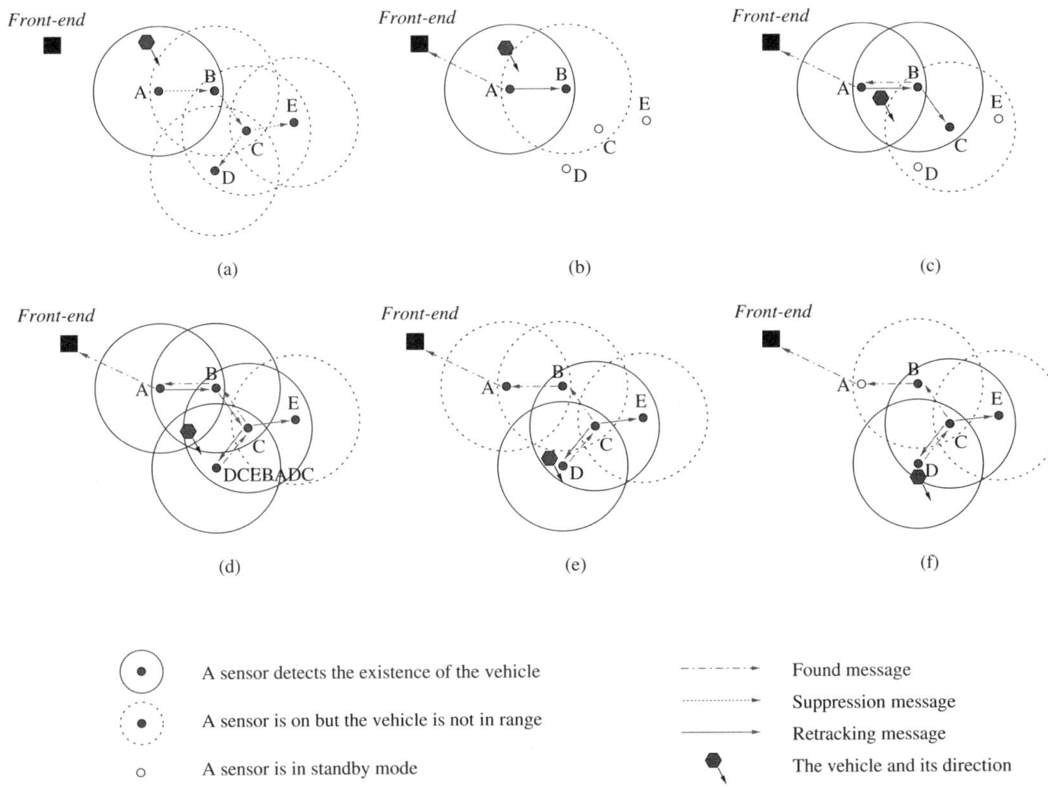

FIGURE 6.7 (a) The incoming vehicle is detected by A; (b) the sensing activities of C, D, and E are suppressed, but B starts tracking again; (c) the vehicle comes into B's area and C restarts its sensor; (d) C and D detect the vehicle and E's sensor is restarted; (e) the vehicle goes out of A's and B's ranges; (f) sensing activity at A stops. (From Shen, et al., *IEEE Personal Commun. Mag.*, 8(4), 52–59, 2001. With permission.)

However, white nodes will become black with some random delay if they receive a request from a gray node. During the delay interval, if white nodes hear any message from other black nodes, they will become gray. Note that all the black and gray nodes ignore all other incoming request messages. After the request has been propagated to the entire network, each node knows its *parent black node*, which is the last black node from which the topology discovery was forwarded to reach it. Each black node also knows the node to which it should forward packets in order to reach its parent black node. By snooping at all incoming request messages, all nodes have their neighborhood information collected.

To respond to the topology discovery message, once a node becomes black, it sets a timer, inversely proportional to the number of hops away from the monitoring node, and waits for responses from its children black nodes. A black node aggregates its own neighborhood list (obtained from snooping) together with neighborhood lists from its children and forwards the aggregated list back to the monitoring node through its default forwarding node.

6.4.2.2 Architectural View of TopDisc

Similar to SINA, TopDisc provides the same set of components described by SNA. First, TopDisc builds a TreC by selecting *distinguished nodes* to become cluster heads. Other nodes then associate with one cluster head. This process has the same functionality as the hierarchical clustering component of SNA. Nodes in TopDisc also perform information aggregation by combining messages obtained from children black nodes. The objective of a TreC and data aggregation is to reduce the number of response messages coming back to the monitoring node. TopDisc also employs attribute-based naming schemes in its data dissemination process. Subsequent requests to the network will be carried over a TreC. Recall that a TreC comprises black (cluster head) and gray (forwarding) nodes. However, only cluster heads will process the requests; gray nodes only forward the requests. This process resembles attribute-based naming. Finally, TopDisc employs location information in one of its proposed applications to schedule sensor nodes' duty cycles.

6.4.2.3 Sample Applications

By using a TreC created by TopDisc, several data dissemination/aggregation applications are possible. The following applications are described in Deb et al. [2]:

- *Retrieving network state.* Connectivity, reachability, and energy maps, as well as a usage model of sensor networks, could be obtained from data collected via TopDisc.
- *Data dissemination and aggregation.* The resulting tree created by TopDisc could also be used in data dissemination and aggregation applications.
- *Duty cycle assignment.* Each pair of closest black nodes can exchange location information of their children. After collecting the complete topology of the surrounding nodes, one of the children may decide to serve as a forwarding node. It then informs other nodes so that they can go into sleep mode. Based on the category presented in Section 6.2, this application can be considered a tasking application.

6.5 Summary

The advent of technology has facilitated development of networked systems of small, low-power devices that combine programmable computing with multiple sensing and wireless communication capability. Already, experimental applications have embedded sensor nodes in the physical environment to facilitate new information-gathering and -processing capabilities. The sheer number of sensor nodes and the dynamics of their operating environments pose unique challenges on how information collected by and stored within a sensor network can be queried and accessed, and how concurrent sensing tasks can be executed internally and programmed by external clients. This chapter described a generic functional architecture for sensor networks by identifying three required functional components: hierarchical clustering, location awareness, and attribute-based naming. Two sample implementation architectures, SINA and TopDisc, were examined in terms of their exploitation of these functional components and the application characteristics they are intended to support.

References

1. Srisathapornphat, C., Jaikaeo, C., and Shen, C.-C., Sensor information networking architecture, in *Proc. 2000 Int. Workshop on Parallel Processing*, 23–30, Toronto, Canada, August 21–24, 2000.
2. Deb, B., Bhatnagar, S., and Nath, B., A topology discovery algorithm for sensor networks with applications to network management, in *IEEE CAS Workshop Wireless Commun. Networking*, Pasadena, CA, September 5–6, 2002.
3. Akyildiz, I.F. et al. Wireless sensor networks: a survey, *Computer Networks*, 38(4), 393–422, 2002.
4. Jaikaeo, C., Srisathapornphat, C., and Shen, C.-C., Diagnosis of sensor networks, in *Proc. IEEE Int. Conf. Commun. (ICC 2001)*, 5, 1627–1632, Helsinki, Finland, June 11–15, 2001.
5. Cerpa, A. et al. Habitat monitoring: application driver for wireless communications technology, in *ACM SIGCOMM Workshop Data Commun. Latin America Caribbean*, 20–41, Costa Rica, April 3–5, 2001.
6. Huang, Q., Lu, C., and Roman, G.-C., Reliable mobicast via face-aware routing, Washington University, St. Louis, MO, Tech. Rep. WUCSE-2003-49, July 2003.
7. Zhang, W. and Cao, G., DCTC: dynamic convoy tree-based collaboration for target tracking in sensor networks, *IEEE Trans. Wireless Communications*, in press.
8. Lin, C., Federspiel, C.C., and Auslander, D.M., Multi-sensor single-actuator control of HVAC systems, in *Proc. Int. Conf. Enhanced Building Operations*, Richardson, TX, October, 2002.

9. Imielinski, T. and Goel, S., DataSpace: querying and monitoring deeply networked collections in physical space, *IEEE Personal Commun.*, 7(5), 4–9, October 2000.
10. Estrin, D., Govindan, R., and Heidemann, J., Embedding the Internet, *Commun. ACM*, 43(5), 39–41, May 2000.
11. Ward, A., Jones, A., and Hopper, A., A new location technique for the active office, *IEEE Personal Commun.*, 4(5), 42–47, October 1997.
12. Bayerdorffer, B.C., Distributed programming with associative broadcast, in *Proc. 28th Hawaii Int. Conf. System Sci.*, 2, 353–362, Hawaii, January 1995.
13. Johnson, D.B., Maltz, D.A., and Broch, J., DSR: the dynamic source routing protocol for multi-hop wireless and ad hoc networks, in *Ad Hoc Networking*, Charles E. Perkins, (Ed.), Addison-Wesley, 2001, chap. 5.
14. Shen, C.-C., Srisathapornphat, C., and Jaikaeo, C., Sensor information networking architecture and applications, *IEEE Personal Commun. Mag.*, 8(4), 52–59, August 2001.
15. Jaikaeo, C., Srisathapornphat, C., and Shen, C.-C., Querying and tasking in sensor networks, in *Proc. SPIE's 14th Annu. Int. Symp. Aerospace/Defense Sensing, Simulation, Control (Digitization of the Battlespace V)*, 4037, 184–197, Orlando, FL, April 24–28 2000.
16. Bonnet, P., Gehrke, J., and Seshadri, P., Querying the physical world, *IEEE Personal Commun.*, 7(5), 10–15, October 2000.
17. Intanagonwiwat, C. et al., Directed diffusion for wireless sensor networking, *ACM/IEEE Trans. Networking*, 11(1), 2–16, February, 2002.

7

A Practical Perspective on Wireless Sensor Networks

7.1 Introduction 7-1
7.2 WSN Applications 7-2
Military Applications • Environment Detection and Monitoring • Disaster Prevention and Relief • Medical Care • Home Intelligence • Scientific Exploration • Interactive Surroundings • Surveillance • Other Applications
7.3 Classification of WSNs 7-6
7.4 Characteristics, Technical Challenges, and Design Directions 7-7
Characteristics • Technical Challenges and Requirements • Design Objectives and Directions
7.5 Technical Approaches 7-11
Hardware Techniques • System Architecture, Protocols, and Algorithms • Software Development
7.6 Conclusions and Considerations for Future Research 7-22

Quanhong Wang
Queen's University

Hossam Hassanein
Queen's University

Kenan Xu
Queen's University

7.1 Introduction

Rapid progress in microelectromechanical system (MEMS) and radio frequency (RF) design has enabled the development of low-power, inexpensive, and network-enabled microsensors. These sensor nodes are capable of capturing various physical information, such as temperature, pressure, motion of an object, etc., as well as mapping such physical characteristics of the environment to quantitative measurements. A typical wireless sensor network (WSN) consists of hundreds to thousands of such sensor nodes linked by a wireless medium.

WSNs have created new paradigms for reliable monitoring. They outperform conventional sensor systems, which use large, expensive macrosensors to be placed and wired accurately to an end user. Detailed discussions of such benefits can be found in the literature [1, 13, 31–33, 43]. Some of these benefits are highlighted as follows:

- *Anywhere and anytime.* The coverage of a traditional macrosensor node is narrowly limited to a certain physical area due to the constraints of cost and manual deployment. In contrast, WSNs may contain a great number of physically separated nodes that do not require human attention. Although the coverage of a single node is small, the densely distributed nodes can work simultaneously and collaboratively so that the coverage of the whole network is extended. Moreover, sensor nodes can be dropped in hazardous regions and can operate in all seasons; thus, their sensing task can be undertaken anytime.

- *Greater fault-tolerance.* This is achieved through the dense deployment of wireless sensor nodes. The correlated data from neighboring nodes in a given area makes WSNs more fault tolerant than single macrosensor systems. If the macrosensor node fails, the system will completely lose its functionality in the given area. On the contrary in a WSN, if a small portion of microsensor nodes fails, the WSN can continue to produce acceptable information because the extracted data are redundant enough. Furthermore, alternative communication routes can be used in case of route failure.
- *Improved accuracy.* Although a single macrosensor node generates more accurate measurement than one microsensor node does, the massively collected data by a large number of tiny nodes may actually reflect more of the real world. Furthermore, after processing by appropriate algorithms, the correlated and/or aggregated data enhance the common signal and reduce uncorrelated noise.
- *Lower cost.* WSNs are expected to be less expensive than their macrosensor system counterparts because of their reduced size and lower price, as well as the ease of their deployment.

In this chapter, Section 7.2 describes diverse applications of WSNs in various domains with examples and Section 7.3 discusses the classifications of the WSNs according to different criteria. Section 7.4 presents the characteristics of WSNs, highlights how they differ from traditional wireless ad hoc networks, and reviews the technique challenges and corresponding design directions. In Section 7.5, various technical approaches with respect to hardware design, system architectures, protocols and algorithms, and software development are illustrated. Finally, Section 7.6 concludes with emphasis on several possible open issues for future research in the area of WSNs.

7.2 WSN Applications

WSNs are able to monitor a wide range of physical conditions, such as [2]:

- Temperature
- Humidity
- Light
- Pressure
- Object motion
- Soil composition
- Noise level
- Presence of a certain object
- Characteristics of an object such as weight, size, moving speed, direction, and its latest position

Due to WSNs' reliability, self-organization, flexibility, and ease of deployment, their existing and potential applications vary widely. As well, they can be applied to almost any environment, especially those in which conventional wired sensor systems are impossible or unavailable, such as in inhospitable terrains, battlefields, outer space, or deep oceans.

7.2.1 Military Applications

WSNs are becoming an integral part of military command, control, communications, computing, intelligence, surveillance, reconnaissance, and targeting (C^4ISRT) systems [2]. In the battlefield, a predictable tendency is that the targets will become smaller and less recognizable/detectable, have higher mobility, and usually move in extremely hostile terrain. To explore the position and strength of the opposing forces, a promising solution lies in dense arrays of sensors to be placed close to the intended targets. Because of their ability to be unattended by humans, ease of deployment, self-organization, and fault tolerance,

WSNs can provide highly redundant and collaborative detected data without the support of friendly forces. Also, WSNs can be mounted on unmanned robotic vehicles, tanks, fighter planes, submarines, missiles, and torpedoes to route them around obstacles, guide them to the exact position and lead them to coordinate with one another to fulfill more effective attacks or defenses. WSNs can also be deployed for remote sensing of nuclear, biological, and chemical weapons, potential terrorist attack detection, and reconnaissance [2, 37]. Obviously, WSNs will take more important roles in the military C^4ISRT tasks and make future attacks and defenses more intelligent, with less human involvement.

7.2.2 Environment Detection and Monitoring

Spreading hundreds to thousands of tiny, cheap, self-configurable wireless sensors in a given geographical region can produce a wide range of applications in collaborative monitoring or control of the environment. This encompasses complex ecosystem monitoring; flood detection; air and sewage monitoring; local climate control in large office buildings; soil composition detection and precise agriculture; wild land fire detection; and exploration of mineral reserves, geophysical studies, etc. [2, 12, 32, 64]. Some representative examples include:

- *Ecosystem monitoring.* WSNs used in ecosystem monitoring represent a class of applications with numerous potential benefits for life science study because WSNs can provide information on several environmental conditions, including soil and air chemistry as well as plant and animal species population and behaviors. It ensures the long-term automatic identification, recording, and analysis of interesting events. These long-term gathered data can help ecosystem scientists to identify, localize, track, and predict species or phenomena in areas of interest [12, 32, 64]. Compared with traditional methods of environment monitoring, WSNs have a number of unique advantages:
    - Noninvasive deployment: unattended wireless sensors can be dropped on remote islands or dangerous places where it would be unsafe, unwise, or even impossible to perform field study repeatedly.
    - Anytime deployment: wireless sensor nodes can be deployed in any selected period, for example, before the producing season of some species of animal or after frozen ground melts.
    - Minimal interference: deploying WSNs for biosystems can eliminate the disturbance impact on the measured objects. For example, some species are very sensitive to the unexpected visits necessary for large-size macrosensor equipment; this can lead to a dramatic increase of mortality in a breeding year.
    - Less cost: deployment of WSNs also leads to a more economical solution to producing long-term observations than human-attended methods do.
    - Higher level of robustness and accuracy: by integrating data aggregation and signal processing within the neighborhood sensors, WSNs become more robust to node failure. Self-configurable WSNs used for biocomplexity mapping are adaptive to the dynamic physical world.
    - Ease of networking: sensor nodes are capable of connecting to the Internet, thus enabling one remote user to control, monitor, and collect data for several different sensed spots or several remote users to gather data for the same spot.

  Mainwaring and colleagues [64] present a real-life experiment of deploying WSN in a natural area — Great Duck Island (44.09N, 68.15W), Maine — to monitor the Leach's Storm Petrel, in terms of short-term cycle (24 to 72 h) of the usage pattern of nesting burrows and long-term (7 months) changes in the burrow and surface environmental parameters. The experiment is intended to guide the reliable environmental monitoring in these previously unaccessible fields.
- *Local climate control in large buildings.* Most people who have worked in large office buildings have experienced that the temperature is seldom proper, i.e., too high or too low; the humidity level is often overly dry or overly wet; too much or too little light is present; or fresh air is lacking. Therefore, local climate monitoring and control systems are highly desirable to ensure healthy and pleasant working places. At present, traditional systems with wired sensors are dominant in such

areas. Distributed WSNs are considered a better solution than their wired counterparts in at least two respects. For one thing, the deployment of a WSN is much more flexible than a wired system. Without the restriction of wire, wireless sensors can sit wherever they are needed; they can also be moved from their original positions to more suitable places. Moreover, WSNs can produce tremendous economical gains compared to wired sensors. According to da Silva et al. [93] and Rabaey et al. [79], for sensing mission, 90% of the total installation cost of a low-cost temperature sensor is due to wiring. Obviously, installation cost can be greatly reduced if wireless sensors are used.

- *Wild land fire detection.* Although significant measures have been exerted, wild land fires still cause extensive loss of lives, property, and resources each year. According to the statistics of the National Interagency Fire Center [71], the 10-year (1992 to 2001) average of wild land fires reached 103,112 and a total of 42,150,890 acres were burned. It costs approximately $1.6 billion (U.S.) on average for fire suppression by federal agencies only. However, because fire weather conditions are predictable, wild land fire prediction is often a possible source of help to support any geographic area before and during periods of high fire danger or fire activity. Because of their ability to be deployed randomly and densely, WSNs are a good choice in wild land fire detection and reporting. By scattering massive numbers of wireless sensors in intended areas, early warning and origin of fires can be caught effectively.

7.2.3 Disaster Prevention and Relief

WSNs may also be effectively deployed in emergency situations and disaster areas [37]. The accurate and prompt location detection provided by the distributed WSNs could be critical in rescue operations, including detection of victims, potential hazards, or sources of the emergency and identification and localization of trapped personnel [83]. For example, microsensors may be embedded/enabled in large-scale buildings during construction, through strategically dropping on the spot at the rescue site, or by automatically triggering standby sensors immediately following the disaster event. The collapse of the walls or ceiling could be predicated and estimated by the stress and motion of buildings. It is also useful to deploy WSNs for long-lasting monitoring tasks, such as detecting and tracking material fatigue, so that the evidence of harmful reaction of the building can be collected continuously and effective measures can be taken before an accident happens. Another example, waterproof sensor arrays, can be automatically triggered to constantly report the location of sunken vessels in the ocean and to provide critically important information for the rescue and salvage operation. Furthermore, wireless sensors can also be used to track fuel, gas, and toxic substances leaked into the neighborhood ocean when a sunken vessel is raised.

7.2.4 Medical Care

WSNs are very helpful in providing prompt and effective health care and will lead to a healthier environment for human beings. Some uses of WSNs in this field include:

- *Remote virus monitoring.* Many widespread disease-ridden regions are impoverished and lack reliable communication. Spreading large number of wireless sensors in such regions could help to collect and transmit crucial ground-based information, such as incident of disease and characteristics of the infected population; to identify features of the area; and to monitor environmental conditions, such as the amount of rainfall and humidity, that support the proliferation of virus-carrying insects. WSNs can also be used to monitor and predict the breakout of some infectious diseases, such as malaria. A project called Health Improvements through Space Technologies and Resources (HI-STAR) proposes development of a global malaria information system [26]. Based on the gathered air and ground-based data via wireless sensors and by integrating and analyzing epidemiological information, this system can generate malaria "risk maps" and provide early warnings about malaria outbreaks. Health officials could also allocate limited disease prevention and treatment resources on a global scale.

- *Integrated patient tracking and monitoring.* Using WSNs to monitor and track possible or suspected patients is a convenient and effective measure to avoid the spread of some infectious diseases. According to a Canadian Broadcast Corporation (CBC) news report in April 2003, discussion was that some people who broke quarantine in Toronto during the period of severe acute respiratory syndrome (SARS) in Spring 2003 could be required to wear a lightweight device with a wireless sensor on their ankles. This device could monitor their movements and report them to the relevant authorities. Moreover, senior citizens without sufficient care could have wireless sensors attached to medical devices to measure their heart rates, blood pressure, etc. In abnormal conditions, an automatic alert reminds the carriers to call their doctors or an automatic notification is directly sent to emergency centers. Furthermore, WSNs can also be used for medical statistics that require data collection from a large number of people or tracing some patients for long period of time.

  Schwiebert and colleagues [88] present a series of applications of WSNs in health care, such as artificial retina; glucose level monitoring for diabetes patients; organ monitoring for organ transplant purposes; and cancer detection for high-risk persons, as well as general health monitoring. WSNs can also be used in drug administration and distribution [2].

7.2.5 Home Intelligence

WSNs can take key roles in providing more convenient and intelligent living environments for human beings. Some predictable examples include:

- *Remote metering.* WSNs can be used in remote reading of utility meters, such as water, gas, or electricity, and then can transmit the readings through wireless connections [37]. Simple attachments of wireless sensors in parking meters can send out warning signals to remind users to recharge the meter remotely before the parking time expires.
- *Smart space.* With recent technological development, it becomes possible to embed various wireless sensors into individual furniture and appliances, which can be connected together to form an autonomous network. For example, a smart refrigerator can understand the family's dietary requirements or doctor's orders and take inventory of refrigerators to relay information to a shopping list on a personal digital assistant [21]. It can also create a menu according to the inventory and transmit the relevant cooking parameters to the smart stove or microwave oven, which will set the desired temperature and cooking time accordingly [46]. Moreover, contents and schedules of TV, VCR, DVD, or CD players can be monitored and operated remotely to satisfy the different requirements of family members.

7.2.6 Scientific Exploration

The effective deployment and operation of self-regulating WSNs is opening novel ways of scientific exploration in higher, further, and deeper environments such as outer space and deep oceans. Hong and colleagues [50] present an example for employing WSNs on the surface of Mars to collect measurements such as seismic, chemical, and temperature and relay the aggregated sensing results to an orbiter. Each distributed sensor node provides time- and position-dependent measurements; via energy-conserved, load-balanced, multihop communications, they can relay the information to the distant base station with prolonged network lifetime. Similarly, WSNs used for underwater exploration may also be possible in the future.

7.2.7 Interactive Surroundings

WSNs produce promising mechanisms for mining information from and reacting to the physical world. By deploying cheap and tiny wireless sensors, monitors and actuators in toys and other children's familiar objects could create "smart kindergartens" to enhance early childhood education [98]. Such a system provides a childhood learning environment with "person–physical world" interaction rather than the conventional "person–computer" or "person–person" communication. Because it allows personalized configuration to each individual child; coordinated activities of children groups; adaptation to the dynamics in children's activities; and constant and unobtrusive data collection in children's actions and learning processes, it provides effective and comprehensive problem-solving strategies in young children's education. Rabaey et al. [79] described WSNs in the real world in an interactive museum in San Francisco's Exploratorium, where children can participate actively in the experiments and get feedback to their touch and speech from the sensor-equipped objects. Yarvis and colleagues [106] present another interactive ad hoc sensor network as a voting platform in San Francisco's Moscone Convention Center.

7.2.8 Surveillance

Instant and remote surveillance inspires significant applications of WSNs. For example, a large number of networked acoustic sensors can be used to detect and track desired targets in a deterministic security area [68, 83, 109]. WSNs can be deployed in buildings, residential areas, airports, railway stations, etc. to identify intruders and report to a command center immediately so that tracking actions can be initiated promptly [62]. Similarly, installing smoke sensor nodes in strategically selected positions at homes, office buildings, or factories is critical to preventing disasters of fires and tracing the spread of fire [37, 65].

7.2.9 Other Applications

Self-configurable WSNs can be used in many other areas, such as robot control and factory instrumentation, automatic warehouse inventory tracking, chemical process control, traffic monitoring and control of smart roads, etc.

7.3 Classification of WSNs

As discussed in Section 7.2, WSNs represent a variety of applications in which environment and technical requirements may greatly differ. Therefore, the design of a WSN is usually application oriented. As a result, the architectures, protocols, and algorithms of WSNs vary case by case. However, different WSNs have some common properties in a broad point of view [100]. They can generally be classified into categories based on several important criteria.

According to the distance of sensor nodes to the base station, WSNs can be single-hop (also known as nonpropagating) or multihop (propagating) systems. In a single-hop WSN, all sensor nodes transmit the data directly to the base station, while in a multihop WSN, some nodes can only deliver their data to the base station via intermediate nodes. In these cases, the intermediate nodes execute the routing function and relay the data along the routing path. Also, data aggregation (or fusion) is an optional function for those intermediate nodes. Single-hop networks have much simpler structure and control and fit into the applications of small sensing areas; multihop networks promise wider applications at the cost of higher complexity.

Based on the sensor node density and data dependency, WSNs can be classified as aggregating and nonaggregating. In nonaggregating systems, all data from each individual node will be sent to the destination "as is." The computational load at intermediate nodes is relatively small and the system can reach high accuracy. However, the total traffic load in the entire system may increase rapidly with the enlargement of the network size, more energy will be consumed for communications, and more collisions and/or congestions will occur, leading to high latency. Therefore, the nonaggregating scheme is suitable for systems that have less node density, sufficient capacity, and/or in which extremely high accuracy is demanded by end users.

While in densely distributed networks, a sensor node is usually located close to its neighboring nodes. Thus, information from multiple sources could be highly correlated and aggregating functions may be executed at the intermediate nodes to eliminate data redundancy. In this way, the traffic load in the system could be reduced considerably, and significant energy savings due to communications can be obtained. However, the intermediate nodes will perform computational functions, which may require

TABLE 7.1 Classification of WSNs according to Different Factors

Factors	Distinct Groups
Distance to base station/processing center	Single hop vs. multihop
Data dependency	Nonaggregating vs. aggregating
Distribution of sensors	Deterministic vs. dynamic
Control scheme	Non self-configurable vs. self-configurable
Application domain	Many

the larger memory size. Therefore, the aggregating scheme is an appropriate option in large-scale systems with massively and densely distributed sensor nodes. It should be noted that end users are only interested in the collective information with moderate accuracy.

WSNs can be deterministic or dynamic according to distribution of the sensor nodes. In deterministic systems, the positions of sensor nodes are fixed or preplanned. The control of this system is simpler and its implementation is easier. However, this scheme can only be used in limited kinds of systems where the information of the sensor node placement could be obtained and planned in advance. However, in many cases, the locations of sensor nodes are not available *a priori*, such as those dropped randomly in remote areas. So, the sensor nodes must work in a distributed dynamic manner. The dynamic scheme is more scalable and flexible, but requires more complex control algorithms.

Moreover, based on the control scheme, WSNs can be non-self-configurable or self-configurable. In the former mechanism, the sensor nodes are not able to organize on their own, but rely on a central controller to offer command to and collect information from them. This scheme can only be used in small-scale networks. However, in most WSNs, the sensor nodes can autonomously establish and maintain connectivity by themselves and collaboratively fulfill sensing and control tasks. This self-configurable scheme fits better in large-scale systems to perform complicated monitoring tasks and information collection and dissemination.

The categories described here may overlap, i.e., a specific WSN may have the characteristics of different domains. For instance, WSNs in a large parking lot are self-configurable, deterministic, nonaggregating, and multihop. A classification of WSNs is shown in Table 7.1.

Although self-configurable systems are more complicated than non-self-configurable ones, they are more practical for deployment in the real world, especially when the network size becomes very large. However, they raise numerous challenges and open issues to be explored further. The remainder of this chapter concentrates mainly on self-configurable systems.

7.4 Characteristics, Technical Challenges, and Design Directions

WSNs aim to bridge the gap between the physical and computational worlds. The salient features of WSNs and their differences from other wireless networks have been discussed by a number of researchers [1, 13, 32, 33, 37, 43, 93, 97, 111, 112]. Some of these features are discussed next.

7.4.1 Characteristics

Most WSNs use the network architecture of wireless ad hoc networks, which are collections of wireless, possibly mobile, nodes that are self-configurable to form a network without the aid of any established infrastructure. The mobile nodes handle the necessary control and networking tasks in a distributed manner. The ad hoc architecture is highly appealing to sensor networks for many reasons [33]:

- Ad hoc architecture overcomes the difficulties raised by the predetermined infrastructure settings of the other families of wireless networks. WSNs can be randomly and rapidly deployed and reconfigured — new nodes can be added on demand to replace failed or powered-off ones and existing nodes can withdraw or depart from the systems without affecting the functionality of other nodes.

TABLE 7.2 Differences between WSNs and Conventional Wireless Ad Hoc Networks

	WSNs	Conventional Wireless Ad Hoc
Number of nodes	Large; hundreds to thousands or even more	Small to moderate
Node density	High	Relatively low
Data redundancy	High	Low
Power supply	Non-rechargeable; irreplaceable batteries	Rechargeable and/or replaceable batteries
Data rate	Low; 1–100kb/s	High
Mobility of nodes	Low	Can have high mobility
Direction of flows	Predominantly unidirectional; sensor nodes → sink	Bidirectional; end-to-end flows
Packet forwarding	Many to one; data centric	End-to-end address centric
Query nature	Attribute based	Node based
Query dissemination	Broadcast	Hop by hop or broadcast
Addressing	No globally unique ID	Globally unique ID
Active duty cycle	Could be as low as 1%	High

- Ad hoc networks can be easily tailored to specific applications.
- This architecture is highly robust to single node failures and provides a high level of fault tolerance because of node redundancy and its distributed nature.
- Energy efficiency can be achieved through multihop routing communication. As reported in Rappoport [82], large-scale propagation follows as exponential law to the transmitting distance (usually with exponent 2 to 4 depending on the transmission environment). It is not difficult to show that power consumption due to signal transmission can be saved in orders of magnitude by using multihop routing with short distance of each hop instead of single-hop routing with a long range of distance for the same destination.
- Ad hoc networks have the advantage of bandwidth reuse, which also benefits from dividing the single long-range hop to multihops; each hop has a considerable short distance. In this case, the communication is local and within a small range.

It is not surprising to see that the majority of existing WSN literature is based on multihop ad hoc architectures. However, because of unique application requirements, WSNs greatly differ from conventional wireless ad hoc networks [56, 93]. As a result, existing ad hoc network architectures and protocols are not directly suitable for or extendible to WSNs. Therefore, new approaches should be developed so as to satisfy the specific requirements of WSNs; numerous research issues remain to be explored. Table 7.2 summarizes the main differences between these two types of networks. These differences raise many technical challenges on system design and implementation. Next, these technical challenges are explored in detail; the corresponding design objective and directions will follow as well.

7.4.2 Technical Challenges and Requirements

WSN design is motivated and influenced by one or more of the following technical challenges [1, 32, 69]:

- *Massive and random deployment.* Most WSNs contain a large number of sensor nodes (hundreds to thousands or even more), which might be spread randomly over the intended areas or are dropped densely in inaccessible terrains or hazardous regions. The system must execute self-configuration before the normal sensing routine can take off.
- *Data redundancy.* The dense deployment of sensor nodes leads to high correlation of the data sensed by the nodes in the neighborhood.
- *Limited resources.* WSN design and implementation are constrained by four types of resources: energy, computation, memory, and bandwidth. Constrained by the limited physical size, microsensors could only be attached with bounded battery energy supply. Moreover, WSNs usually operate in an untethered manner, so their batteries are nonrechargeable and/or irreplaceable. At the same time, their memories are limited and can perform only restricted computational functionality. The bandwidth in the wireless medium is significantly low as well.

- *Ad hoc architecture and unattended operation.* The attributes of no fixed infrastructure and human-unattended operation of such networks require the system to establish connections and maintain connectivity autonomously.
- *Dynamic topologies and environment.* On the one hand, the topology and connectivity of WSNs may frequently vary due to the unreliability of the individual wireless microsensors. For example, a node may fail to function because of exhaustion of power at any time without notification to other nodes in advance. As well, new nodes may be added randomly in an area without prior notification of existing nodes. On the other hand, the environment that the WSNs are monitoring can also change dramatically, which may cause a portion of sensor nodes to malfunction or render the information they gather obsolete.
- *Error-prone wireless medium.* Sensor nodes are linked by the wireless medium, which incurs more errors than their wired counterpart. In some applications, the communication environment is actually noisy and can cause severe signal attenuation.
- *Diverse applications.* As described in Section 7.2, WSNs could be used to perform various tasks, such as target detection and tracking, environment monitoring, remote sensing, military surveillance, etc. Requirements for the different applications may vary significantly.
- *Safety and privacy.* Safety and privacy should be an essential consideration in the design of WSNs because many of them are used for military or surveillance purposes. Denial of service attacks against these networks may cause severe damage to the function of WSNs. However, security seems to be a significantly difficult problem to solve in WSNs because of the inevitable dilemma: WSNs are resource limited and security solutions are resource hungry. Indeed, most existing communication protocols for WSNs do not address security and are susceptible to adversaries [104].
- *QoS concerns.* The quality provided by WSNs refers to the accuracy with which the data reported match what is actually occurring in their environment. Different from others, accuracy in WSNs emphasizes the characteristic of the aggregated data of all sources instead of individual flows. One way to measure accuracy is the amount of data. Another aspect of QoS is latency. Data collected by WSNs are typically time sensitive, e.g., early warning of fires. It is therefore important to receive the data at the destination/control center in a timely manner. Data with long latency due to processing or communication may be outdated and lead to wrong reactions.

7.4.3 Design Objectives and Directions

The following objectives and directions are identified in the design of WSNs so as to deal with the challenges and satisfy the various application requirements [1, 13, 32, 33, 40, 43, 55, 69, 78, 97]:

- *Small microsensor devices.* Affordable and compact sensor units are essential factors to massive and random deployment of WSNs. For a large-scale WSN application, the cost of individual sensor devices would contribute to the major part of the total expense. Besides, the smaller the sensor is, the lower interference the sensor would have on the observed objects and the easier the deployment would be.
- *Scalable and flexible architectures and protocols.* In addition to the requirement on individual sensor devices, the system should be scalable and flexible to the enlargement of the network scale. The approaches to scalability and flexibility include clustering, multihop delivery, and localization of computation and protocols.
- *Localized processing and data fusion.* To eliminate data redundancy, collaborative efforts should be made among the sensor nodes performing a variety of localized processing. Instead of sending the raw data to the destination directly, sensor nodes might locally filter the data according to the requirements, carry out simple computation, process the data, and transmit only the processed data. Some intermediate nodes may also perform data fusion in order to reach high efficiency.
- *Resource efficiency design.* In WSNs, resource efficiency is extremely critical and is desirable regardless of its complexity. Above all, energy-efficient protocols are in high demand in order to extend the lifetime of the system. Indeed, power saving should be achieved in every component

of the network by integrating the corresponding mechanisms, such as power-saving mode on MAC layer, power-aware routing on network layer, etc. In addition, efforts should be made to increase efficiency for the utilization of other resources. For example, using algorithms with low complexity will reduce the computation time and thus save power; it also decreases the latency of data delivery. Bandwidth-efficient architectures and protocols can accelerate data delivery as well.

It should be noted that it is difficult to issue a unique definition of system lifetime for all applications or cases. The system can be declared dead when the first node exhausts its energy, when a certain fraction of nodes dies, or even when all nodes die. Using one or the other definition depends on the particular application. On the other hand, system lifetime can also be measured using application-specific parameters, such as the time until the system can no longer provide acceptable results.

- *Self-configuration.* Naturally, randomly and massively deployed sensor nodes have to execute self-configuration in order to set up the network connection and commence routine operation. WSNs are highly dynamic during the lifetime of the network. Sensor nodes transit among the states of off, sleep, startup, idle, transmitting, receiving, and failure* for the purpose of energy conservation. Thus, WSN protocols should have the capability of forming connections autonomously — regardless of the condition of sensor nodes. New links should be accommodated in case of node failure or link congestion, and the transmitting power or signaling rate may be adjusted actively to reduce energy consumption based on up-to-date topology information. As well, packets could be rerouted through some subsets of the network in which nodes have more residual energy so as to realize an equal dissipation of energy among nodes over the entire network.
- *Adaptability.* To cope with dynamic/varying conditions, WSNs should adapt to changing connectivity and system stimuli over time. To detect the nondeterministic phenomena with disturbance caused by communication noise and sensor diversity, adaptive fidelity signal processing at individual sensor nodes is also desired to make trade-offs among resources, accuracy, and latency requirements.
- *Reliability and fault tolerance.* For many WSN applications, data must be delivered reliably over the noisy, error-prone, and time-varying wireless channel. In such cases, data verification and correction on each layer of the network are critical to provide accurate results. Additionally, sensor nodes are expected to perform self-testing, self-calibrating, self-repair, and self-recovery procedures during their lifetime.
- *Application-specific design.* Because no unique protocol satisfies all applications of WSNs, the design of WSNs is in many cases application specific.
- *Security design.* Data privacy and safe communications are of utmost importance. Wood and Stankovic [104] argue that the best way to ensure successful network deployment is to take security issues into consideration at the design stage of WSNs.
- *QoS design with resource constraints.* As stated previously, the two measures of QoS in WSNs are accuracy and timely delivery of information. Accuracy reflects the basic value of the information. In general, the amount of data determines the level of accuracy. Data should be delivered in a timely manner. It is essential to make a trade-off between these two aspects because large amounts of data consume a large portion of bandwidth and cause more contention during transmission. As a result, the latency would be increased with higher accuracy requirement. Furthermore, it is critical to realize the trade-off between QoS and resource consumption. High accuracy requires large amounts of data delivery, thus leading to more power and bandwidth consumption. Local computation is helpful to eliminate the amount of data transmitted, but complex and memory costly computation will cause long latency. At the same time, more complex computation reduces power efficiency.

*Note that nodes in the same network may be in different states.

TABLE 7.3 Summary of Technical Challenges and Design Objectives in WSNs

Technical Challenges and/or Requirements	Design Objectives and Directions
Massive and random deployment	Cheap and small sensor node; scalable and flexible architecture and protocols
Data redundancy	Localized processing and data fusion
Limited resources	Resource efficiency design
Ad hoc architecture and unattended operation	Self-configuration and coordination
Dynamic surrounding	Adaptability
Error-prone medium	Reliability and fault tolerance
Diverse applications	Application-specific design
Safety and privacy	Security
QoS concerns	QoS design with resource constraint; localization; attribute-based naming and data-centric routing

- *Other attributes.* In addition to the preceding objectives and directions, WSN design should accommodate the following objectives:
  - *Locality of information.* The reported data from a sensor are only meaningful when associated with exact knowledge of the sensor's location. This can significantly simplify the network discovery and maintenance efforts. The data-centric query should be forwarded directly and efficiently to targeted areas of interest.
  - *Attribute-based naming and data centric routing.* When deploying WSNs, users are more interested in querying the property of the interested phenomenon, rather than a specific node. For example, "the temperature in room 717" or "the areas where the temperature is over 50°C" are more common than the query of "the temperature read by a certain sensor node."

It is impractical to achieve all objectives in a single network. Most WSN designs are application specific and have different stress on some of the objectives described previously. Thus, the protocols should be designed to satisfy the unique quality demands of each individual network and trade-offs should be made among the different parameters when designing protocols and algorithms for WSNs. Table 7.3 summarizes the technical challenges and corresponding design objectives and directions.

7.5 Technical Approaches

In many cases, it is very challenging to design and implement a resource-efficient and QoS-enabled WSN. This is usually constrained by many factors and has several objectives to meet at the same time; often such factors and objectives are contradictory to each other. Nevertheless, research on WSNs have achieved significant progress. Emphasizing on one or two aspects of the constrained factors or objectives, these research efforts take diverse approaches. Here, they are broadly grouped into three categories: hardware techniques; system architecture, protocols, and algorithms; and software development.

7.5.1 Hardware Techniques

7.5.1.1 Cheap, Compact, Low-Power Wireless Sensor Nodes

A WSN node integrates sensing, signal processing, data collection and storage, computation, and wireless communications, along with attached power supply on a single chip. The system architecture of a typical microsensor node is shown in Figure 7.1 [81, 95]. Generally, each node is composed of four components: (1) a power supply unit that is usually an attached battery with desirable output voltage to drive all other components in the system; (2) a sensing unit consisting of the embedded sensor and actuator as well as an analog-digital converter that links the sensor node to the physical world; (3) a computing/processing unit that is a microcontroller unit (MCU) or microprocessor with memory and provides intelligence to the sensor node (widely used MCUs include Intel's Strong ARM microprocessor and Atmel's AVR microcontroller); and (4) a communication unit consisting of a short-range RF circuit and performing

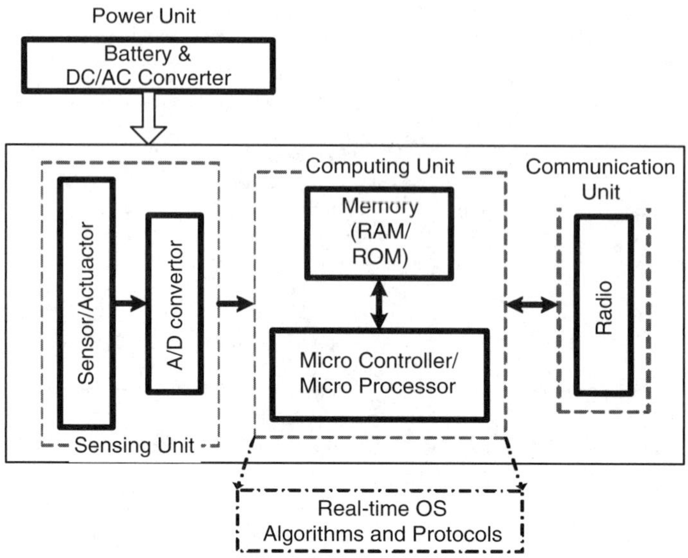

FIGURE 7.1 System architecture of a typical microsensor node.

data transmission and reception. Moreover, a real-time micro-operating system controls and operates the sensing, computing, and communication units through microdevice drivers and decides which parts to turn off and on.

Advances in microelectromechanical systems (MEMS) and continuous developments in wireless communications are spurring more intelligent, less expensive, much smaller sensor nodes to be embedded into the physical world. For example, *piconodes* in the PicoRadio project are a promising "system-on-chip" implementation to provide ubiquitous distribution of computation and communications for sensor/monitor networks. Each PicoRadio node has a small size of less than 0.10 to 0.15 in.^{3}, consumes less than 10 mW, and costs less than $1 [79, 80, 103].

Another system, called *WINS* (wireless integrated network sensors), integrates multiple functions including sensing, signal processing, decision making, and wireless networking capability in a compact, low-power device. These intelligent sensors are tiny and powerful in establishing low-cost and robust self-organizing networks for continuous sensing and event detection and identification [4, 75, 76].

A project called *μAMPS* (microadaptive multidomain power-aware sensors) [67] has the objective of implementing a microsensor system on a chip of 1 cm^3, with the integration of MEMS sensors, A/D, data and protocol processing, and a radio transceiver on a single die. Moreover, the *Smart Dust* project aims to explore the limits on size and power consumption of self-organizing sensor nodes that are not more than a few cubic millimeters in size, i.e., small enough to float in the air detecting and communicating for hours or days [54, 110–112]. For information on other experimental systems, refer to Hill et al. [47, 48], Mainwaring et al. [64], and Yarvis et al. [106].

7.5.1.2 Low Duty Cycle Electronics

Because the detected environment would not vary frequently or rapidly, the sensor node and its components should operate in alternating active and inactive modes for the purpose of power conservation. As the major contributors of the power consumption in a sensor node, data processing and radio subsystems have been under extensive study [13, 19, 92]. The energy consumed by the static CMOS-based microprocessor unit in a typical sensor node can be modeled as follows [92, 94]:

$$E_{total} = E_{switch} + E_{leakage} = C_{total} V_{dd}^2 + (V_{dd} t) I_0 e^{\frac{V_{dd}}{n' V_T}} \tag{7.1}$$

Total power consumption is composed of two parts: switching power and leakage power. Switching power is determined by supply voltage, V_{dd}, and the total capacitance switched by executing software, C_{total}. The leakage power refers to the energy consumption while no computation is conducted. Here, V_T is the thermal voltage. An effective way to reduce the energy consumption in the processor is to minimize the power wasted while no useful work is done, i.e., the leakage power part.

For the radio module, a possible scheme of power conservation is to turn off the radio electronics (such as frequency synthesizers, mixers, etc.) during periods of inactivity and to wake them up when interesting events occur [13, 92]. The average power consumed by the radio is modeled as [92]:

$$P_{radio\text{–}ave} = N_{tx}\,[P_{tx}\,(T_{tx\text{–}on} + T_{start}) + P_{out}T_{tx\text{–}on}] + N_{rx}\,[P_{rx}\,(T_{rx\text{–}on} + T_{start})] \quad (7.2)$$

where $N_{tx/rx}$ is the average number of times per second that the transmitter/receiver is active; $P_{tx/rx}$ is the power consumed by the transmitter/receiver; P_{out} is the output transmit power; T_{start} is the transceiver startup time; and $T_{tx/rx\text{–}on}$ is the actual data transmitting/receiving time equal to L/R, where L is the packet length in bits and R is the data rate in bits per second. Obviously, it is natural to turn off the radio as long as no work is to be done in order to reduce power consumption. However, significant overhead in terms of time and energy dissipation will be raised when switching the electrics from the inactive to the active state. Optimal schemes are necessary to estimate the traffic dynamics and make the switching decision accordingly.

7.5.2 System Architecture, Protocols, and Algorithms

7.5.2.1 Sensor Deployment Strategies

Sensor deployment is a fundamental issue for WSNs. The objective of a sensor deployment plan is to achieve desirable coverage with a minimum number of sensor nodes while complying with constraints of QoS, cost, reliability, and scalability of a certain application.

In WSNs, coverage has a twofold meaning: range and spatial localization. Range refers to the geometric area of a designated sensing mission, while spatial localization emphasizes the relative spatial positions of sensor nodes and targets so as to extract accurate measurements. Meguerdichian and colleagues [65] interpret the coverage problem in terms of deterministic vs. statistical and worst vs. best cases in WSNs, and propose an optimal polynomial-time algorithm for coverage calculation by combining computational geometry (specifically, Voronoi diagrams) and graph search algorithm. Mehta and coworkers [66] describe several algorithms that quickly and interactively compute the optimal coverage paths in WSNs. With greatly diverse applications, sensor deployment strategies and mechanisms vary significantly from case to case. In general, four methods of sensor deployment exist: predetermined, self-regulated, randomly undetermined, or biased distribution [24, 101].

Predetermined strategy applies to two situations: (1) knowledge about the environment or the possible targets is sufficient, as described in Musman et al. [70]; (2) sensor nodes can be regularly placed in some grid-based topology in which the sensing site is spatially modeled as a grid-based distribution, i.e., the two- or three-dimensional space is represented by point coordinates. The granularity of the grid (distance between adjunctive grid points) is determined by the desired accuracy [24]. Salhieh [87] and Schwiebert and colleagues [88] illustrate several examples of placing sensor nodes in some preplanned geometric topologies for medical care purposes. Using code identification, Chakrabarty and coworkers [14] describe methods for determining the placement of sensor nodes for unique target location and provide code-theoretic bounds on the number of sensors. Chakrabarty et al. [15] developed an integer linear programming (ILP) model for optimistically minimizing the cost of sensor deployment under the constraint of complete coverage of the sensor field. In general, predetermined strategy can provide an optimal solution for desirable coverage and obtain high QoS and cost efficiency at the same time. However, the first situation is often impractical in the real world because knowledge of the environment and targets is often not available *a priori*. A regular grid-based approach has better adaptation to the variation of the conditions, although it experiences some drawbacks as well. For one thing, the computational complexity makes the schemes not scalable to large-scale networks. However, the grid coverage relies on accurate sensor detection, although, in reality, sensor detection is often uncertain.

To overcome the difficulties of the predetermined approach, self-regulated strategy is developed. Howard and colleagues [51] propose a potential field-based method to deploy sensor nodes automatically in an unknown environment. Because the sensing fields are established in a manner in which each sensor node is repelled by obstacles and by other nodes, the entire network is self-spread throughout the environment and can reach the maximum coverage. Clouqueur et al. [20] present a scheme to deploy sensor nodes sequentially in steps by introducing path exposure as a metric of goodness. With the strategy of properly choosing the number of sensors in each step, the cost of deployment can be minimized to achieve the desired detection performance. Self-regulated methods are scalable to increasing the number of sensor nodes, but the computational expense may become prohibitive.

Randomly undermined strategy is more realistic for a large-scale WSN application, such as unknown battlefields or hostile terrains. With methods of this approach, sensor nodes are generally spread uniformly in a given area [42–44, 60, 61, 101]. This strategy is preferable because of easy placement of nodes and therefore low cost. Although sensing devices can be randomly deployed in two- or three-dimensional spaces, the coverage might not be uniform due to obstacles or other sources of noise in an environment. Based on an initial random distribution, Zou and Chakrabarty [109] introduced a practical virtual force algorithm (VFA) to reposition the sensors in order to enlarge coverage to the desired optimal results, thus dealing with cases of high- and low-detection accuracy while considering energy constraints.

Furthermore, in some contexts, the uniform deployment of sensor nodes may not always satisfy the design requirements and biased deployment can then be a viable option. Willig and coworkers [103] illustrate an example of biased placement of sensors in a large-scale office building in which the density of sensor nodes close to the windows is much higher than that in the middle of the room. Some comparisons of different deployment strategies by means of simulations have been presented by Tilak et al. [101].

Most research on sensor deployment discussed here has an implicit assumption that every sensor node operates in a reliable manner; however, because this is not always true in reality, some proposals have been introduced to handle unreliable conditions. Considering the uncertainty of sensor detection, a statistical optimization framework is presented in Dhillon et al. [24]. Assuming a given set of detection probabilities in a sensor field, it optimizes the number of sensors and determines their position so as to achieve sufficient grid coverage. Guibas [116] discusses the coverage and connectivity for WSNs with unreliable sensor nodes, deriving the necessary and sufficient conditions to cover a unit square region by a random grid network and maintain connectivity. These authors also formulate the sufficient conditions for connectivity between active nodes.

The framework described in Ray et al. [83] allows the sensor coverage areas to overlap so that each resolvable position is covered by a unique set of sensors. Using novel identification codes and based on a polynomial-time algorithm, it not only requires fewer sensors than existing proximity-based schemes in order to achieve required coverage, but also is robust against sensor failure or physical damage to the system. An alternate approach to achieving desirable and reliable coverage is by means of hardware redundancy, i.e., to deploy a greater density of sensor nodes in a sensing region and exploit redundancy to extend the overall system lifetime by operating distinct subsets that are, in turn, based on local density and local demand [32]. This is effective when the cost of deploying a node during the initial placement is much smaller than the cost of adding a new node at a later time.

7.5.2.2 Dynamic Power Optimization at the Nodal Level

Energy consumption at sensor node level has been described in Raghunathan et al. [81], Shih et al. [92], and Sinha and Chandrakasan [95]. From a functionality perspective, energy is consumed for sensing, computation, and communications. Power conservation can be achieved in any of these functions.

First, it should be noted that workload in WSNs typically has the characteristic of burstiness [10, 96]. Therefore, some nodes or certain components of nodes should switch to power-saving states between consecutive bursts while the functionality and QoS are still maintained. Dynamic power management (DPM) [9, 14, 81] is an example of this approach. As listed in Table 7.4, a particular combination of

TABLE 7.4 States of the Sensor Node and Its Components

No.	Node State	MCU	Memory	Sensor and A/D	Radio
S0	Transmitting	Active	Active	On	Tx
S1	Receiving	Active	Active	On	Rx
S2	Ready	Idle	Sleep	On	Rx
S3	Observing	Sleep	Sleep	On	Rx
S4	Standby	Sleep	Sleep	On	Off
S5	Sleep	Sleep	Sleep	Off	Off
S6	Off	Off	Off	Off	Off

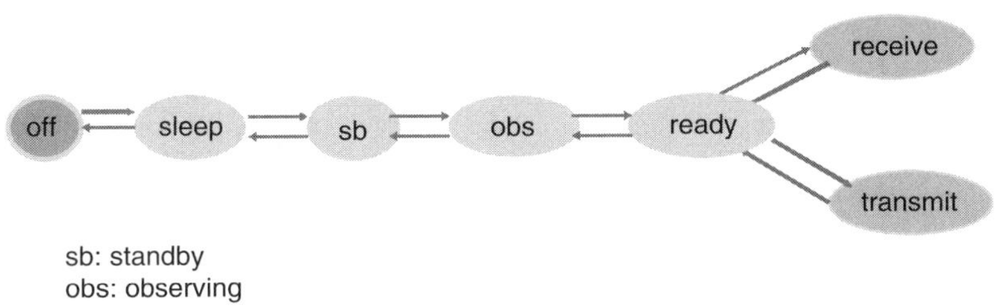

FIGURE 7.2 State transition diagram of a sensor node.

component states will determine a specific node state [92, 95]. For a sensor node, the states in decreasing order of power consumption are: transmitting, receiving, ready, observing, standby, sleep, and off. The state transition diagram of a sensor node is shown in Figure 7.2. For detailed numerical analysis of power consumption, see Raghunathan and colleagues [81]. However, transitions among states have power consumption and latency costs. Specifically, some transitions, for example, from "off" to "sleep," might cost much more energy than others, such as from "sleep" to "active." As a result, well-designed control algorithms are needed to achieve the trade-off between power saving and latency, power consumption, and state transitions.

Second, adaptively adjusting the operating voltage and frequency to meet the dynamically changing workload without degrading performance is a method of energy saving on computation. The rationale behind this technique is that the computational workload of MCU in WSNs is usually time varying and peak system performance is not always demanded. Dynamic voltage scaling (DVS) [14, 39, 73, 81] is an example of this approach. However, this scheme needs to predict the microprocessor's workload so as to adjust the power supply and operating frequency. A workload prediction strategy in WSNs is described in Chakrabarty et al. [14]. More accurate prediction can lead to higher power efficiency with less degradation to the system's performance. Nevertheless, workloads in current and future WSNs are mostly nondeterministic, so accurately modeling the workload is an open issue.

Another approach is to optimize the transmission power of sensor nodes. The change in transmission power has great impact on many aspects of WSN communication, including one-hop communication radius; network topology and hierarchy; retransmission rate; routing path selection; etc. Researches of this approach can be further divided into two types, depending on whether the node has the power control.

According to [113], an optimal transmission range, or transmission power in terms of energy efficiency, exists in certain ad hoc networks. The optimal value is mainly affected by propagation environment and device parameters. Contrary to intuition, [114] discovered that small transmission power might cause excessive power consumption due to a combined effect of increased number of hops and larger retransmission probabilities. Both researches were conducted in a flat, symmetric, multihop ad hoc network with no power control for individual nodes. Further research with various network and nodal conditions is strongly desired in the future.

Some other research assumes the power control capability on individual nodes. In such case, a large amount of communication energy can be saved through dynamically adjusting the transmission power based on the estimation of transmitting distance of each transmission. Proposed in [115], ROAD is a new MAC scheme for variable-radius multihop networks.

7.5.2.3 Optimal Schemes at System Level

7.5.2.3.1 Topology Management

As discussed earlier, dense deployment of sensors ensures the required coverage and sufficient precision of detection. Meanwhile, the redundant data generated by densely deployed nodes can be treated as backups for each other, so as to ensure the reliable function of the network. In the process of system operation, some node may operate in low duty cycles by transiting the hardware to sleep or off states to conserve energy. In these states, the sensor nodes are unable to communicate and forward packets. The nodes would then need to be awakened in certain situations, such as when it is time to collect data or neighboring nodes are depleted. Therefore, the active topology of the network changes over time. This leads to the critical issue of how to arrange sleep state transitions while ensuring robust, undegraded information collection [81].

A typical approach is to rotate the node functionality periodically to achieve balanced energy consumption among nodes. The protocol SPAN, proposed in Chen et al. [17], is an example of this approach for wireless ad hoc networks. Randomly, a limited number of nodes are self-selected as coordinators to construct the backbone in a peer-to-peer fashion within the network for traffic forwarding, while others can make local decisions to transit to a sleep state or keep active. The geographical adaptive fidelity (GAF) algorithm proposed in Xu and colleagues [105] is another way to rotate the active nodes within the network. Identified equivalent nodes, based on geographic locations on a virtual grid, can substitute each other directly and transparently without affecting the routing topology. Considering the fact that a WSN is only sensing its environment and waiting for an interesting event to happen, a new technique — sparse topology and energy management (STEM) described in Schurgers and coworkers [89, 90] — claims to improve beyond SPAN and GAF in terms of obtaining higher energy savings so as to prolong the system lifetime by trading off an increased latency to establish a multihop path.

7.5.2.3.2 Clustering and Hierarchical Architectures

It is reported that the energy consumed by communication is much higher than that for sensing and computation; in fact, this actually dominates the total energy consumption in WSNs. Experiments show that the ratio of communicating 1 bit over the wireless medium to that of processing the same bit could be in the range of 1,000 to 10,000 [108]. Furthermore, in most WSNs, power for transmission contributes to a majority of the total energy consumed for communication and the required transmission power grows exponentially with the increase of transmission distance. Therefore, reducing the amount of traffic and distance of communications can greatly prolong the system's lifetime.

On the other hand, a WSN usually contains a large number of sensor nodes in a wide area, and the base station may be far from the wireless sensors. Thus, dividing the entire system into distinct clusters replaces the one-hop long-distance transmission by multihop short-distance data forwarding. This would reduce the energy consumed for data communications and also has the advantages of load balancing, and scalability when the network size grows. Challenges faced by such clustering-based approach include how to select the cluster heads and how to organize the clusters. The clustering strategy could be single-hop cluster or multihop cluster, based on the distance between the cluster heads and their members, as shown in Figure 7.3(a) and Figure 7.3(b), respectively [38]. According to the hierarchy of clusters, the clustering strategies can also be grouped into single-level or multilevel clustering. Figure 7.4 illustrates the system architecture of multilevel hierarchical clustering [7].

Various clustering approaches for wireless ad hoc and/or sensor networks have been proposed in the literature [6–8, 16, 30, 36, 38, 42–44, 59, 72, 84, 87]. Heinzelman et al. [42] propose a distributed low-energy adaptive clustering hierarchy (LEACH). At the beginning, each node self-selects itself as a cluster head with a predetermined probability; the cluster head then advertises its decision to the other nodes, which decide to join a specific cluster that requires minimum communication energy. In order to ensure the balanced energy dissipation among all nodes, LEACH invokes the rotation of the cluster head by calling the self-selection and cluster formation procedure periodically. Moreover, the analytical and

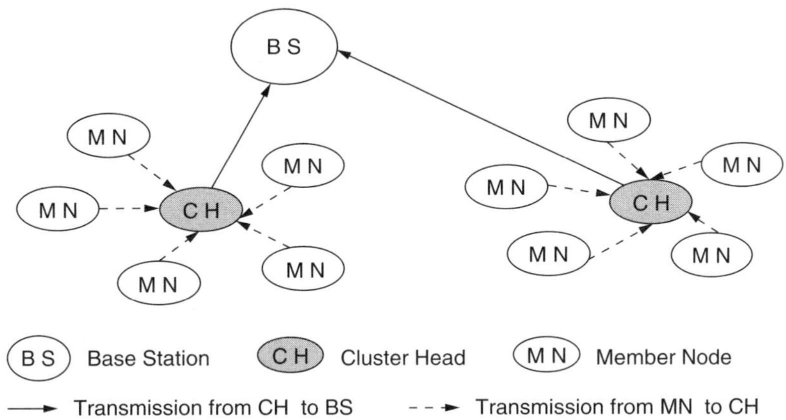

FIGURE 7.3(a) Single-hop clustering architecture.

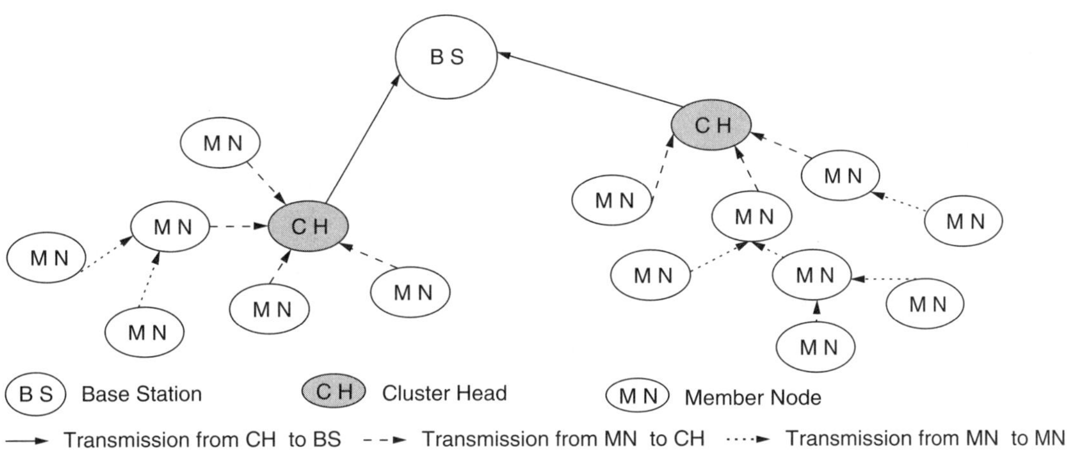

FIGURE 7.3(b) Multihop clustering architecture.

simulation results show that there is an optimal number of cluster heads that minimize the energy consumption.

Chiasserini et al. [18] attempt to solve the optimal problem of the balanced k-clustering, where k denotes the number of cluster heads in the system. Based on minimum weight matching, the algorithm attempts to realize load balancing among different clusters by partitioning the nodes into groups so that each cluster has a similar number of nodes. It achieves minimum energy consumption by optimizing the total spatial distance between the cluster members and the cluster heads. The power-aware virtual base stations (PA-VBS) protocol proposed by Safwat and colleagues [84, 86] is a first attempt to use the residual power capacity to select cluster heads in mobile ad hoc networks. It is attractive to WSNs because of its characteristics of load balancing and scalability to the growth of network size. In Gupta and Younis [38], a load-balanced clustering approach is introduced for heterogeneous sensor networks. The gateway nodes (cluster heads) with high energy manage the cluster member nodes and forward the data collected from the cluster member to a faraway base station. However, all the preceding schemes are single-hop cluster head formation algorithms, which may result in a large number of clusters. Therefore, they are only suitable for networks with a small to medium number of nodes.

In a large-scale network, multihop clusters and multilevel clustering hierarchy are highly in demand in order to decrease the communication distance further. Amis et al. [3] propose the max–min d-cluster to generate d-hop clusters, which can achieve better load balancing among clusters with fewer clusters than the

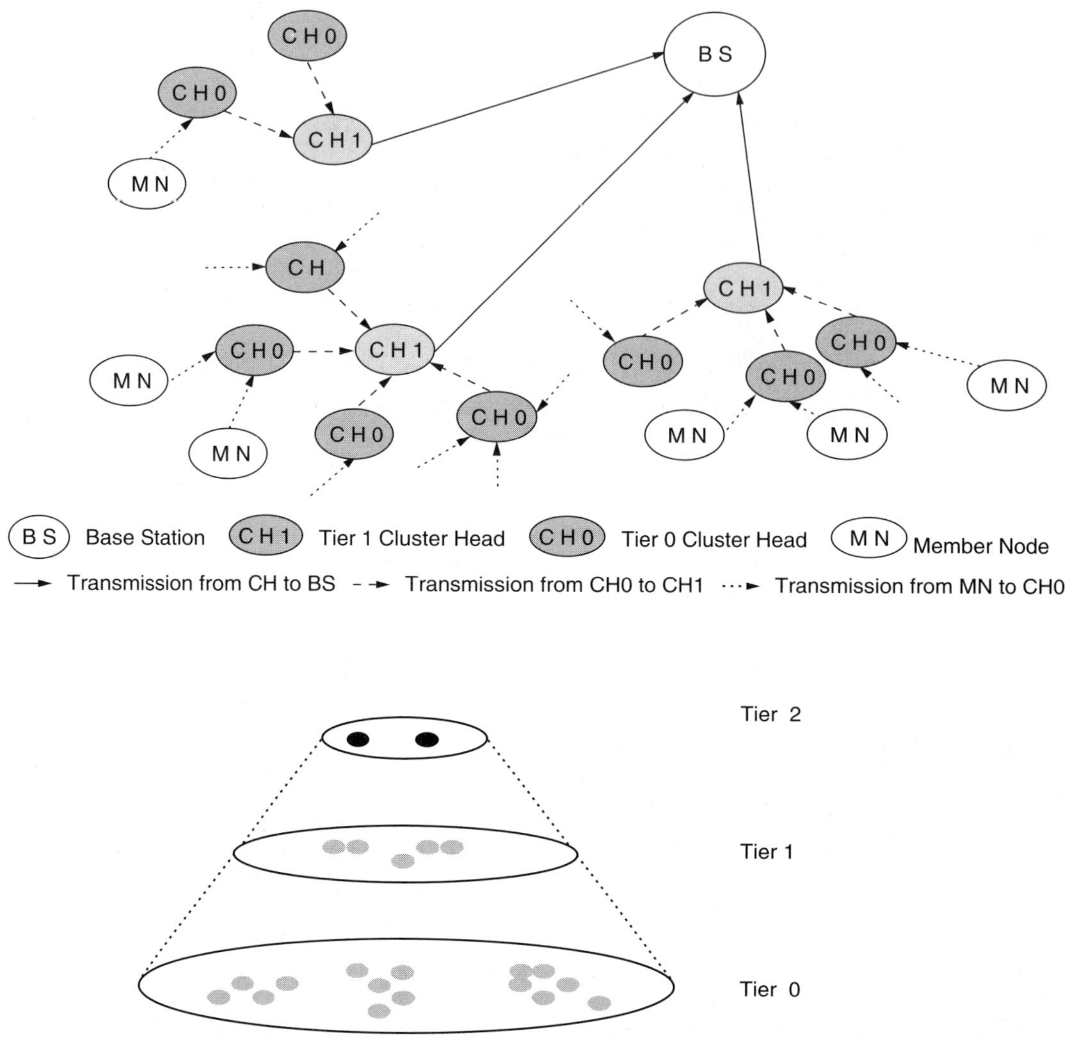

FIGURE 7.4 Multilevel hierarchical clustering architecture.

single-hop clustering schemes can [6, 30]. In Chiasserini et al. [18], a clustering algorithm is described to maximize the system lifetime through optimizing cluster size and assignment of nodes to each cluster head. However, this requires predetermining the number and locations of cluster heads, and each node should have knowledge of global network topology, which is impractical in WSNs. A chain-based protocol called power-efficient gathering in sensor information systems (PEGASIS) is presented in Lindsey and Raghavendra [60] and Lindsey et al. [61]. Instead of sending data packets directly to the cluster heads as shown in the LEACH protocol, each node forwards its packets to the destination through its closest neighbors.

Inheriting the feature of randomized creation and rotation of cluster heads as proposed in LEACH, as well as the advantages of a multihop clustering algorithm, Bandyopadhyay and Coyle [7] introduce a new energy-efficient, single-level, multihop clustering algorithm; these authors also provide the formulation for finding optimal parameter values to minimize the energy consumption. Moreover, based on the results of Foss and Zuyev [35] and Baccelli and Zuyev [5], Bandyopadhyay and Coyle [7] also provide a novel energy-efficient hierarchical clustering algorithm with a total of h levels (i.e., some of the cluster heads in level $k-1$ select themselves as kth level cluster heads, and the remaining level $k-1$ cluster heads are cluster members in level k). They derive optimal parameters to achieve minimum energy consumption within the whole system. Experimental results for up to 10,000 nodes have been reported.

7.5.2.3.3 Traffic Distribution and System Partitioning

Due to the limited resources in WSNs, one key element of traffic forwarding is the selection of an energy-efficient path from the source to the destination. Some algorithms have been proposed to select a route that minimizes total energy consumption within the entire network. However, this is not always the case in order to maximize the overall system lifetime. Because the nodes on such route are overused, their batteries are more likely to be exhausted. This can result in discontinuity of the network, as well as unavailability of sensing in the corresponding regions. Therefore, taking the point of view of the system's overall availability and longevity, it is preferable to avoid continuously forwarding traffic through the same route, even though it always consumes the minimum energy from source to destination. Thus, it is desirable to distribute the traffic more evenly within the whole system [81].

It is also possible to introduce the concept of system partitioning [13] to reduce power dissipation in the sensor nodes by removing some intensive computation to remote base stations that are not energy constrained, or spreading some of the complex energy-consuming computation among more sensor nodes instead of overloading several centralized processing elements. Chandrakasan et al. [13], Min et al. [68], and Wang and Chandrakasan [102] describe examples of implementing system partitioning.

7.5.2.3.4 Collaborative Signal and Information Processing (CSIP) and Data Aggregation

In addition to the approaches described in previous subsections, local processing of raw data before direct forwarding will effectively reduce the amount of communication and improve the efficiency (information per bit transmitted). CSIP and data aggregation are two typical localized paradigms for the purpose of data processing in WSNs.

With the combination of interdisciplinary techniques, such as low-power communication and computation, space-time signal processing, distributed and fault-tolerant algorithms, adaptive systems, and sensor fusion and decision theory, CSIP is expected to provide solutions to many challenges, including dense spatial sampling of interested events; distributed asynchronous processing; progressive accuracy; optimized processing and communication; data fusion; and querying and routing tasks [58]. CSIP can be implemented through coherent signal processing on a small number of nodes in a cluster or through noncoherent processing across a larger number of nodes when synchronization is not a strict requirement [32]. CSIP algorithms can be classified [78] as information-driven schemes [107, 108], mobile agent-based schemes [77], or relation-based schemes [116].

Data aggregation or fusion [45, 52, 56] is another efficient data processing approach in WSNs. It tries to minimize traffic load (in terms of number and/or length of packets) through eliminating redundancy. Specifically, when an intermediate node receives data from multiple source nodes, instead of forwarding all of them directly, it checks the contents of incoming data and then combines them by eliminating redundant information under some accuracy constraints. It applies a novel data-centric approach to replace the traditional address-centric approach in data forwarding [56]. The examples depicted in Figure 7.5(a) and Figure 7.5(b) demonstrate the difference in these two approaches. In an address-centric approach, the intermediate node, M, must forward all the packets received from different source nodes, e.g., S1, S2, to the destination D. However, in a data-centric approach, node M first fuses the data from S1 and S2 by eliminating the redundant information, then relays the processed data to D. This leads to higher efficiency and more energy savings.

Several data aggregation algorithms have been reported in the literature. The most straightforward is duplicate suppression, i.e., if multiple sources send the same data, the intermediate node will only forward one of them. Maximum or minimum functions are also very simple approaches. Heinzelman and colleagues [41] and Julik and coworkers [57] propose a scheme named sensor protocols for information via negotiation (SPIN) to realize traffic reduction for information dissemination. It introduces metadata negotiations between sensors to avoid redundant and/or unnecessary data through the network. Proposed in Intanagonwiwat et al. [52], directed diffusion is a data distribution scheme that incorporates in-network data aggregation, data caching, and data-centric dissemination, while enforcing adaptation to the empirically best path. It aims to establish efficient n-way communication from single or multiple sources to sinks. Heidemann and colleagues [45] present a physical implementation of directed diffusion

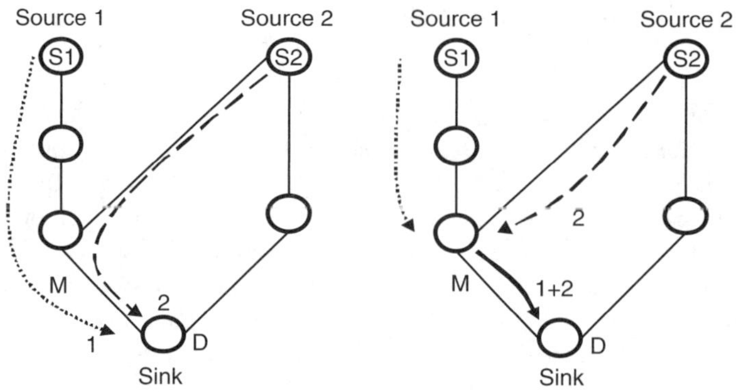

FIGURE 7.5 (a) Example of address-centric data forwarding. (b) Example of data-centric data forwarding.

with a wireless sensor test bed and shows that the traffic can be reduced by up to 42% when deploying a duplicate suppression data aggregation scheme.

The greedy aggregation approach proposed in Intanagonwiwat et al. [53] can improve path sharing and attain significant energy savings when the network has higher nodal density compared with the opportunistic approach. Krishnamachari and coworkers [56] describe the impact of source-destination placement on the energy costs and delay associated with data aggregation; they also investigate the complexity of optimal data aggregation. In [117], a polynomial-time algorithm for near-optimal maximum lifetime data aggregation (MLDA) is described for data collection in WSNs. The scheme is superior to others in terms of system lifetime, but has a high computational expense for large sensor networks. In Dasgupta et al. [22], a simple and efficient clustering-based heuristic for maximum lifetime data aggregation (CMLDA) is proposed for small- and large-scale sensor networks.

7.5.2.3.5 Cross-Layer Design

Traditional design of wireless ad hoc network protocols is mainly based on the layered stack as shown in Figure 7.6(a). This layered model makes a significant contribution to simplifying network design. Consequently, the layer structure leads to robust and scalable protocols. However, the design and operation of each layer in the stack are isolated, and the interface between layers is static and independent of the individual network constraints and applications. Therefore, inheriting such a stack will lead to poor WSN performance in which resources, especially energy, bandwidth, memory size, and CPU speed are greatly constrained. Many WSNs are dedicated for real-time data collection and strict delay bounds and high bandwidth demands could occur. Thus, new approaches are desirable to break the traditional border between the adjunct layers and create cross-layer paradigms. A possible cross-layer stack architecture is depicted in Figure 7.6(b) [37].

Goldsmith and Wicker [37] discuss not only the principles and strategies of cross-layer design in wireless ad hoc networks, but also the functionality of the individual layers and interactions among the different layers. Cross-layer design has become an attractive and active research topic in protocol designs of WSNs in recent years. Although some efforts have been made in literature, such as Heinzelman et al. [42, 43] and Safwat et al. [85], numerous open issues — how to understand and apply this design principle, how to deal with problems of information exchange across stack layers, and how to realize a specific application requirement with global system constrains — remain to be explored.

7.5.3 Software Development

Because of severe resource constraints, the software environment of WSNs is very different from those other distributed and parallel computing systems. Issues such as energy efficiency, scalability, and reliability are fundamental factors in software development for WSNs [13, 47, 49, 67, 81, 94, 99].

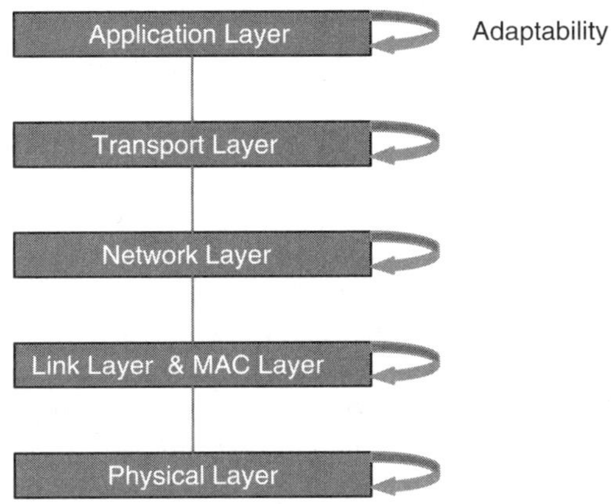

FIGURE 7.6(a) Traditional layered protocol stack for ad hoc networks.

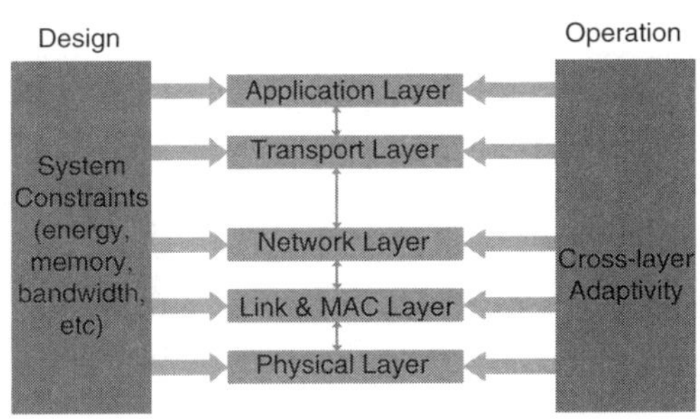

FIGURE 7.6(b) Cross-layer protocol stack for WSNs.

7.5.3.1 Single Node Level

System support on the lowest level begins at each single node. The concept of energy-aware software is introduced in Sinha and Chandrakasan [95]; who also illustrate the energy model of a typical microprocessor used for microsensors. With the proper operating systems, DPM and DVS can be deployed to reduce system power consumption at the node level. Described in Hill et al. [47, 49], TinyOS is one of the earliest operating systems dedicated for tiny sensor nodes; this system is event driven and uses only 178 bytes of memory, but supports communication, multitasking, and code modularity. Min and colleagues [67] present the concept of energy-scalable software, which is claimed to balance the trade-off between energy and quality characteristics.

7.5.3.2 Middleware

The middleware in WSNs abstracts the system as a collection of massively distributed objects and enables sensor applications to originate queries and tasks, gather responses and results, and monitor the changes within the network [91]. Sensor information networking architecture (SINA), proposed in Shen et al. [91], provides a middleware implementation of the general abstraction; these authors also describe sensor query and tasking language (SQTL), the sensor programming language used to implement such middleware architecture.

7.5.3.3 Application Programming Interface (API)

Considerable operation complexity exists in a WSN. However, with proper API implementation, the underlying system complexity can be transparent to end users who are experts in their specific application domain, but not necessarily experts in WSNs. The detailed functionalities of API in WSNs have been discussed in Shen and colleagues [91]. Stankovic et al. [99] consider other issues and advances in WSN software development.

7.6 Conclusions and Considerations for Future Research

A wireless sensor network consists of a large number of sensor nodes performing various distributed sensing and control tasks that are linked by a wireless medium. In general, a sensor is a device capable of capturing physical information, such as temperature, pressure, motion of an object, and mapping such physical characteristics of the environment to quantitative measurements. WSNs are evolving from simple networks with a small number of sensor nodes into diverse forms containing rapidly growing numbers of distributed nodes with enriched functions. These networks exhibit many benefits over their conventional wired counterparts and have been turning impossible monitoring and detection tasks into reality. Because of their ease of deployment, self-organization, reliability, versatility, scalability, and flexibility, WSNs have revealed significant potential in providing safer and healthier environments for human beings and thus have attracted much attention from academia as well as industry over the past few years.

This chapter presented an overview of WSNs and their evolution, describing numerous applications of self-configurable WSNs for target monitoring, detection, localization, and tracking in distinct military and civil domains. A discussion on technical challenges and design requirements was provided. Also highlighted were the state-of-the-art technical approaches in three aspects: hardware design; systems architectures, protocols, and algorithms; and software development.

Despite of the great progress on development of WSNs, quite a few issues still need to be explored in the future:

- *Tiny hardware components and sensor nodes with high efficiency* are still to be developed.
- *Protocols and algorithms for WSNs with heterogeneous sensor nodes.* Currently, many WSN protocols/algorithms are based on homogeneous sensor networks. However, sensors with different power capacities, sensing and transmitting range, and computing/processing abilities are usually more practical for constructing highly reliable networks [55, 63].
- *Combination of data-centric and address-centric operations.* As a long-term goal, WSNs are designated to be the first-class candidates in ubiquitous networks [118]. However, end-to-end communication fashion in traditional networks may not be suitable for the collective fashion in sensor networks. Combining WSNs' data-centric operation with the address-centric operation in traditional networks will lead to numerous open issues.
- *Security issues.* Most existing WSN communication protocols have not addressed security and are susceptible to attacks by adversaries. The issue of integrating security at the design stage in a resources-constrained WSN is a serious technical challenge.
- *Analytical modeling.* More accurate and expeditious implementation of WSNs in the real world is highly dependent on the ability to devise analytical models to evaluate and predict WSNs' performance characteristics, such as efficiency for information gathering, delay properties, granularity, and energy consumption. However, due to the diverse forms of applications and massive number of nodes in a single network, many technical problems remain to be solved in modeling the behavior of WSNs.
- *Clock synchronization.* Large numbers of sensor nodes in a WSN need to collaborate to fulfill the sensing task and the collected data are time sensitive in most cases. Thus, clock synchronization is a key requirement for algorithm and protocol design. However, due to resource and size limitation and lack of a fixed infrastructure, as well as the dynamic topology, existing time synchronization strategies designed for other traditional wired and wireless networks are not suitable for WSNs. Although Elson and Estrin [27] and Elson et al. [29] propose some synchronization proposals for WSNs, and some design principles are given in Elson and Romer [28], quite a few open issues still need to be explored in the future.

- *Other issues.* Optimal sensor node selection and allocation, discovery, localization, and network diagnoses are other open issues in this direction. Many software issues remain open as well. These include the design of distributed control and coordination algorithms to ensure balanced load assignment and energy consumption; efficient techniques for sensor data storage; and protocols with mobility consideration and dynamic group communications.

The issues discussed in this chapter are not exhaustive: many open issues remain to be explored so as to enable WSNs to achieve desirable connectivity, availability, reliability, and survivability in an energy-efficient fashion.

References

1. J. Agre and L. Clare, An integrated architecture for cooperative sensing networks, *Computer Mag.*, 33(5), 106–108, 2000.
2. I.F. Akyildiz, W. Su, Y. Sankarasubramaniam, and E. Cayirci, Wireless sensor network: a survey, *Computer Networks*, 38(4), 393–422, March 2002.
3. A.D. Amis et al., Max–min D-cluster formation in wireless ad hoc networks, IEEE *INFOCOM 2000*, 1, 32–41, Tel Aviv, March 2000.
4. G. Asada et al., Wireless integrated network sensors: low power systems on a chip, *24th IEEE Eur. Solid-State Circuits Conf.*, 9–12, The Hague, the Netherlands, 1998.
5. F. Baccelli and S. Zuyev, Poisson Voronoi spanning trees with applications to the optimization of communication networks, *Operations Res.*, 47(4), 619–631, 1999.
6. D.J. Baker and A. Ephremides, The architectural organization of a mobile radio network via a distributed algorithm, *IEEE Trans. Commun.*, 29(11), 1694–1701, November 1981.
7. S. Bandyopadhyay and E.J. Coyle, An energy efficient hierarchical clustering algorithm for wireless sensor networks, *IEEE INFOCOM 2003*, 3, 1713–1723, San Francisco, March–April 2003.
8. S. Basagni, Distributed clustering for ad hoc networks, *Int. Symp. Parallel Architectures, Algorithms Networks*, 310–315, Freemantle, Australia, June 1999.
9. L. Benini and G.D. Micheli, *Dynamic Power Management: Design Techniques and CAD Tools*, Kluwer Academic Pub., Norwell, MA, 1998.
10. L. Benini et al., A discrete-time battery model for high-level power estimation, *Design, Automation Test in Eur. Conf.*, 35–41, Paris, March 2000.
11. D.W. Carman, P.S. Kruus, and B.J. Matt, Constraints and approaches for distributed sensor network security, NAI labs Technical Report 00-010, September 2000.
12. A. Cerpa, J. Elson, D. Estrin, L. Girod, M. Hamilton, and J. Zhao, Habitat monitoring: application driver for wireless communications technology, *1st ACM SIGCOMM Workshop Data Commun. Latin Am. Caribbean*, San Jose, Costa Rica, 31(2), 20–41, 2001.
13. A. Chandrakasan et al. Design considerations for distributed microsensor systems, *IEEE 1999 Custom Integrated Circuits Conf.*, 279–286, San Diego, May 1999.
14. K. Chakrabarty, S.S. Iyengar, H. Qi, and E. Cho, Coding theory framework for target location in distributed sensor networks, *Int. Conf. Inf. Technol.: Coding and Computing*, 130–134, Las Vegas, April 2001.
15. K. Chakrabarty, S.S. Iyengar, H. Qi, and E. Cho, Grid coverage for surveillance and target location in distributed sensor networks, *IEEE Trans. Computers*, 51, 1448–1453, 2002.
16. M. Chatterjee, S.K. Das, and D. Turgut, WCA: a weighted clustering algorithm for mobile ad hoc networks, *J. Cluster Computing*, special issue on mobile ad hoc networking, 5, 193–204, 2002.
17. B. Chen et al., SPAN: An energy-efficient coordination algorithm for topology maintenance in ad hoc wireless networks, *ACM/IEEE MOBICOM* 2001, 85–96, Rome, July 2001.
18. C.F. Chiasserini et al., Energy efficient design of wireless ad hoc networks, *Proc. Networking 2002*, 387–398, Pisa, May 2002.

19. S. Cho and A.P. Chandrakasan, Energy efficient protocols for low duty cycle wireless microsensor networks, *ICASSP'01*, 4, 2041–2044, Salt Lake City, May 2001.
20. T. Clouqueur, V. Phipatanasuphorn, P. Ramanathan, and K.K. Saluja, Sensor deployment strategy for target detection, *ACM WSNA*, 42–48, Atlanta, 2002.
21. Collaborative Sensor Networks, Internet article, http://wwwhome.cs.utwente.nl/~havinga/sensor.html, 2003.
22. K. Dasgupta, K. Kalpakis, and P. Namjoshi, An efficient clustering-based heuristic for data gathering and aggregation in sensor networks, *IEEE WCNC'03*, 3, 1948–1953, New Orleans, 2003.
23. V. De and S. Borkar, Technology and design challenges for low power and high performance, *ISLPED 1999*, 163–168, San Diego, August 1999.
24. S.S. Dhillon, K. Chakrabarty, and S.S. Iyengar, Sensor placement for grid coverage under imprecise detections, *Int. Conf. Inf. Fusion (FUSION) 2002*, 2, 1581–1587, Annapolis, 2002.
25. S. Dulman et al., Collaborative communication protocols for wireless sensor networks, *Eur. Res. Middleware Architectures for Complex Embedded Syst. Workshop*, Pisa, 2003.
26. M. Easton, Using space technology to fight malaria, *Queen's Gazette*, 13, April 7, 2003.
27. J. Elson and D. Estrin, Time synchronization for wireless sensor networks, in *2001 Int. Parallel Distributed Processing Symp.* (IPDPS), *Workshop Parallel Distributed Computing Issues Wireless Networks Mobile Computing*, 1965–1970, April 2001.
28. J. Elson and K. Romer, Wireless sensor networks: a new regime for time synchronization, *Workshop Hot Topics Networks (HotNets-I)*, Princeton, NJ, October 2002.
29. J. Elson, L. Girod, and D. Estrin, Fine-grained network synchronization using reference broadcasts, *Symp. Operating Syst. Design Implementation (OSDI 2002)*, Boston, MA. December 2002. UCLA technical report 020008.
30. A. Ephremides, J.E. Wiesethier, and D.J. Baker, A design concept for reliable mobile radio networks with frequency hopping signaling, *Proc. IEEE*, 75(1), 56–73, 1987.
31. D. Estrin, R. Govindan, J. Heidemann, and S. Kumar, Next century challenges: scalable coordination in sensor networks, *ACM/IEEE MOBICOM '99*, 263–270, Seattle, August 1999.
32. D. Estrin, L. Girod, G. Pottie, and M. Srivastava, Instrumenting the world with wireless sensor networks, *IEEE ICASSP 2001*, 4, 2033–2036, 2001.
33. J. Feng, F. Koushanfar, and M. Potkonjak, System-architectures for sensor networks issues, alternatives, and directions, *IEEE ICCD'02: VLSI Computers Processors*, 226–231, Freiburg, Germany, 2002.
34. J. Feng and M. Potkonjak, Power minimization by separation of control and data radios, short paper, *IEEE CAS Workshop Wireless Commun. Networking*, Pasadena, September 2002.
35. S.G. Foss and S.A. Zuyev, On a Voronoi aggregative process related to a bivariate Poisson process, *Adv. Appl. Probability*, 28(4), 1981, 965–981.
36. S. Ghiasi et al., Optimal energy aware clustering in sensor networks, *Sensors Mag.*, 19(2), 258–269, 2002.
37. A.J. Goldsmith and S.B. Wicker, Design challenges for energy-constrained ad hoc wireless networks, *IEEE Wireless Commun.*, 8–27, August 2002.
38. G. Gupta and M. Younis, Load-balanced clustering of wireless sensor networks, *IEEE ICC 2003*, 3, 1848–1852, May 2003.
39. V. Gutnik and A.P. Chandrakasan, An embedded power supply for low-power DSP, *IEEE Trans. VLSI Syst.*, 5(4), 425–435, December 1997.
40. P.J.M. Havinga and G.J.M. Smit, Energy-efficient wireless networking for multimedia applications, *Wireless Communications and Mobile Computing*, John Wiley & Sons, New York, 2001, 165–184.
41. W. Heinzelman, J. Kulik, and H. Balakrishnan, Adaptive protocols for information dissemination in wireless sensor networks, *ACM/IEEE MOBICOM '99*, 174–185, Seattle, August 1999.
42. W. Heinzelman, A. Chandrakasan, and H. Balakrishnan, Energy-efficient communication protocol for wireless microsensor networks, *HICSS 2000*, 8020–8029, Maui, January 2000.

43. W. Heinzelman, Application-specific protocol architecture for wireless networks, Ph.D. dissertation, Massachusetts Institute of Technology, June 2000.
44. W.B. Heinzelman, A.P. Chandrakasan, and H. Balakrishnan, An application-specific protocol architecture for wireless microsensor networks, *IEEE Trans. Wireless Commun.*, 1(4), 660–670, October 2002.
45. J. Heidemann et al., Building efficient wireless sensor networks with low-level naming, *18th ACM Symp. Operating Syst. Principles*, 146–159, October 2001.
46. C. Herring and S. Kaplan, Component-based software systems for smart environments, *IEEE Personal Commun.*, 60–61, October 2000.
47. J. Hill et al., System architecture directions for networked sensor networks, *9th Int. Conf. Architectural Support Programming Languages Operating Syst.*, 28(5), 93–104, Cambridge, MA, November 2000.
48. J. Hill and D. Culler, A wireless embedded sensor architecture for system level optimization, University of California Berkeley technical report, 2002.
49. J. Hill et al., TinyOS: operating system for sensor networks, hppt://tinyos.millennium.berkeley.edu, 2003.
50. X. Hong, M. Gerla, H. Wang, and L. Clare, Load balanced, energy-aware communications for Mars sensor networks, *IEEE Aerospace Conf.*, 3, 1109–1115, 2002.
51. A. Howard, M.J. Mataric, and G.S. Sukhatme, Mobile sensor network deployment using potential fields: a distributed, scalable, solution to the area coverage problem, *6th Int. Symp. Distributed Autonomous Robotics Syst.* (DAR02) 299–308, Fukuoka, Japan, June 2002.
52. C. Intanagonwiwat, R. Govindan, and D. Estrin, Directed diffusion: a scalable and robust communication paradigm for sensor networks, *ACM/IEEE MOBICOM 2000,* 56–67, Boston, August 2000.
53. C. Intanagonwiwat, D. Estrin, and R. Govindan, Impact of network density on data aggregation in wireless sensor networks, Technical report 01-750, University of Southern California, November 2001.
54. J.M. Kahn, R.H. Katz, and K.S.J. Pister, Next century challenges: mobile networking for Smart Dust, *ACM/IEEE MOBICOM,* 271–278, Seattle, 1999.
55. F. Koushanfar, M. Potkonjak, and A. Sangiovanni–Vincentelli, Fault-tolerance techniques for sensor networks, *IEEE Sensors Mag.*, 2, 1491–1496, 2002.
56. B. Krishnamachari, D. Estrin, and S. Wicker, Impact of data aggregation in wireless sensor networks, *Int. Workshop Data Aggregation Wireless Sensor Networks*, 575–578, Vienna, Austria, July 2002.
57. J. Julik, W. Heinzelman, and H. Balakrishnan, Negotiation-based protocols for disseminating information in wireless sensor networks, *Wireless Networks*, 8, 169–185, 2002.
58. S. Kumar, F. Zhao, and D. Shepherd, Collaborative signal and information processing in microsensor networks, *IEEE Signal Processing Mag.*, 13–14, March 2002.
59. C.R. Lin and M. Gerla, Adaptive clustering for mobile wireless networks, *J. Selected Areas Commun.*, 15(9), 1265–1275, September 1997.
60. S. Lindsey and C.S. Raghavendra, PEGASIS: power-efficient gathering in sensor information systems, *IEEE Aerospace Conf. 2002*, 3, 1125–1130, March 2002.
61. S. Lindsey, C. Raghavendra, and K.M. Sivalingam, Data gathering algorithms in sensor networks using energy metrics, *IEEE Trans. Parallel Distributed Syst.*, 13(9), 924–935, September 2002.
62. C. Lu et al., RAP: A real-time communication architecture for large-scale wireless sensor networks, *8th IEEE Real-Time Embedded Technol. Applications Symp.* (RTAS), 55–66, San Jose, CA, 2002.
63. Y. Ma et al., ROP: A resource oriented protocol for heterogeneous sensor networks, *2003 Virginia Tech Symp. Wireless Commun.*, Blacksburg, VA, 2003.
64. A. Mainwaring, J. Polastre, R. Szewczyk, D. Culler, and J. Anderson, Wireless sensor networks for habitat monitoring, *ACM WSNA'02*, 88–97, Atlanta, September, 2002.
65. S. Meguerdichian, F. Koushanfar, M. Potkonjak, and M.B. Srivastava, Coverage problems in wireless ad-hoc sensor networks, *IEEE INFOCOM*, 3, 1380–1387, Anchorage, 2001.

66. D.P. Mehta, M.A. Lopez, and L. Lin, Optimal coverage paths in ad-hoc sensor networks, *IEEE ICC*, 1, 507–511, Anchorage, 2003.
67. R. Min et al., Low-power wireless sensor networks, *IEEE VLSID 2001*, 205–210, India, 2001.
68. R. Min et al., Energy-centric enabling technologies for wireless sensor networks, *IEEE Wireless Commun.*, 28–39, August 2002.
69. J. Mirkovic, G.P. Venkataramani, S. Lu, and L. Zhang, A self-organizing approach to data forwarding in large-scale sensor networks, *IEEE ICC*, 5, 1357–1361, St. Petersburg, Russia, 2001.
70. S. Musman, P.E. Lehner, and C. Elsaesser, Sensor planning for elusive targets, *J. Computer Math. Modeling*, 25(3), 103–115, 1997.
71. National Interagency Fire Center, http://www.nifc.gov.
72. A.K. Parekh, Selecting routers in ad-hoc wireless networks, *Proc. ITS*, 1994.
73. T.A. Pering, T.D. Burd, and R.W. Brodersen, The simulation and evaluation of dynamic voltage scaling algorithms, *Int. Symp. Low Power Electron. Design* (ISLPED), 1998.
74. E.M. Petriu et al., Sensor-based information appliances, *IEEE Instrumentation Measurement Mag.*, 31–35, December 2000.
75. G.J. Pottie and L.P. Clare, Wireless integrated network sensors: towards low cost and robust self-organizing security networks, *Proc. of SPIE* 1998, 3577, 86–95, 1999.
76. G.J. Pottie and W.J. Kaiser, Wireless integrated network sensors, *Commun. ACM* 2000, 43(5), 51–58, 2000.
77. H. Qi, S.S. Iyengar, and K. Chakrabarty, Multi-resolution data integration using mobile agents in distributed sensor networks, *IEEE Trans. Syst., Man Cybernetics (part C)*, 31, 383–391, August 2001.
78. H. Qi, P.T. Kuruganti, and Y. Xu, The development of localized algorithm in wireless sensor networks, *Sensors Mag.*, 2, 286–293, 2002.
79. J.M. Rabaey et al., PicoRadio supports ad hoc ultra-low power wireless networking, *IEEE Computer Mag.*, 33(7), 42–48, July 2000.
80. J.M. Rabaey, PicoRadio communication/computation piconodes for sensor networks year one report, 2001, EECS Department, University of California at Berkeley.
81. V. Raghunathan, C. Schurgers, S. Park, and M.B. Srivastava, Energy-aware wireless microsensor networks, *IEEE Signal Process. Mag.*, 40–50, March 2002.
82. T.S. Rappoport, *Wireless Communications, Principles and Practice*, Prentice Hall, Upper Saddle River, NJ, 1996.
83. S. Ray et al., Robust location detection in emergency sensor networks, *IEEE INFOCOM 2003*, 2, 1044–1053, March 2003.
84. A. Safwat, H. Hassanein, and H.T. Mouftah, Power-aware fair infrastructure formation for wireless mobile ad hoc communications, *IEEE GLOBECOM 2001*, 5, 2832–2836, November 2001.
85. A. Safwat, H. Hassanein, and H. Mouftah, Optimal cross-layer designs for energy-efficient wireless ad hoc and sensor networks, *22nd IEEE Int. Performance, Computing, Commun.* Conf., (IPCCC 2003), 123–128, April 2003.
86. A. Safwat, H. Hassanein, and H.T. Mouftah, Power-aware virtual base stations (PW-VBS) for wireless mobile ad hoc communications, *J. Computer Networks*, 41(3), 331–346, 2003.
87. A. Salhieh et al., Power efficient topologies for wireless sensor network, *Int. Conf. Parallel Processing*, 156–163, Spain, 2001.
88. L. Schwiebert, S.K.S. Gupta, and J. Weinamann, Research challenges in wireless networks of biomedical sensors, *ACM SIGMOBILE 2001*, 151–165, Rome, July 2001.
89. C. Schurgers, V. Tsiatsis, and M. Srivastava, STEM: topology management for energy efficient sensor networks, *2002 IEEE Aerospace Conf.*, 3, 1099–1108, March 2002.
90. C. Schurgers, V. Tsiatsis, S. Ganeriwal, and M. Srivastava, Optimizing sensor networks in the energy–latency–density design space, *IEEE Trans. Mobile Computing*, 1(1), 70–80, January–March 2002.

91. C-C. Shen, C. Srisathapornphat, and C. Jaikaeo, Sensor information networking architecture and applications, *IEEE Personal Commun.*, 52–59, August 2001.
92. E. Shih et al., Physical layer driven protocol and algorithm design for energy-efficient wireless sensor networks, *ACM/IEEE MOBICOM'01*, 272–287, Italy, July 2001.
93. J.L. da Silva Jr. et al. Design methodology for PicoRadio networks, *Design, Automation Test Eur.*, 314–325, Germany, March 2001.
94. A. Sinha and A. Chandrakasan, Energy aware software, *VLSID'00*, 50–57, Calcutta, January 2000.
95. A. Sinha and A. Chandrakasan, Dynamic power management in wireless sensor networks, *IEEE Design Test Computers*, 18(2), 62–74, 2001.
96. S. Slijepcevic and M. Potkonjak, Power efficient organization of wireless sensor networks, *IEEE ICC*, 2, 472–476, St. Petersburg, Russia, 2001.
97. K. Sohrabi, J. Gao, V. Ailawadhi, and G.J. Pottie, Protocols for self-organization of a wireless sensor network, *IEEE Personal Commun.*, 16–27, October 2000.
98. M. Srivastava, R. Muntz, and M. Potkonjak, Smart kindergarten: sensor-based wireless networks for smart developmental problem-solving environments, *ACM MOBICOM 2001*, 132–138, Italy, July 2001.
99. J.A. Stankovic et al., Real-time communication and coordination in embedded sensor networks, *Proc. IEEE*, 91(7), 1022–1032, 2003.
100. L. Subramanian and R.H. Katz, An architecture for building self-configurable systems, *MOBIHOC 2000*, 63–73, Boston, 2000.
101. S. Tilak, N.B. Abu–Ghazaleh, and W. Heinzelman, Infrastructure trade-offs for sensor networks, *ACM WSNA'02*, 49–58, September 2002.
102. A. Wang and A. Chandrakasan, Energy efficient system partitioning for distributed wireless sensor networks, *2001 IEEE Int. Conf. Acoustics, Speech, Signal Processing*, 2, 905–908, Salt Lake City, 2001.
103. A. Willig, R. Shah, J. Rabaey, and A. Wolisz, Altruists in the PicoRadio sensor network, *4th IEEE Int. Workshop Factory Commun. Syst.*, 175–184, Sweden, August 2002.
104. A.D. Wood and J.A. Stankovic, Denial of service in sensor networks, *IEEE Computer*, 35(10), 48–56, October 2002.
105. Y. Xu, J. Heidemann, and D. Estrin, Geography-informed energy conservation for ad hoc routing, *MOBICOM 2001*, 70–84, Italy, 2002.
106. M.D. Yarvis et al., Real-world experiences with an interactive ad hoc sensor network, *Int. Conf. Parallel Processing Workshops (ICPPW'02)*, 143–151, 2002.
107. F. Zhao, J. Shin, and J. Reich, Information-driven dynamic sensor collaboration for tracking applications, *IEEE Signal Process. Mag.*, 19(2), 61–72, March 2002.
108. F. Zhao et al., Collaborative signal and information processing: an information directed approach, *Proc. IEEE*, 91(8), 1199–1209, 2003.
109. Y. Zou and K. Chakrabarty, Sensor deployment and target localization based on virtual forces, *IEEE INFOCOM 2003*, 2, 1293–1303, San Francisco, 2003.
110. J.M. Kahn, R.H. Katz, and K.S.J. Pister, Emerging challenges: mobile networking for "Smart Dust," *Journal of Communications and Networks*, 2(3), 188–196, 2000.
111. K.S.J. Pister, SMART DUST — Autonomous sensing and communication in a cubic millimeter, *Internet article*, http//www-bsac.eecs.berkeley.edu/~pister/SmartDust/.
112. K.S.J. Pister, My view of sensor networks in 2010, *Internet article*, http://robotics.eecs.berkeley.edu/~pister/SmartDust/in2010.
113. P. Chen, B. O'Dea, and E. Callaway, Energy efficient system design with optimum transmission range for wireless ad hoc networks, *IEEE ICC 2002*, 2, 945–952, New York, May 2002.
114. S. Bansal et al., Energy efficiency and throughput for TCP traffic in multi-hop wireless networks, *IEEE INFOCOM 2002*, 1, 210–219, 2002.
115. C.-H. Yeh, ROAD: A variable-radius MAC protocol for ad hoc wireless networks, *IEEE VTC 2002 (Spring)*, 1, 399–403, 2002.

116. L.J. Guibas, Sensing, tracking, and reasoning with relations, *IEEE Signal Processing Magazine*, 19(2), 73–85, March 2002.
117. K. Kalpakis, K. Dasgupta, and P. Namjoshi, Efficient algorithms for maximum lifetime data gathering and aggregation in wireless sensor networks, Technical Report UMBC-TR-02-13, 2002, Computer Science and Electrical Engineering Department, University of Maryland, Baltimore, County.
118. A. Köpke, V. Handziski, and H. Karl, Making sensor networks intelligent, *7th Wireless World Research Forum (WWRF)*, the Netherlands, 2002.

8

Sensor Network Architecture

Jessica Feng
University of California at Los Angeles

Farinaz Koushanfar
University of California at Berkeley

Miodrag Potkonjak
University of California at Los Angeles

8.1 Overview 8-1
8.2 Motivation and Objectives 8-1
8.3 SNs — Global View and Requirements 8-3
8.4 Individual Components of SN Nodes 8-4
Processor • Storage • Power Supply • Sensors • Radio
8.5 Sensor Network Node 8-8
Berkeley Mote Node • UCLA Medusa MK-2 Node • BWRC PicoNode • Sensor-Centric Design: Light Compass
8.6 Wireless SNs as Embedded Systems 8-13
8.7 Summary 8-16

8.1 Overview

Emergence of the concept of multihop ad hoc wireless networks, low-power electronics, low-power, short-range wireless communication radios, and intelligent sensors is considered the major technological enabler for deployment of sensor networks (SNs). The goal in this survey is to identify key architectural and design issues related to sensor networks, critically evaluate the proposed solutions, and outline the most challenging research directions. The evaluation has three levels of abstraction:

- Individual components on SN nodes (processor, communication, storage, sensors and/or actuators, and power supply)
- Node level
- Distributed networked system level

Special emphasis is placed on architecture, system software, to some extent, and new challenges related to using new types of components in networked systems. The evaluation is guided by anticipated technology trends and current and future applications. The main conclusion of the analysis is that the architectural and synthesis emphasis will be shifted from computation and, to some extent communication, components to sensors, actuators, and different types of sensors and applications that require distinctly different architectures at all three levels of abstraction.

8.2 Motivation and Objectives

Embedded wireless SNs are systems consisting of a large number of nodes, each equipped with a certain amount of computational, communication, storage, sensing, and actuation resources [20]. SNs aim to provide efficient and effective connection between physical and computational worlds and are also widely considered the new big frontier for the Internet. Furthermore, they have high potential economic impact in many fields, including military, education, monitoring, retail, and science. At the same time, SNs pose numerous new research and development challenges, including the need for the next generation of low power; low cost; small size; error and fault resiliency; flexibility; conceptually new security and privacy; and a need for new types of input/output (I/O) operations.

However, before any of these challenges can be properly addressed, one must have the sensor network in place; the network must be designed and implemented and the need for flexible mechanisms and means for efficient and convenient use must be realized. In addition to algorithms, hardware and software architecture will decide to a significant extent the effectiveness of SNs. Furthermore, SN design methodology will have primary impact on the cost and performance of SNs. The third aspect with major potential impact — algorithms and modeling techniques for SN — is mainly out of the scope of this survey. Comprehensive surveys on SNs include Estrin et al. [20], Pottie and Kaiser [50], and Akyildiz et al. [2].

The overall strategic goal is to summarize current state of the art with respect to architecture and synthesis techniques for SNs and to provide a starting point and impetus for research and development of new architectures and synthesis tools for SNs. More specifically, the emphases are on:

- *Identifying requirements for typical SN application.* Traditionally, design of new computer architectures has been based on comprehensive and representative benchmark suites for typical target applications. It is of exceptional importance to create such benchmarks for sensor networks. In addition, it is important to predict the nature of future SN applications. However, even before the benchmarks are available, qualitative analysis of representative application can greatly facilitate identification of more accurate design goals.
- *Identifying relevant technological trends.* It is well known that many electronics and optical systems follow exponential performance growth rates. SN systems are heterogeneous and complex; therefore, it is important to anticipate which design and cost bottlenecks are intrinsic and which will be resolved due to technological progress. Importance of technological trends is well illustrated during power optimization. Depending on future ratios of computation, communication, and storage cost, very different types of algorithms will be best suited for SNs.
- *Balanced design.* In order to achieve a balanced design, the first instinct could be to optimize each and every component to the maximum extent. From a research and economic point of view, it is important to identify where to put the main optimization effort. In addition, new computational models are needed, but one must keep in mind that they are not the ultimate goal per se.
- *Techniques for design and the use of the design components.* The six components of SN node can be grouped in two categories according to their maturity. Power supplies, and in particular storage and power supply, are considered mature technologies. On the other hand, ultralow power wireless communication, sensors, and actuators are technologies waiting for major technological revolutions. It is important to identify which techniques, architectures, and tools can be reused, and where the new design effort is required.
- *Overall node architecture and trade-offs.* One can envision a number of possible trade-offs. For example, the TinyOS approach [27] advocates aggressive communication strategy in order to reduce complexity of computation and storage at local sensor nodes. On the other hand, the sensor-centered approach [22] advocates aggressive sensor data processing, filtering, and compression in order to reduce communication.
- *Survey of state-of-the-art technology, components, and sensor network nodes.* Special emphasis is placed on providing qualitative and quantitative analysis. In addition, several state-of-the-art sensor nodes are surveyed and decisions that influenced their structure are critically evaluated.

8.3 SNs — Global View and Requirements

It is well known that characteristics of computing or communication systems are direct consequences of targeted applications. A number of characteristics of sensor networks that have direct impact on architectural and design decisions have been identified. These characteristics rise naturally from a confluence of typical application requirements and technology limitations. Typical SN applications include contaminant transport monitoring; marine microorganisms analysis; habitat sensing; and seismic and home monitoring [9]. These applications show a great deal of diversity. Nevertheless, a number of general characteristics are shared among the majority of SN applications, regardless of the specific types of sensors and application objectives. These characteristics include low cost; small size; low power consumption; robustness; flexibility; resiliency on errors and faults; autonomous mode of operation; and privacy and security.

Sensor network nodes typically consist of six components: processor; radio; local storage; sensors and/or actuators; and power supply. A number of relevant technology trends need to be considered. For example, a huge variety of powerful low-power, low-cost processors, and low-cost memory technologies are widely accessible. Also, memory and processor technologies are growing more and more powerfully according to Moore's law, and wireless bandwidth has increased by a factor of more than 100 in the last 7 years; the capacity of batteries is growing at a rate as low as 3% per year. The cost of application-specific designs is growing rapidly: only masks cost $1 million and keep increasing by the factor of two every 2 years. Sensors and actuators are relatively young industrial fields and predictions are still uncertain.

Because of these application requirements and technology constraints, the following architectural and design objectives are most relevant:

- *Small physical size.* Reducing physical size has always been one of the key design issues. Therefore, the goal is to provide powerful processor, memory, radio, and other components while keeping a reasonably small size, dictated by a specific application.
- *Low power consumption.* The capability, lifetime, and performance of the sensors are all constrained by energy. The sensors should be able to be active for a reasonably long time without recharging the battery because maintenance is expensive.
- *Concurrency-intensive operation.* In order to achieve the overall performance, the sensor data must be captured from the sensor, processed, compressed, and then sent to the network simultaneously in pipelined processing mode, instead of sequential action. Two conceptual approaches address this requirement: (1) partitioning the processor into multiple units in which each is assigned responsibility for a specific task; and (2) reduction of the context switching time.
- *Diversity in design and usage.* Because each node should be small in size, low on power consumption, and have limited physical parallelism, the sensor nodes tend to be application specific. However, different sensors have different requirements. For example, cameras and simple thermometers are two extremes in terms of functionality and complexity. Therefore, the design should facilitate trade-offs among reuse, cost, and efficiency.
- *Robust operations.* Because sensors will be deployed over a large and sometimes hostile environment (forests, military usage, human body), they must be tolerant of fault and error. Therefore, sensor nodes need abilities to self-test, self-calibrate, and self-repair [33].
- *Security and privacy.* Each sensor node should have sufficient security mechanisms in order to prevent unauthorized access, attacks, and unintentional damage of the information inside the SN node. Furthermore, additional privacy mechanisms must be included.
- *Compatibility.* The cost to develop software dominates the cost of the overall system. In particular, it is important to be able to reuse the legacy code through binary compatibility or binary translation.
- *Flexibility.* It is necessary to accommodate functional and timing changes. Flexibility can be achieved through two means: (i) programmability (by employing programmable processors such as microprocessors, DSP processors, and microcontrollers); and (2) reconfiguration (by using FPGA-based platforms). Flexibility will be mainly achieved by programmability and use of specialized ASIC and coprocessors due to low power consumption.

8.4 Individual Components of SN Nodes

SN nodes generally are composed of six components: processor; storage unit; power supply; sensors and/or actuators; and, finally, communication (radio) subsystems. It is apparent that standard processors, possibly augmented with DSP, and other coprocessors and some ASIC units will provide adequate processing capabilities at acceptable low-energy rates. Also the state of the art of the actuators is such that they are still not used in the current generation of SN nodes. Therefore, the focus is on the other five components. For the sake of completeness, the discussion begins by presenting a processor specifically designed for sensor networks.

8.4.1 Processor

Berkeley BWRC research group has designed and implemented a prototype processor; its main target areas include voice processing and related applications for wireless devices. For example, the processor can be used in museums to provide better interaction between visitors and displayed items. The Maia processor [63] is built around an ARM8 core with 21 coprocessors. These 21 processors include: two MACs; two ALUs; eight address generators; eight embedded memories; and an embedded low-energy FPGA [24]. The goal is to provide enough parallelism at low energy levels. ARM8 core configures the memory-mapped satellites using a 32b configurable bus and also communicates data with the satellite coprocessors using two pairs of I/O interface ports by applying direct memory reads/writes. The interactions between the ARM8 and coprocessor satellites are carried out through an interface control unit.

A two-level, hierarchical, mesh-structured, reconfigurable interconnect network is used to establish the connections between all satellites. This network provides a favorable trade-off between bandwidth and low area (cost) and low power requirements. This 210-pin chip contains 1.2 M transistors and measures 5.2×6.7 mm^2 in 0.25-μm, six-metal CMOS. In order to minimize the overall energy consumption, the embedded ARM8 core is additionally optimized and can operate under variable supply voltages [8]. In addition, the dualstage pipelined media access control (MAC) and the ALU are configurable. The address generators and embedded memories provide multiple concurrent data streams to the computational components. The embedded FPGA has a 4×8 array of five-input, three-output CLBs. It can be optimized for tasks such as arithmetic operations and data-flow control functions. The interface control unit interacts and coordinates the synchronization and communication between the synchronous ARM8 core and the asynchronous reconfigurable data paths. It also enables the ARM8 core to reconfigure the satellites. The overall targeted computation model is globally asynchronous, locally synchronous computation and supports multirate operation.

8.4.2 Storage

Depending on the overall sensor network structure, the requirements for storage in terms of fast and nonvolatile memory at each node can be sharply different. For example, if one follows the architecture model in which all information is instantaneously sent to the central node, there is very little need for local storage on individual nodes. However, in a more likely scenario in which the goal is to minimize the amount of communication and conduct a significant part of computation at each individual node, there will be significant requirement for local storage. At least two alternatives exist for storing data in a local node. In addition, in the case in which the node is physically larger, one can store the data in microdisks [17].

The first option is to use flash memory, which is very attractive in terms of cost and storage capacity. However, it has relatively severe limitations in terms of how many times it can be used for storing different data in the same physical locations [28]. The second option is to use nanoelectronics-based MRAM [56]. It is expected that MRAM will soon be able to support significant numbers of applications in a number of areas.

It is important to note that historically, nonvolatile semiconductor and disk storage capacity has been growing at a rate higher than that indicated in Moore's law. At least two major challenges for the use of nonvolatile memory in sensor nodes are: (1) partitioning for power reduction and (2) developing memory structures that will fit short, word-length data produced by sensors. Note that a significant percentage of network control and sensor data will have low entropy. Therefore, it is likely that aggressive compression techniques will be used to reduce the amount of data that must be stored or transferred [14].

8.4.3 Power Supply

A wide consensus is that energy will be one of the main technological constraints for SN nodes [46, 57]. For example, the current generation of smart badges and motes enables continuous operations for only a few hours. Energy supply can be addressed in at least two conceptually different ways. The first is to equip each sensor node with a (rechargeable) source of energy. Two main options for this approach exist. Currently, the dominant option is to use high-density battery cells [23, 37]; the other alternative is to use full cells. Full cells provide exceptionally high density and a clean source of energy. However, they are not currently available in a physical format appropriate for SN nodes.

The second conceptual alternative is to harvest energy available in the environment [52]. In addition to solar cells, which are already widely used for mobile appliances such as calculators, a number of proposals concern converting vibration to electric energy [45]. An interesting solution for a power source is introduced in Douseki et al. [18]. A battery-less wireless system that harvests ambient heat is used instead of adopting traditional batteries as the power source. The main component of the system is a switched-capacitor DC–DC converter; a microthermoelectric module makes such a system possible. The chip is fabricated in a 0.8-μm fully depleted SOI process and its effectiveness has been demonstrated.

8.4.4 Sensors

The importance of sensors cannot be overstated. The purpose of SN nodes is not to compute or to communicate, but rather to sense. The sensing component of SN nodes is the current technology bottleneck; these technologies currently are not progressing as fast as semiconductors. Conceptual limitations are significantly stricter for sensors than for processors or storage. For example, sensors interface to the real physical world, while computing and communicating units are dealing with a greatly controlled environment of a single chip. Transducers are front-end components in sensor nodes that are being used to transform one form of energy into another. Design of transducers is considered out of a system architect's scope. In addition, sensors may have four other components: analog, A/D, digital, and microcontroller. The simplest design option includes only the transducer; however, because the current trend is to put more "smartness" into sensor network nodes, significant processing and computing abilities are being added to sensor nodes [41].

One of the main challenges of SNs is to select the type and quantity of sensors and determine their placement. This task is difficult because of the numerous types of sensors with different properties such as resolution, cost, accuracy, size, and power consumption. In addition, often more than one sensor type is needed to ensure the correctness of operation and data from different sensors that can be combined. For example, in the Cricket Compass [51, 65], the orientation and the movement of the studying object can be obtained by measuring the distance between several fixed-location referencing sensors; therefore the location of the sensor is crucial to minimize error [65].

Another challenge is to select the correct types of sensors and the way to operate them. The source of difficulty is sensor interactions. For example, consider determining distance using audio sensors. Because the speed of sound depends greatly on temperature and humidity of the environment, it is necessary to take both measurements into account in order to get the accurate distance.

Several other design tasks are associated with sensors, including fault tolerance, error control, calibration, and time synchronization [33]. There are a large number of different sensor technologies [46, 60]; as an example, consider Kulah et al. [35] and Luo et al. [39]. The accelerometer is one of the most popular

MEMS-based sensors. A state-of-the-art capacitive accelerometer was recently reported by the MEMS group at the University of Michigan. It uses a two-element sensor array in two Σ (sigma-delta) loops to improve accuracy by a factor more than two times in comparison with a traditional second-order Σ modulator. The design is clocked at 1 MHz and provides 1 V/pF sensitivity. It has dynamic range of more than 120 dB and consumes less than 12 mW. Another state-of-the-art accelerometer has been designed at Carnegie Mellon University. The design combines lateral accelerometer and vertical gyroscopes with signal processing circuits.

8.4.5 Radio

Short-range radios as communication components are exceptionally important because the part of the energy budget dedicated to sending and receiving messages usually dominates the overall energy budget [52]. During the design and the selection of radios, one must considers at least three different abstraction layers: physical, MAC, and network. The physical layer is responsible for establishing physical links between a transceiver and one or more receivers. The main tasks at this level involve signal modulation and encoding of data in order to maintain communication in the presence of channel noise and signal interferences. In order to use the bandwidth efficiently and reduce the development cost to some extent, the standard practice is that several radios share the same interconnect medium. The sharing of media (e.g., time or frequency) is facilitated by the MAC layer. Finally, the network layer is responsible for establishing the path that a message must travel through the network in order to be transferred from its source to the destination.

Design of power and bandwidth efficient radios is one of the main research and development tasks. It is important to realize that radio architecture is a function of the employed network structure and protocols. The main trade-off is between the relative energy cost of transmission and reception; the key observation is that listening to the channel is expensive. Therefore, it is necessary to develop schemes that will enable long periods of sleep mode for receivers. For example, one option is to use coordinated policy for deciding which node will go to sleep while the connectivity in the node is maintained [53]. The other option is to use two radios; one of them is responsible for data reception and is power hungry. It is used only when the other ultralow power radio invokes it. The ultralow power radio is only used to detect if one wants to transmit data to this node.

Table 8.1 surveys the state-of-the-art radio design alternatives from ISSCC 2001 [29] and ISSCC 2002 [30]; several notable radio designs are briefly outlined. One radio design alternative is the fully integrated GPS radio described in Behbahani et al. [4]. The low-IF architecture of the radio enables a high level of integration and low power consumption simultaneously. The integrated radio measures a 9.5-mm^2 chip area. It can operate under a various range of voltage and temperature, namely, from 2.2 to 3.6 V and from –40 to +85°C and consumes 27 mW from a 2.2-V supply.

Another notable design is the IEEE 802.11a wireless local area network (WLAN) transceivers presented in Xargari et al. [62]. A 0.25-µm CMOS technology is used to integrate a 5-GHz transceiver compressing the RF and analog circuits of an IEEE 802.11a-compliant wireless local area network (WLAN). The integrated circuit has 22 dBm maximum transmitted power; 8 dB overall receive-chain noise figure; and –112 dBc/Hz synthesizer phase noise at 1 MHz frequency offset.

Other state-of-the-art radio designs have been developed [7, 11, 32, 64]. Chien and colleagues [11] introduced a fully integrated 2.4-GHz transceiver in 0.25-µm CMOS and its associated baseband processor in 0.15-µm CMOS. Kluge and coworkers [32] have recently designed advanced microdevices — a 2.4-GHz CMOS radio for 802.11b wireless LAN. They used 0.25-µm feature size to design 10-mm^2 integrated circuits that consume 86 mA in receiver mode and 73 mA in transceiver mode from a 2.5-V supply. The receiver has a short settling time and is equipped with a separate receiver channel filter and transceiver pulse-shaping filter. In addition, it provides filter calibration circuitry. Bouros and colleagues [7] introduced a digitally calibrated transceiver in 0.18-µm CMOS that occupies 18.5 mm^2. The integrated phase noise can be minimized to less than –37.4 dBc using the fully integrated VCO and synthesizer.

TABLE 8.1 Comparison of State-of-the-Art Radio Design Alternatives

	Technology	Silicon Area (mm^2)	ICC_RX (mA)	ICC_TX (mA)	VCC (V)
Alcatel (RF+BB) ISSCC 2001-13.1	0.25-μm CMOS	40	41	52	2.5
IME + OKI (RF) ISSCC 2001-13.2	0.35-μm CMOS	18	66	47	2.7–3.3
Broadcom (RF) ISSCC 2001-13.3	0.35-μm CMOS	20?	46	47	2.7–3.3
Conexant (RF) ISSCC 2001-13.4	0.35-μm SiGE BiCMOS	12?	16	12	1.6–3.0
SiliconWave (RF) ISSCC 2002-5.2	0.35-μm SOI BiCMOS	19.5	39	37	2.7
Transilica (RF) ISSCC 2002-5.3	0.25-μm CMOS	13.3	45	36	3.0
Hitachi (RF) ISSCC 2002-5.5	0.35-μm BiCMOS	11.2	45	35	2.7
Bluetooth (RF) ISSCC 2002-5.1	0.18-μm CMOS	5.5 (4.0)	30	35	2.5–3.0

Source: Zeijl, P. et al., *IEEE J. Solid-State Circuits*, 37, Dec. 2002. With permission.

TABLE 8.2 MCU Comparison

	AT91FR4081	ATMega128L
Datapath	16/32 b	8 b
Clock speed (MHz)	40	4
MIPS/MHz	(ARM 0.9); (THUMB 0.7)	1
Power @ 3 V (mW)	75	15
MIPS/W	480	242

Source: Savvides, A. and Srivastava, M.B., in *Proc. Int. Conf. Computer Design*, 2002. With permission.

TABLE 8.3 Current Drawn by Node Components

Component	Active (mA)	Sleep (mA)
ATMega128L	5.5	1
RFM	2.9	5
AT91FR4081	25	10
RS-485	3	1
RS-232	3	10
Total	39.4	27

Source: Savvides, A. and Srivastava, M.B., in *Proc. Int. Conf. Computer Design*, 2002. With permission.

Chien et al. [11] have developed a 2.4-GHz radio for 802.15.4 WPANs using 0.18-μm CMOS technology that consumes 21 and 30 mW at 1.8-V supply in receiving and transmitting mode, respectively. It incorporates a poly-phase filter and applies transistor linearization techniques to achieve a low-IF architecture. Other alternatives are also available [13, 16].

8.5 Sensor Network Node

This section addresses the key issues related to the architecture and synthesis of an individual SN node. Architecture aspects are discussed along three lines: hardware, software, and middleware; design issues are presented from synthesis and analysis points of view.

The architecture of SN nodes has been addressed in at least three main directions. The first group of initial efforts comprises a number of designs of individual sensor nodes and badges [1, 3, 38, 40, 45, 50, 59]. The emphasis in this class has been placed on ensuring creation of working prototypes and, in some cases, pushing the state of the art of an individual component (e.g., radio, low power, energy harvesting). The second group was represented by the Mote/TinyOS development team at UC Berkeley [15, 27], who made the first effort to address the trade-offs between various components of the node by developing a new architecture and operating system (OS). The main characteristic of the last effort is sensor centered. The emphasis is on exploiting relatively inexpensive off-the-shelf components in terms of cost and energy as a basis for exploring qualitative and quantitative trade-offs between node components and, in particular, sensors.

It is difficult to anticipate technological trends, but one can easily identify at least some high-impact trends and required solutions. For example, it is apparent that overall energy consumption-balanced architectures are needed. Another high-impact research topic concerns sensor organization and development of the interface between components. Finally, due to privacy, security, and authentication needs, techniques such as unique ID for CPU and other components that facilitate privacy will be in high demand.

In the software domain, main emphasis will be on RTOS (real time operating system) [36]. Ultra-aggressive low-power management is needed because of energy constraints and comprehensive resource accounting is desired due to demands for privacy and security. In a number of cases, support for mobility functions (e.g., location discovery) is also needed. Middleware will be in even stronger demand in order to enable rapid development and deployment of new applications. Tasks such as sensor data filtering; compression; sensor data fusion, sensor data searching and profiling; exposure coverage; and tracking will be ubiquitous.

Synthesis of sensor nodes will pose a number of new problems in the CAD world. It is obvious that new types of models, abstractions, and tasks will be defined and solved. Sensor allocation and selection, sensor positioning, sensor assignment, and efficient techniques for sensor data storage are typical examples of pending synthesis tasks. Development of conceptually simple and clean, yet inexpressive, models of computation is of prime importance as a starting point for synthesis of modern computing systems. Sensor nodes will require new models of computations as well as new models of the physical world. One such example is standard Euclidian space with classical physical laws (e.g., Newton's law, thermodynamics law).

It is well known that modern design flow, debugging, and verification are the most expensive and time-consuming components. Due to the heterogeneous nature of and complex interactions between components, the same scenario is expected in the case of sensor nodes. In particular, techniques for error and fault discovery, testing, and calibration will be of prime importance. In the rest of this section, four representative SN nodes designs are described: Berkeley mote; piconode node; UCLA Medusa II; and light compass node.

8.5.1 Berkeley Mote Node

The starting point for designing modern computer systems is a comprehensive set of benchmarks that are representative for common users. Unfortunately, such a set of benchmarks currently is not available to designers of SN nodes. The starting point for designing mote wireless sensor network nodes was the set of qualitative observations about the requirements of wireless sensor networks. Special emphasis was placed on small physical size and low energy consumption. In addition, attempts were made to facilitate concurrency intensive operations to provide control hierarchy and take advantage of limited physical parallelism. Furthermore, the design decisions were driven by robust operations' ability to be retargeted, at least at the network level.

The design went though several iterations and until recently was leveraging on the availability of standard off-the-shelf components. Generally speaking, the design is radio centric in the sense that all main decisions are made in order to facilitate low-energy communications. The main processor is Atmel 90LS8535 microcontroller that has 8-b Harvard architecture with 16-b addresses. It achieves a speed of 4 MHz at 3 W. The system has a rather minimal amount of memory that consists of 8 kbytes of flash for program memory and 512 bytes SRAM for data memory. Therefore, the system can be integrated only with low-frequency sampling sensors and must communicate frequently.

The processor integrates a system of timers and counters and can be placed in four energy modes: active, idle, power down, and power save. In the idle mode, the processor is completely shut off. In the power-down mode, only the watchdog and asynchronous interrupt logic are awake. Finally, in the power-save mode, in addition to watchdog and interrupt logic, the asynchronous timer is also active. The system also has a coprocessor Atmel 90LS2343 microcontroller that has 2 kbytes flash instruction memory and 128 bytes of SRAM and EEPROM memory. The coprocessor can be used to reprogram the main microcontroller.

The authors consider the RF Monolithic 916.50 transceiver as the central part of the design. The radio is equipped with an antenna and a system of discrete components that can be used to alter characteristics of the physical layer such as signal strength. The radio operates at a speed of 19.2 kbytes/sec. The transceiver can operate in three modes: transmission, reception, and power off. The system can have up to eight sensors; the two most widely used are photoelectric optical sensor and temperature sensor. Each sensor is placed on the bus that is controlled using software.

It is instructive to consider power characteristics of the design. MCU core consumes between 2.5 to 6.5 mA; radio consumes between 5 to 12 mA. Optical sensor and temperature sensor consume 0.3 and 1 mA, respectively, and the coprocessor consumes 1 to 2.4 mA. Finally, EEPROM consumes 1 to 3 mA. In particular, it is instructive to compare energy spent for bit transmission and bit processing. The system spends about 1 mJ to send, and 0.5 mJ to receive, 1 b. At the same time, the system can execute approximately 120 instructions for each millijoule spent. The system does provide for energy reduction using variable voltage; therefore, energy is saved mainly by turning the system off. The core of the system software for the design is an exceptionally compact microthreading operating system (TinyOS).

The Berkeley design team concluded that the new application domain requires a new OS; therefore, they decided not to adopt any great variety of RTOS 8-b controllers. Although this decision certainly resulted in higher power efficiency and more interesting system software architecture, it also created additional demands and constraints in programming already highly constrained hardware. Nevertheless, the system has been highly popular in the research community. Several thousand copies of the motes in several versions have been used by more than 200 research teams. The greatest strength in the system is its small size and low power. Probably the most serious disadvantages are related to the development of real applications. Although motes have been tremendously popular in research communities, it is still unclear how well they are suited for applications in which more complex systems of sensors are needed.

8.5.2 UCLA Medusa MK-2 Node

The Medusa MK-2 node is a representative of the state-of-the-art design of more powerful sensor nodes [55]. The computational unit of Medusa MK-2 nodes consists of two microcontrollers. The first is an 8-b Atmel STMega128L MCU with 4 MHz that has 32 K of flash memory and 4 KB of RAM. This processor serves as an interface between sensors and radio base band processing. The second microcontroller is an ATMEL ARM THUMB processor enclosed within a 120-ball BGA package. It has significantly more processing power and 40 MHz. It includes 136 KB of RAM and 1 MB of on-chip FLASH memory.

The communication unit of Medusa MK-2 nodes is a combination of a TR 1000 low-power radio from RF Monolithics for wireless and an RS-485 serial bus transceiver for wireline communication. The sensing unit has two components: a MEMS accelerometer and a temperature sensor. It can also be augmented with other types of sensors. Medusa nodes also incorporate a variety of interfaces, including eight 10-b ADC inputs, serial ports, and numerous general purpose I/O ports. An ultrasonic ranging unit is implemented on an accessory board using 40-kHz transducers. Ultrasonic measurements are coordinated with RF measurements in order to calculate internode distances and therefore enable localization of nodes. Localization is conducted using iterative linearized multilateration.

The nodes also have two external connectors. The first is used to communicate with a PC to download and debug software. It also provides the necessary wiring requirements for connecting to an external GPS module. The second connector serves as an expansion slot for attaching add-on boards carrying different sensors because it has a set of ADC and GPIO. Finally, Medusa nodes also have two pushbuttons that serve as a user interface. They are mainly used for triggering events and executing different tests during experiments.

It is interesting to take a closer look at the computational unit of Medusa Mk-2 nodes. According to the computation requirements, the computational tasks are classified into two broad categories: low-demand tasks and high-demand, low-frequency processes. The low-demand tasks are the periodic processes such as base band processing for the radio while listening for new packets, sensor samplings, handling of sensor events, and power management. Even though these tasks usually require a high concurrency, they are not particularly demanding in terms of computational resource requirements and therefore can be easily handled by an 8-b microcontroller. The Medusa-MK-2 nodes use a low-power AVRMega128L microcontroller.

The second category — the low-frequency, high-demanding tasks — is related to the processing of acquired sensor data in order to produce user-requested information. For example, in the case of a fine-grained localization problem, a sensor node is expected to compute an estimate of its location based on a set of distance measurements to known beacons or neighbors. In order to avoid error propagation, a node must perform a set of high-precision operations. If an 8-b processor were used to conduct this type of computation, it would result in high latencies and lower precision. Therefore, a high-end processor is a more adequate solution. More specifically, Medusa adopts the 40-MHz ARM THUMB processor to perform this type of operation.

Another advantage is that the node can use existing standard applications and libraries. The THUMB microcontroller also has sufficient resources to support shelf-embedded operating systems such as Red Hat eCos and uCLinux. The inclusion of the THUMB processor is also justified by a comparison of the two processors made from a power/latency perspective conducted by the UCLA group. The THUMB processor executes instructions at the rate of 0.9 MIPS per megahertz at 40 MHz while consuming 25 mA with a 3-V supply, which has a performance of 480 MIPS/W. On the other hand, the ATMega128L only provides a 242-MIPS/W performance when operating at 4 MHz and consumes 5 mA at 3-V supply.

Communication between the two processors is handled by a pair of interrupt lines — one for each microcontroller — and an SPI bus. The two nodes remain in sleep mode until an interrupt indicating the need for data exchange occurs. The communication takes place over the 1-Mbs SPI bus.

Medusa MK-2 nodes are capable of two types of communications: wired and wireless. All nodes are equipped with a wired and a wireless link. The wireless link is a low-power TR1000 radio from RF Monolithics. This radio has transmitting power of 0.75 mW at maximum and has an approximate transmission range of 20 m. Two modulation schemes are supported by this radio: of-off keying (OOK) and amplitude shift keying (ASK). Selection of the appropriate modulation can be done in software. On a Medusa MK-2 node, the base band processing for the radio is done by an ATMega128L microcontroller. This also allows the node to run the low power S-MAC [61] protocol on the ATMega128L processor. In addition to the wireless link, Medusa nodes also incorporate an RS-485 serial bus interface for wireline communication. Attaching a low-power RS-485 transceiver to one of the RS-232 ports of the THUMB processor allows the node to connect to an RS-485 network using an RJ-11 connector and regular telephone wire. A single RS-485 has occupancy up to 32 nodes that span over a total wire length distance of 1000 ft.

The power unit of Medusa MK-2 nodes consists of two main components: the power supply and the power management and tracking unit (PMTU) [12]. The power supply consists of a 540-mAh lithium-ion rechargeable battery and an up–down DC–DC converter with a 3.3-V output that can reach up to 300 mA of current from the battery. The power supply is designed in such a way that power-additional sensors can be attached later on as accessory boards because the node only requires less than 50 mA with no sensors attached. In a typical SN setting, putting the ARM THUMB processor together with the RS-485 and RS-232 transceivers in sleep mode most of the time, yields an 80% reduction of the overall node power consumption. Comprehensive energy consumption comparisons between Medusa MK-2 nodes and other SN nodes designs can be found in Savvides and Srivastava [55].

8.5.3 BWRC PicoNode

Another communication-centered sensor node design is the PicoNode [52]. The main overall objective of this design is to provide flexibility and low energy consumption simultaneously. It consists of four main modules. The first two units are processors: an embedded processor unit and configurable satellite units. The embedded processor is dedicated mainly for application and protocol-stack layers that require higher flexibility but have relatively low computational complexity and are infrequently requested. Configurable processing modules are targeted for the more frequent tasks with higher computational requirements. Two other modules are dedicated to communication tasks — a parameterized and configurable digital physical layer and a simple direct-down conversion RF front end.

These modules are interconnected by a flexible and low-power consumption interconnect scheme. The authors claim that a dynamic matching between application and architecture leads to a significant energy savings for signal-processing applications while maintaining implementation flexibility. One of the main premises of the design is the observation that the processor implementation is three orders of magnitude more expensive in terms of energy consumption than the implementations of the dedicated hardware. However, a trade-off occurs between flexibility and programmability (software on programmable platforms) and energy consumption (ASIC hardware).

The traditional approach is to design the wireless transceiver using only RF and analog circuit modules. More recently, a primarily digitalized design approach has emerged. This is inspired by the insight that digital circuits can improve exponentially with the scaling of technology, while analog circuits get linearly worse because of reduction of the supply voltage. Therefore, it is beneficial to incorporate a small, noncritical analog front end and use digital back-end processing to balance the limitations.

Many design challenges are related to the physical layer. They are mostly related to the low-energy targets and variable demands from the network. Therefore, in order to satisfy various demands from the network, the PicoNode physical layer can be made into parameters. These parameters include power control modes, modulation scheme, and bit rate.

In order to meet the low-energy requirement, the physical layer must meet two mutually exclusive criteria: fast signal acquisition and low standby power. The first criterion refers to the process of requiring least amount of time to wake up, receive bursts of data, and immediately go back to sleeping mode after data acquisition. The second criterion emphasizes consuming the least amount of energy while sleeping. The reason that they are usually mutually exclusive is that an inverse proportional relationship exists between the depth of sleep (i.e., energy consumed) during standby and the time required to wake up.

PicoNode is designed so that it does not require an interval power supply. It is self-constrained and self-powered using energy extracted from the environment. The two major constraints for harvesting ambient energy from the environment are: applicability within the environment and the size of the node (Berkeley group targets the 1-cm^3 design). PicoNodes harvest energy from light and vibrations [52].

8.5.4 Sensor-Centric Design: Light Compass

The final sensor node design alternative for overviewed is the light compass node [66]. The emphasis in this approach is completely shifted from computation, communication, and storage to sensors. The first three functions are provided by a standard laptop or PDA. The rationale is that this type of design will progress on its own to become a viable platform for SN nodes. Even the interface toward sensor is built using off-the-shelf components. The focus is placed on sensors and how to select and place them in such a way that sensor data fusion is facilitated. In addition, special emphasis is placed on how to rapidly develop sensor data fusion software that can be retargeted and how to develop systematic procedures for design of sensor nodes.

Figure 8.1 shows the used light sensor components. The smallest device (on the left) is a miniature silicon solar cell used for converting light impulses directly to electrical charges (photovoltaic). It generates its own power and therefore does not require any external bias. This silicon cell is further mounted on a 0.78 × 0.58 × 0.18 cm thick plastic carrier that generates roughly 400 mV in moderate light (most typical rooms). A significantly larger sensor (on the right), measuring 2.54 × 2.15 cm, also can be referred to as a photoconductor and can be surrounded by a 0.18-cm thick plastic encapsulated ceramic package. In strong light, its resistance measures 20 Ω and 5 kΩ in complete darkness. These components are very economically viable (roughly $0.30 each) and they can be easily purchased in large quantities.

These sensors can be used in multiple prototypes, such as the ones shown in Figure 8.2. On the left side of Figure 8.2, the six-sided cut-pyramid structure has a base length of 3 cm and a top edge length of 1 cm with a 60° slope. Sensors can be attached to each side of the structure depending on the application and purpose. The structure on the right is a cube with 2-cm edges; therefore, it can incorporate up to six sensors with one on each surface.

In this light sensor platform, the heart component is an eight-channel analog to digital converter (ADC) module. It is used to read the sensor values through the parallel port of a standard PC laptop. This ADC component comprises a Maxim MAX186 ADC, which has an internal analog multiplexer that can be configured for eight single-ended, or four differential, inputs at a 12-b resolution; the conversion time is under 10 s. This component is pictured on the left in Figure 8.3. In addition, some of the other components of the circuit include: several resistors to protect the analog inputs; capacitors to filter noise; an external 4.096-V voltage regulator, and an 8-b digital latch required for parallel port communications. The overall design flow of a sensor appliance is presented in detail in Figure 8.4.

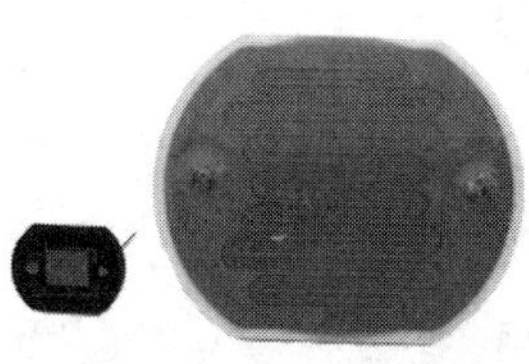

FIGURE 8.1 Light sensor components. (From Wang, J., et al., *40th IEEE/ACM Design Automation Conf.*, pp. 66–71, 2003. With permission.)

FIGURE 8.2 Light appliance prototypes: 60° six-sided, cut pyramid and cube. (From Wang, J., et al., *40th IEEE/ACM Design Automation Conf.*, pp. 66–71, 2003. With permission.)

FIGURE 8.3 Light appliance platform. (From Wang, J., et al., *40th IEEE/ACM Design Automation Conf.*, pp. 66–71, 2003. With permission.)

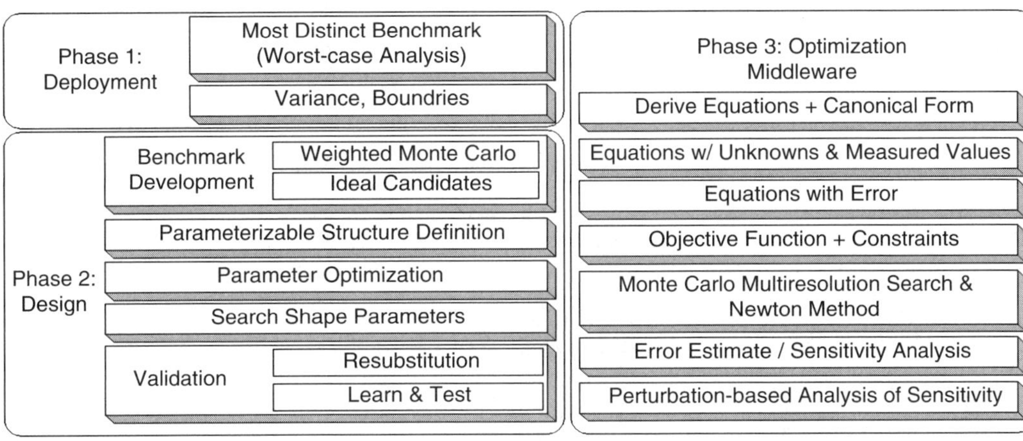

FIGURE 8.4 Overall design flow of a sensor appliance. (From Wang, J., et al., *40th IEEE/ACM Design Automation Conf.*, pp. 66–71, 2003. With permission.)

The main goal of this design was to achieve low power consumption while maintaining a tolerable level of coverage. Figure 8.5 through Figure 8.7 depict the results obtained from four different sensor structures: a four-sensor pyramid (square base); a four-sensor cut pyramid (triangular-based pyramid with a flat sensor on top); a five-sensor pyramid (pentagonal base); and a five-sensor cut pyramid (square-based pyramid with a flat sensor on top). In all cases, the objective was to estimate the positions of 5000 randomly placed light instances.

8.6 Wireless SNs as Embedded Systems

The architecture of wireless SNs at the network level is briefly surveyed in this section. For the networking of the wireless devices and appliances, several communication schemes have been proposed, such as satellite, WLAN, cellular, and ad hoc multihop architectures [25, 26, 48, 49, 58]. Based on the different architectures, the communication between the nodes can be all low power (ranges in meters), high power (ranges in megameters), or medium power (ranges in kilometers).

For example, wireless SNs are the widely used cellular wireless networks. In this architecture, a number of base stations are already deployed within the field. Each base station forms a cell around itself that covers part of the area. Mobile wireless nodes and other appliances can communicate wirelessly as long as they are at least within the area covered by one cell. An example of such a network is shown in Figure 8.8. The communication requires medium power, although the fixed and immobile base stations are consuming a large amount of power to cover a large area and to communicate to and from the lower power mobile wireless nodes. However, cellular wireless architecture has the drawback that it must be

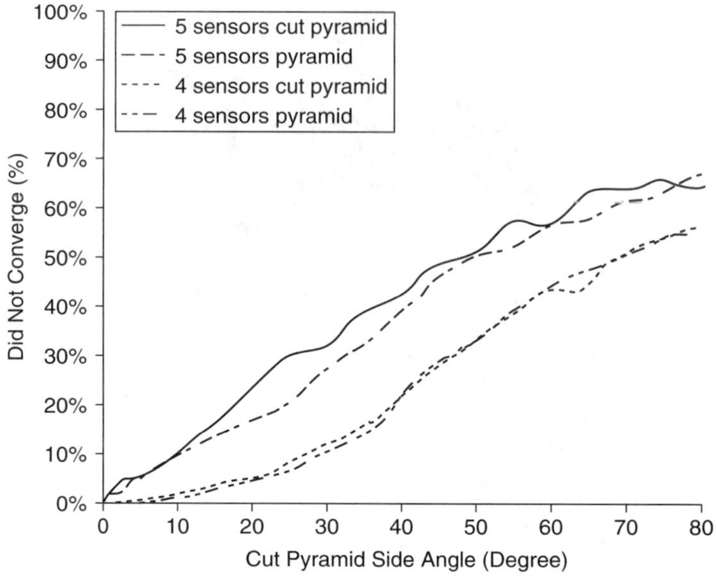

FIGURE 8.5 Fraction of failure convergence vs. sensor angles. (From Wang, J., et al., *40th IEEE/ACM Design Automation Conf.*, pp. 66–71, 2003. With permission.)

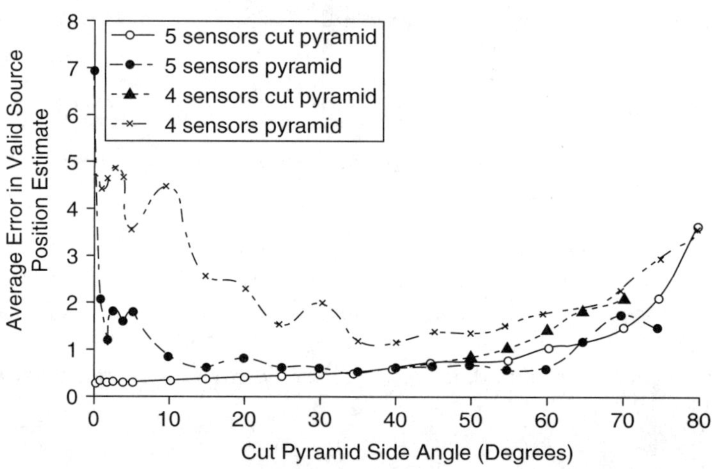

FIGURE 8.6 Fraction of valid solutions vs. sensor angles. (From Wang, J., et al., *40th IEEE/ACM Design Automation Conf.*, pp. 66–71, 2003. With permission.)

implanted in the field; also, cells should be carefully designed to have full coverage and transparency with respect to the cells.

The WLAN is built for high-frequency radio waves. The WLAN also needs its own infrastructure within the designated local area. It is very well suited for local private areas, such as offices, campuses, and buildings. In some of the applications of the sensor network, such as smart buildings, connecting the sensor networks to the WLAN implanted within the area is very suitable. The power consumption in LAN is also medium, although the fixed part of the infrastructure is naturally higher powered.

In order to overcome the difficulties caused by the infrastructure settings for wireless satellites, WLAN, and cellular networks, a new generation of wireless networks architecture has emerged — the wireless multihop ad hoc networks. In such networks, the infrastructure architecture is not needed and the nodes

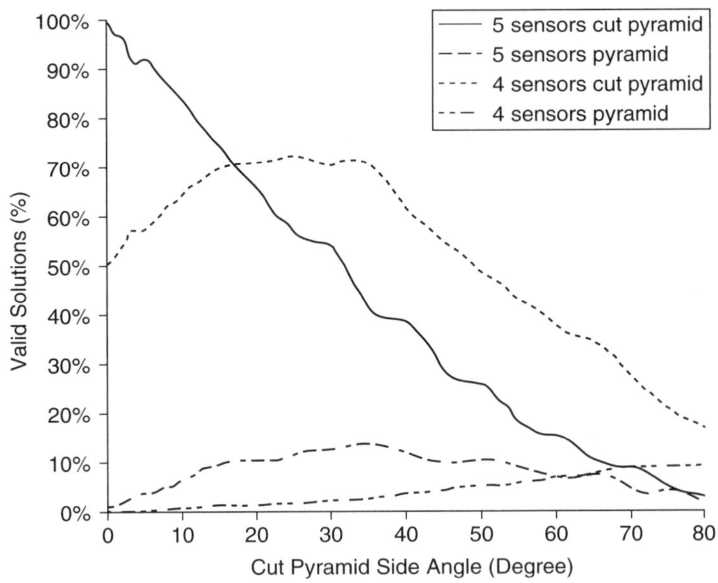

FIGURE 8.7 Average error in positions vs. sensor angles. (From Wang, J., et al., *40th IEEE/ACM Design Automation Conf.*, pp. 66–71, 2003. With permission.)

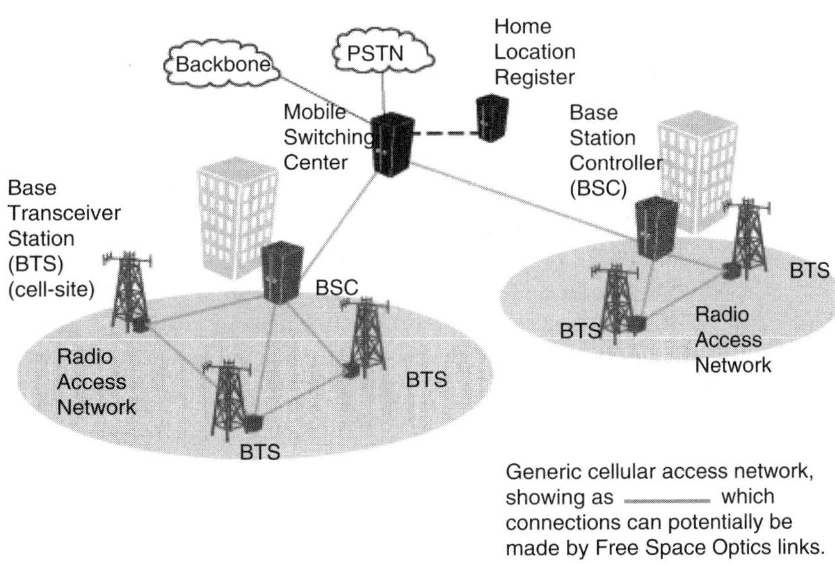

FIGURE 8.8 Wireless cellular network architecture. (From: http://w.w.w.holoplex.com/technology_backhaul. html.)

can configure to communicate to other nodes within their communication range on the fly. The nodes are short range and therefore all of the communications are low power. If two nodes that are not within each other's range need to communicate to each other, they use the intermediate nodes as the relays. The multihop ad hoc wireless SN architecture appears as an attractive alternative to the WLAN and cellular technologies for at least four reasons:

- On-demand formation of the network does not require predeployed architecture.
- Multihop routing can save orders of magnitude of power consumption when compared to long-range routing for the same distance [52].

- Because communications between the nodes are short range and local, the bandwidth is reusable, as opposed to that in long-range communications.
- The fourth reason is the fault tolerance [10]. SNs are envisioned to have a lot of inexpensive nodes embedded in the environment. The ad hoc multihop architecture supports the advent of the new nodes and departure or failure of the old ones.

Most of the current SN literature has been advocating ad hoc multihop architecture [2, 20, 27, 34, 52, 61]. Nevertheless, there are no indications that this architecture would be the best architecture for all of the sensor network applications. Because of the quantity of the radios and the number of the packets flowing in the network, a natural asymmetry is present in the multihop ad hoc implementation. In fact, for some applications, such as smart buildings or scientific experiments in which the network does not change over the space, having a number of static components in the network is a natural solution. The static parts would be connected to the constant power supply, so wireless parts could use low power to communicate to them and nodes could go into the standby mode from time to time.

Another important issue related to sensor networks is the topology of the network [10]. The question is how to distribute the nodes within the field to achieve the best range and coverage from the sensors. This question is a variation of the well-known art gallery problem [47], in which the new constraints on the nodes are that they are short communication range. The other big issue in topology consideration is that not all of the nodes should be uniformly distributed, as is the assumption in the current literature and simulations for SNs. Furthermore, network architecture should address the concerns of various layers of the network.

Better components are still needed in the physical layer [31], power control, and MAC layer [61]; routing protocols [20] are needed at the network layer. The only proposed OS for the sensor network is TinyOS, which is an operating system at the node level [27]. There is a need for a more complex network operating system (NOS) that can (1) facilitate the autonomous mode for ad hoc multihop architecture; (2) address privacy and security concerns; and (3) provide efficient execution of localized algorithms.

This section concludes with a very brief overview of three industrial wireless networks standards: IEEE 802.11b; Bluetooth; and HomeRF. IEEE 802.11b, or WiFi, primarily targets computer communication. Although its main target is indoor connectivity at speeds of 11 Mbps within 150 m, it is expected that it will provide the same level of service outdoors within a 300-m range. With specially equipped radios (amplifiers and special antenna) it may establish connectivity within a range of 30+ km. It can operate in several modes, including peer–peer and infrastructure access point. The wired equivalent privacy (WEP) standard ensures data protection using 40- and 128-b RC4-based encryption. Bluetooth mainly targets personal area networks on very short distances and applications such as audio, video, and multimedia. IEEE802.11b and Bluetooth use 2.4-GHz ISM band for unlicensed radio communication. HomeRF provides inexpensive residential-oriented wireless connectivity.

8.7 Summary

This chapter surveyed the architectural and synthesis issues related to SNs. The analysis has been conducted at three levels of abstraction: subsystem, individual node, and network. The main design objectives and current trends, as well as their relative advantages and limitations, were identified. Furthermore, several architecture and design case studies have been conducted. Special emphasis was placed on formulating the highest impact architectural and synthesis challenges.

Acknowledgment

This material is based upon work supported in part by the National Science Foundation under Grant No. ANI-0085773 and NSF CENS Grant.

References

1. Agre, J.R. et al., Development platform for self-organizing wireless sensor networks, in *Proc. SPIE Int. Soc. Optical Eng.*, 3713, 257, 1999.
2. Akyildiz, I.F. et al., Wireless sensor networks: a survey, *Computer Networks*, 38, 393, 2002.
3. Asada, G. el at., Wireless integrated network sensors: low power systems on a chip, in *Proc. 24th Eur. Solid-State Circuits Conf.*, 9, 1998.
4. Behbahani, F. et al., A fully integrated low-IF CMOS GPS radio with on-chip analog image rejection, *IEEE J. Solid-State Circuits*, 37, Dec 2002.
5. Beneden, B.V., Examining Windows CE 3.0 real time capabilities, *Dr. Dobb's J.*, 26, 66, 2001.
6. Bridges, S., The R380s — the first smartphone from the Ericsson–Symbian partnership, *Ericsson Rev.*, 78, 44, 2001.
7. Bouras, I. et al., A digitally calibrated 5.15 — 5.825-GHz transceiver for 802.11a wireless LANs in 0.18-µm CMOS, in *Proc. ISSCC*, 2003.
8. Burd, T. et al., A dynamic voltage scaled microprocessor system, *Dig. Tech. Papers ISSCC*, 2000.
9. Cerpa, A. et al., Habitat monitoring: application driver for wireless communications technology, in *Proc. ACM SIGCOMM Workshop Data Commun. Latin Am. Caribbean*, 2001.
10. Cerpa, A. and Estrin. D., ASCENT: adaptive self-configuring sensor networks topologies, in *Proc. INFOCOM 2002*, 2002.
11. Chien, G. et al., A 2.4G-Hz CMOS transceiver and base band processor chipest for 802.11b wireless LAN application, in *Proc. ISSCC*, 2003.
12. Chen, A. et al., A support infrastructure for the smart kindergarten, *IEEE Pervasive Computing Mag.*, 1, 49, 2002.
13. Cojocaru, C. et al., A 43-mW Bluetooth transceiver with –91 dBm sensitivity, in *Proc. ISSCC*, 2003.
14. Drinic, M., Kirovski, D. and Potkonjak, M. *Model-based compression in wireless ad hoc networks*, in *Proc. of Sensys*, 2003.
15. Culler, D.E. et al., EMSOFT 2001: network-centric approach to embedded software for tiny devices, in *Proc. Workshop Embedded Software*, 2001.
16. Darabi, H. et al., A dual-mode 802.11b/Bluetooth radio in 0.35-µm CMOS, in *Proc. ISSCC*, 2003.
17. Dietzel, A. and Berger, R., Trends in hard disk drive technology, in *Proc. VDE World Microtechnol. Cong.*, 1, 2000.
18. Douseki, T. et al., A batteryless wireless system uses ambient heat with a reversible-power-source compatible CMOS/SOI DC-DC converter, in *Proc. ISSCC*, 2003.
19. Doyle, M., Fuller, T.F., and Newman, J. Modeling of galvanostatic charge and discharge of the lithium/polymer/insertion cell, *J. Electrochem. Soc.*, 140, 1526, June 1993.
20. Estrin, D. et al., Next century challenges: scalable coordination in sensor networks, in *Proc. MOBICOM*, 262, 1999.
21. Faroque, M. and Maru, H.C., Fuel cells — the clean and efficient power generators, *in Proc. IEEE*, 89, 1819, Dec 2001.
22. Feng, J. et al., Sensor networks: quantitative approach to architecture and synthesis, UCLA, technical Report, 2002.
23. Fuller, T.F., Doyle, M., and Newman, J. Simulation and optimization of the dual lithium ion insertion cell, *J. Electrochem. Soc.*, 141, 1, Jan 1994.
24. George, V. et al., The design of a low-energy FPGA, in *Proc. ISLPED*, 1999.
25. Gupta, P. and Kumar, P.R., Internets in the sky: capacity of 3D wireless networks, in *Proc. IEEE Conf. Decision Control*, 3, 2290, 2000.
26. Hamburgen, W.R. et al., Itsy: stretching the bounds of mobile computing, *Computer*, 34, 28, April 2001.
27. Hill, J. et al., System architecture directions for networked sensors, *ASPLOS*, 93, 2000.
28. Ishii, T. el al., A 126.6-mm/sup 2/AND-type 512-Mb flash memory with 1.8-V power supply, *IEEE J. Solid-State Circuits*, 36, 1707, Nov 2001.

29. Session 13 on Bluetooth transceivers, *ISSCC Dig. Tech. Papers*, Feb 2001.
30. Session 5 on Bluetooth transceivers, *ISSCC Dig. Tech. Papers*, Feb 2002.
31. Kahn, J.M., Katz, R.H., and Pister, K.S. Next century challenges: mobile networking for "Smart Dust," in *Proc. MobiCom*, 271, 1999.
32. Kluge, W. et al., A 2.4GHz CMOS transceiver for 802.11b wireless LANs, in *Proc. ISSCC*, 2003.
33. Koushanfar, F., Potkonjak, M., and Sangiovanni–Vincentelli, A., Fault-tolerance techniques for sensor networks, in *Proc. IEEE Sensors*, 49, 2002.
34. Koushanfar, F. et al., Processors for mobile applications, in *Proc. Int. Conf. Computer Design*, 603, 2000.
35. Kulah, H., Yazdi, N., and Najafi, K., A multi-step electromechanical ΣΔ converter for micro-g capacitive accelerometers, in *Proc. ISSCC*, 2003.
36. Li, Y., Potkonjak, M., and Wolf, W., Real-time operating systems for embedded computing, in *Proc. Int. Conf. Computer Design*, 388, 1997.
37. Linden, D., *Handbook of Batteries*, 2nd ed., New York: McGraw-Hill, 1995.
38. Locher, I. et al., System design of iBadge for smart kindergarten, unpublished manuscript.
39. Luo, H., Fedder, G., and Carley, L., Integrated multiple-device IMU systems with continuous-time sensing circuitry, in *Proc. ISSCC*, 2003.
40. Maguire, G.Q. et al., Smartbadges: a wearable computer and communication system, in *Proc. 6th Int. Workshop Hardware/Software Codesign*, 1998.
41. Mason, A. et al., A generic multielement microsystem for portable wireless applications, *IEEE*, 86, 1733, Aug 1998.
42. Meguerdichian, S. et al., Coverage problems in wireless ad hoc sensor networks, in *Proc. IEEE INFOCOM*, 3, 1380, 2001.
43. Meguerdichian, S. et al., Localized algorithms in wireless ad hoc networks: location discovery and sensor exposure, in *Proc. MOBIHOC*, 106, 2001.
44. Meng, T.H. and McFarland, B., Wireless LAN revolution: from silicon to systems, in *Proc. IEEE Radio Frequency Integrated Circuits (RFIC) Symp.*, 3, 2001.
45. Meninger, S. et al., Vibration-to-electric energy conversion, in *Proc. IEEE Trans. VLSI Syst.*, 9, 64, Feb 2001.
46. Min, R. et al., An architecture for a power-aware distributed microsensor node, in *Proc. IEEE Workshop Signal Process. Syst.*, 581, 2000.
47. O'Rourke, J., *Art Gallery Theorems and Algorithms,* Oxford University Press, 1987.
48. Pehrson, S., WAP — the catalyst of the mobile Internet, *Ericsson Rev.*, 77, 14, 2000.
49. Perkins, C.E., *Ad Hoc Networking*, Boston: Addison-Wesley, 2001.
50. Pottie, G.J. and Kaiser, W.J. Wireless integrated network sensors, in *Proc. Commun. ACM*, 43, 51, May 2000.
51. Priyantha, N.B., Chakraborty, A., and Balakrishnan, H., The Cricket location-support system, in *Proc. MobiCom*, 32, 2000.
52. Rabaey, J.M. et al., PicoRodio supports ad hoc ultra-low power wireless networking, *Computer*, 33, 42, July 2000.
53. Rozovsky, R. and Kumar, P.R., SEEDEX: a MAC protocol for ad hoc networks, in *Proc. MOBIHOC*, 67, 2001.
54. Savvides, A., Han, C.C., and Srivastava, M., Dynamic fine-grained localization in ad-hoc networks of sensors, in *Proc. ACM SIGMOBILE 7th Annu. Int. Conf. Mobile Computing Networking*, 2001.
55. Savvides, A. and Srivastava, M.B., A distributed computation platform for wireless embedded sensing, *Proc. Int. Conf. Computer Design*, 2002.
56. Slaughter, J.M. et al., Fundamentals of MRAM technology, *J. Superconductivity*, 15, 19, Feb 2002.
57. Slijepcevic, S. and Potkonjak, M., Power efficient organization of wireless sensor networks, in *Proc. IEEE Int. Conf. Commun.*, 472, 2001.
58. Sohrabi, K. et al., Protocols for self-organization of a wireless sensor network, *IEEE Personal Commun.*, 16, Oct 2000.

59. Want, R. et al., The active badge location system, *ACM Trans. Inf. Syst.*, 10, 91, 1992.
60. Yazdi, N., Ayazi, F., and Najafi, K., Micromachined inertial sensors, *IEEE*, 86, 1640, Aug 1998.
61. Ye, W., Heidemann, J., and Estrin, D., An energy-efficient MAC protocol for wireless sensor networks, in *Proc. INFOCOM*, 2002.
62. Xargari, M. et al., A 5-GHz CMOS transceiver for IEEE 802.11a wireless LAN system, *IEEE J. Solid-State Circuits*, 37, Dec 2002.
63. Zhang, H. et al., A 1-V Heterogeneous reconfigurable processor IC for base band wireless applications, in *IEEE Int. Solid-State Circuits Conf.*, 2000.
64. Zeijl, P. et al., A Bluetooth radio in 0.18-μm CMOS, *IEEE J. Solid-State Circuits*, 37, Dec 2002.
65. Nissanka, P.B., Min, A.K.L., Balakrishnan, H., and Teller, S., The Cricket Compass for context-aware mobile applications, *Proc. 7th Ann. Intl. Conf. Mobile Computing and Networking (MobiComm 2001)*, pp. 1, 2001.
66. Wong, J., Megerian, S., and Potkonjak, M., Design techniques for sensor appliances: foundations and light compass case study, *40th IEEE/ACM Design Automation Conf.*, pp. 66–71, June 2003.

9 Power-Efficient Topologies for Wireless Sensor Networks*

9.1 Motivation 9-1
9.2 Background 9-2
9.3 Issues for Topology Design 9-3
Three-Neighbors WSN • Four-Neighbors WSN • Five-Neighbors WSN • Six-Neighbors WSN • Seven-Neighbors WSN • Eight-Neighbors WSN • Six-Neighbors for Three Dimensions
9.4 Assumptions 9-8
Calculation of Power Usage for Each Path
9.5 Analysis of Power Usage 9-10
Two-Dimensional Analysis • Three-Dimensional Analysis
9.6 Directional Source-Aware Routing Protocol (DSAP) 9-13
9.7 DSAP Analysis 9-15
Two-Dimension Analysis • Three-Dimension Analysis
9.8 Summary 9-19

Ayad Salhieh
Wayne State University

Loren Schwiebert
Wayne State University

9.1 Motivation

This chapter examines the relationship between power usage and the number of neighbors in a wireless sensor network. The study of wireless network topology must be approached from a point of view different from that for wired networks. In a wired network, one examines how nodes are physically connected and the resulting available routing paths. In a wireless sensor network (WSN), the definition of the network topology is derived from the physical neighborhood and transmission power, so it is necessary to determine which topology gives the optimal number of neighbors that a node can handle to transmit or receive. Many of the topologies proposed for wired networks cannot be used for wireless networks because, in wired networks, a higher dimension can be implemented by connecting the nodes in some fashion to simulate higher dimensions. In WSNs, however, one is dealing with three dimensions in the physical world and thus restricted in choice of topologies. Therefore, this chapter concentrates on two-dimensional and three-dimensional mesh topologies.

In this chapter, performance issues associated with different network topologies are analyzed. The question to answer concerns the best topology for a wireless network of sensors, assuming that one can control the placement of these sensors and the sensor locations are fixed relative to each other. Because

*This research was supported in part by National Science Foundation Grants DGE-9870720 and ANI-0086020.

control over the placement of these sensing nodes is assumed and mobility of the sensors relative to each other is not required, the research problem changes. Instead of considering self-organization of the sensor nodes into a network, efficient placement of fixed nodes is addressed.

Some of these networks can be installed in a building to monitor the building or in an assembly, where the use of regular topology will have an advantage over mobile. In a fixed topology, nodes can be placed so that they can give better coverage. Also, in the use of regular topology or mesh topologies, a node can also function as a router and can relay messages for its neighbors. These networks offer multiple redundant communication paths throughout the network. If one node dies or fails, other nodes can be used to reroute the message. Also, regular topologies enhance the overall reliability of the network.

This chapter does not consider the effects of communication with a base station. Because the topology is fixed and known, it is assumed that the base station can be placed at an appropriate place for each topology. Thus, the power requirements for communicating with the base station should be essentially independent of the topology. This enables one to concentrate on the effects of the topology on the communication among the network nodes only.

9.2 Background

Much of the related research addresses WSNs that are mobile and battery powered. Because of these requirements, most of the literature is concentrated on finding solutions at various levels of the communication protocol, including being extremely energy efficient. Energy efficiency is often gained by accepting a reduction in network performance [7]. Although one does not wish to waste energy, this system does have a constant, renewable energy source. However, a very low-power dissipation allowance offers constraint, which fits nicely with an energy-efficient scheme. Popular power-saving ideas include specialized nodes, negotiation, and data fusion.

Low-energy adaptive clustering hierarchy (LEACH) [2, 13] is a new communication protocol that tries to distribute the energy load evenly among the network nodes by randomly rotating the cluster head among the sensors. This assumes a finite amount of power and aims at conserving as much as possible despite a dynamic network. LEACH uses localized coordination to enable scalability and robustness for dynamic networks, as well as data compression to reduce the amount of data that must be transmitted to a base station. Performing some calculations and using data fusion locally conserves much energy at each node.

Sensor protocols for information via negotiation (SPIN) [3, 5] is a unique set of protocols for energy-efficient communication among wireless sensors. The authors propose solutions to traditional wireless communication issues such as network implosion caused by flooding, overlapping transmission ranges, and power conservation. The SPIN protocols incorporate two key ideas to overcome implosion, overlap, and resource blindness: negotiation and resource adaptation. Using very small metadata packets to negotiate, SPIN efficiently communicates with fewer redundancies than in traditional approaches, dealing with implosion and overlap. The metadata are application specific — they could be used to describe the amount of power dissipated, for instance. To solve the resource blindness issue, each node has an individual resource manager, allowing the node to limit activity when power is low.

Pottie has studied design issues and trade-offs that need to be considered for power-constrained WSNs with low data-rate links [8] and advocates "aggressive power management at all levels," noting that the communication protocol is more helpful in reducing the power consumption than is optimizing the hardware. Local processing of information is key to reducing the amount of communication between nodes and thus reducing the amount of power consumed by the network.

Chen and colleagues have also provided a useful comparison of multiple protocols used for WSNs [1]. Although the authors' main focus is on energy efficiency due to battery power, they provide very useful guidelines for designing access protocols for wireless networks. Specifically, they recommend that "protocols should reduce the number of contentions to improve power conservation," as well as using shorter packet lengths. The receiver usage time, however, tends to be higher for protocols that require the mobile nodes to sense the medium before attempting transmission.

Limited research has been conducted on topology's effect on wireless networking [4, 9, 12]. The concentration, however, has been on mobile networks rather than ones with fixed node placement. Although novel approaches have been devised, none of them would be appropriate, for example, in the biomedical arena, in which a surgeon places the nodes, giving a nominally fixed topology. Although much research has been completed in the area of WSN, nothing has sufficiently answered the question of fixed topology's impact on low-power requirements.

9.3 Issues for Topology Design

This section analyzes the performance issues associated with different network topologies. Unlike previous studies, mobility is not an issue. The question concerns what the best topology for a wireless network of sensors is, assuming placement of these sensors can be controlled and the sensor locations fixed relative to each other. One factor in the choice of topology is the amount of contention for the wireless media. The level of contention will vary with the application because the message pattern and overall message generation rate are functions of the application. However, this study should provide some insights that can be used, along with knowledge of the application, to select an appropriate topology. Again, the goal is not to find a single topology appropriate for all applications, but rather to provide a structured analysis of the options and give guidance on the best choices so that a more informed decision is possible.

Each of the different topologies used in this chapter will be considered as a grid on nodes in two or three dimensions. The vertices of this grid are the nodes that will transmit the packets, and the edges are the neighbors of each node that will receive the transmission. According to the mesh topologies that will be used in this section, the optimal path will be found between a source (S) and a destination (D) or the shortest path between them. We will introduce this optimal path and use it later to show how much power is used in the network using each topology to send a packet from S to D.

The WSN, WSN(m,n), is an $m \times n$ grid, where $m \times n$ represents the number of nodes in the network. Each node is represented as (y,x) for $0 \leq y \leq m - 1$ and $0 \leq x \leq n - 1$. For each of the topologies, the following will be assumed:

- $S = (y_s, x_s)$
- $D = (y_d, x_d)$
- $\Delta y = ||y_s - y_d||$
- $\Delta x = ||x_s - x_d||$

Each network will be defined by identifying the neighbors of each node according to the different number of neighbors (as shown in Figure 9.1) and presenting the optimal number of hops from a source to a destination. Next, identifying whether two nodes are neighbors and the optimal number of hops between a source and a destination will be discussed.

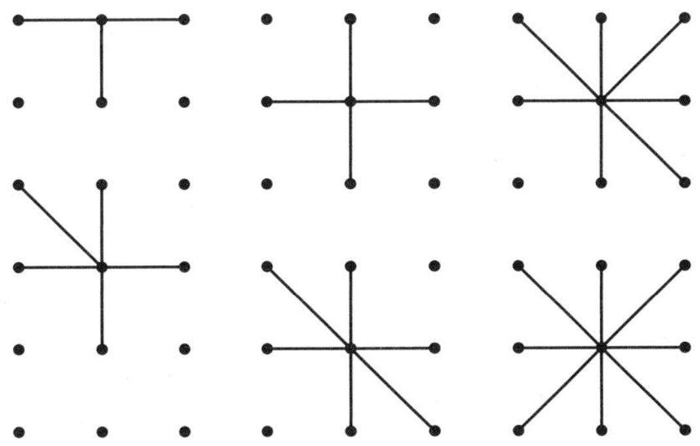

FIGURE 9.1 Possible number of neighbors.

9.3.1 Three-Neighbors WSN

According to Figure 9.2,

- Two nodes are neighbors if:
  - $\langle(y, x), (y, x + 1)\rangle$ for $x < n - 1$
  - $\langle(y, x), (y + 1, x)\rangle$ for even (y, x) and $y < m - 1$
- Two nodes are not neighbors if $\langle(y, x), (y + 1, x)\rangle$ for odd (y, x) and $y < m - 1$
- Optimal number of hops $(s, d) = \begin{cases} \Delta x + \Delta y & \text{if } \Delta x \geq \Delta y \\ 2\Delta y \pm 1 & \text{if } \Delta x < \Delta y \end{cases}$

9.3.2 Four-Neighbors WSN

According to Figure 9.3 note the following:

- Two nodes are neighbors if:
  - $\langle(y, x), (y, x + 1)\rangle$ for $x < n - 1$
  - $\langle(y, x), (y + 1, x)\rangle$ for $y < m - 1$
- Optimal number of hops $(s, d) = \Delta z + \Delta y$.

9.3.3 Five-Neighbors WSN

According to Figure 9.4,

- Two nodes are neighbors if:
  - $\langle(y, x), (y, x + 1)\rangle$ for $x < n - 1$
  - $\langle(y, x), (y + 1, x)\rangle$ for $y < m - 1$
  - $\langle(y, x), (y + 1, x + 1)\rangle$ for even x.
  - $\langle(y, x), (y - 1, x - 1)\rangle$ for odd x.
- Optimal number of hops $(s, d) = \begin{cases} \Delta x + 2 & \text{if } x_s \geq x_d \text{ and } y_s > y_d \\ & \text{or } x_s \leq x_d \text{ and } y_s < y_d \\ \Delta x + \Delta y & \text{Otherwise} \end{cases}$

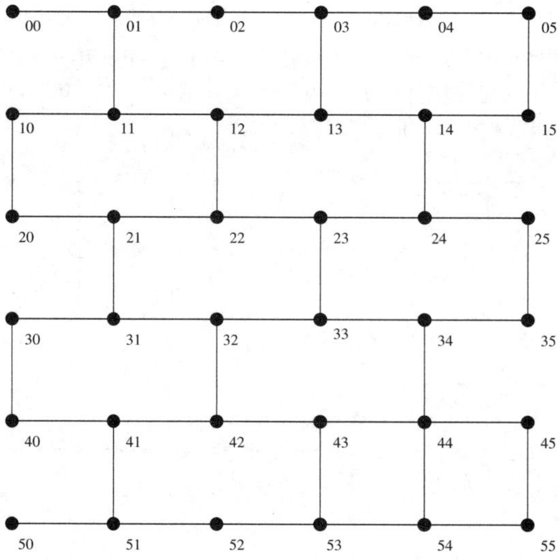

FIGURE 9.2 Two-dimensional topology with up to three neighbors.

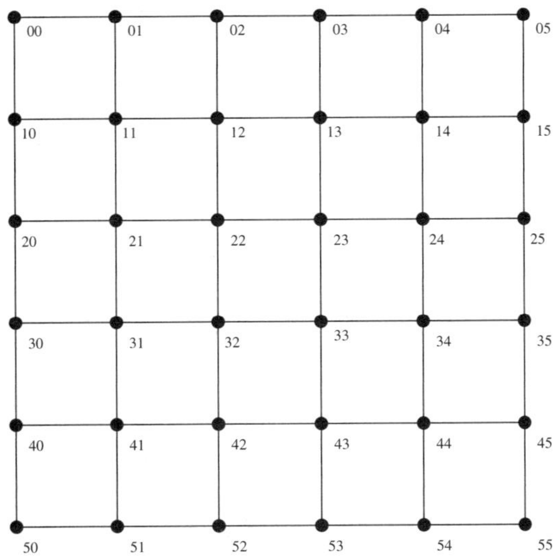

FIGURE 9.3 Two-dimensional topology with up to four neighbors.

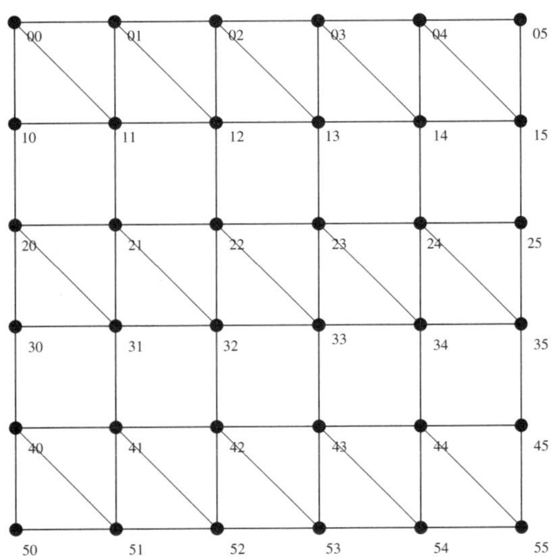

FIGURE 9.4 Two-dimensional topology with up to five neighbors.

9.3.4 Six-Neighbors WSN

According to Figure 9.5,

- Two nodes are neighbors if:
  - $\langle(y, x), (y, x + 1)\rangle$ for $x < n - 1$
  - $\langle(y, x), (y + 1, x)\rangle$ for $y < m - 1$
  - $\langle(y, x), (y + 1, x + 1)\rangle$ for every $y < y + 1$ and $x < x + 1$
  - $\langle(y, x), (y - 1, x - 1)\rangle$ for every $y < y - 1$ and $x < x - 1$
- Two nodes are not neighbors if
- Optimal number of hops $(s, d) = \begin{cases} \Delta x + \Delta y & \text{if } x_s > x_d \text{ and } y_s < y_d \\ & \text{or } x_s < x_d \text{ and } y_s > y_d \\ \max(\Delta x, \Delta y) & \text{Otherwise} \end{cases}$

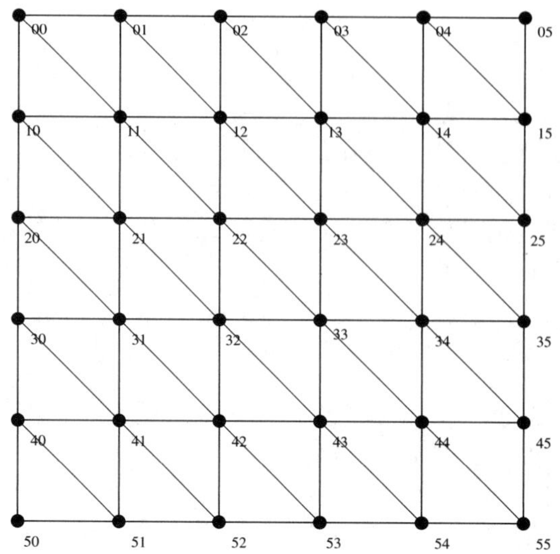

FIGURE 9.5 Two-dimensional topology with up to six neighbors.

9.3.5 Seven-Neighbors WSN

According to Figure 9.6,

- Two nodes are neighbors if:
  - $\langle (y, x), (y, x + 1) \rangle$ for $x < n - 1$
  - $\langle (y, x), (y + 1, x) \rangle$ for $y < m - 1$
  - $\langle (y, x), (y + 1, x - 1) \rangle$ for $x = 0$ or x is even
  - $\langle (y, x), (y - 1, x + 1) \rangle$ for $x = 1$ or x is odd
  - $\langle (y, x), (y + 1, x + 1) \rangle$ for every $y < y + 1$ and $x < x + 1$
  - $\langle (y, x), (y - 1, x - 1) \rangle$ for every $y < y - 1$ and $x < x - 1$
- Optimal number of hops $(s, d) = \begin{cases} \Delta x + 2 & \text{if } x_s > x_d \text{ and } y_s < y_d \\ & \text{or } x_s < x_d \text{ and } y_s > y_d \\ \max(\Delta x, \Delta y) & \text{Otherwise} \end{cases}$

9.3.6 Eight-Neighbors WSN

According to Figure 9.7,

- Two nodes are neighbors if:
  - $\langle (y, x), (y, x + 1) \rangle$
  - $\langle (y, x), (y + 1, x) \rangle$
  - $\langle (y, x), (y + 1, x - 1) \rangle$
  - $\langle (y, x), (y - 1, x + 1) \rangle$
  - $\langle (y, x), (y + 1, x + 1) \rangle$
  - $\langle (y, x), (y - 1, x - 1) \rangle$
- Optimal number of hops $(S, D) = \max(\Delta x, \Delta y)$.

9.3.7 Six-Neighbors for Three Dimensions

The WSN (m, n, k) is an $m \times n \times k$ grid where a node is represented as (y, x, z) for $0 \leq y \leq m - 1$, $0 \leq x \leq n - 1$, and $0 \leq z \leq k - 1$. For three-dimensional topology, assume the following:

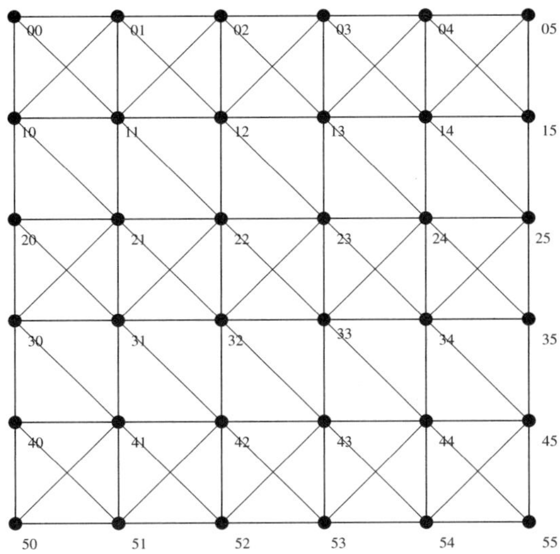

FIGURE 9.6 Two-dimensional topology with up to seven neighbors.

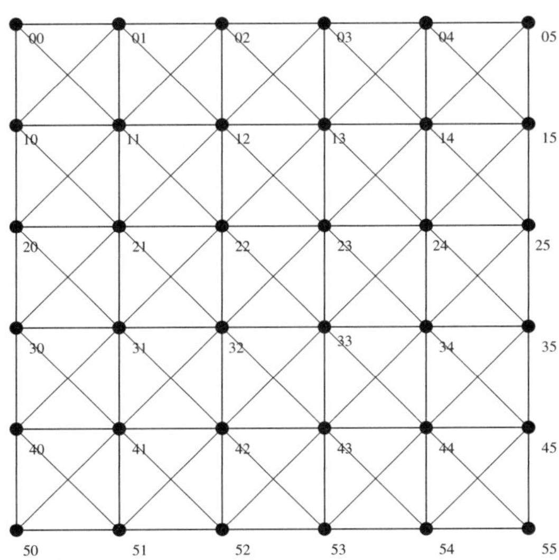

FIGURE 9.7 Two-dimensional topology with up to eight neighbors.

- $S_{3D} = (y_s, x_s, z_s)$
- $D_{3D} = (y_d, x_d, z_d)$
- $\Delta y = ||y_s - y_d||$
- $\Delta x = ||x_s - x_d||$
- $\Delta z = ||z_s - z_d||$

According to Figure 9.8, two nodes are neighbors if:

- $\langle(y, x, z), (y, x + 1, z)\rangle$ for $x < n - 1$
- $\langle(y, x, z), (y + 1, x, z)\rangle$ for $y < m - 1$
- $\langle(y, x, z), (y, x, z + 1)\rangle$ for $z < k - 1$
- Optimal number of hops $(S_{3D}, D_{3D}) = \Delta x + \Delta y + \Delta z$.

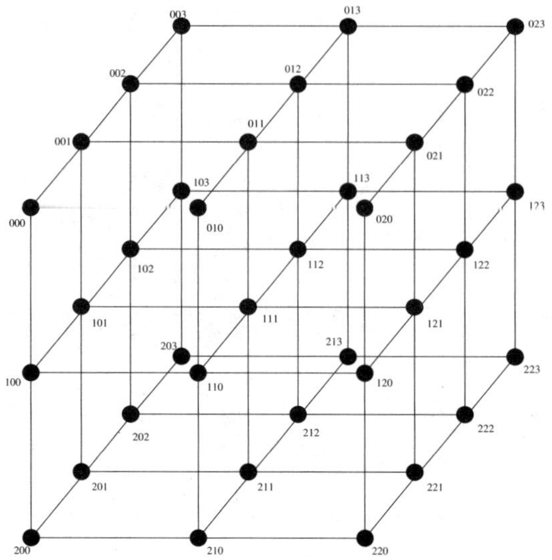

FIGURE 9.8 Three-dimensional topology with up to six neighbors.

9.4 Assumptions

In this work, a simple model is assumed in which the radio dissipates E_{elec} = 50 nJ/b to run the transmitter or receiver circuitry and E_{amp} = 100 pJ/b/m^2 for the transmit amplifier to achieve an acceptable E_b/N_0 (see Figure 9.9 and Table 9.1) [2]. To transmit a k-b message a distance of d meters using this radio model, the radio expends:

$$\begin{aligned} E_{Tx}(k,d) &= E_{Tx-elec}(k) + E_{Tx-amp}(k.d) \\ &= E_{elec} * k + E_{amp} * k * d^2 \end{aligned} \quad (9.1)$$

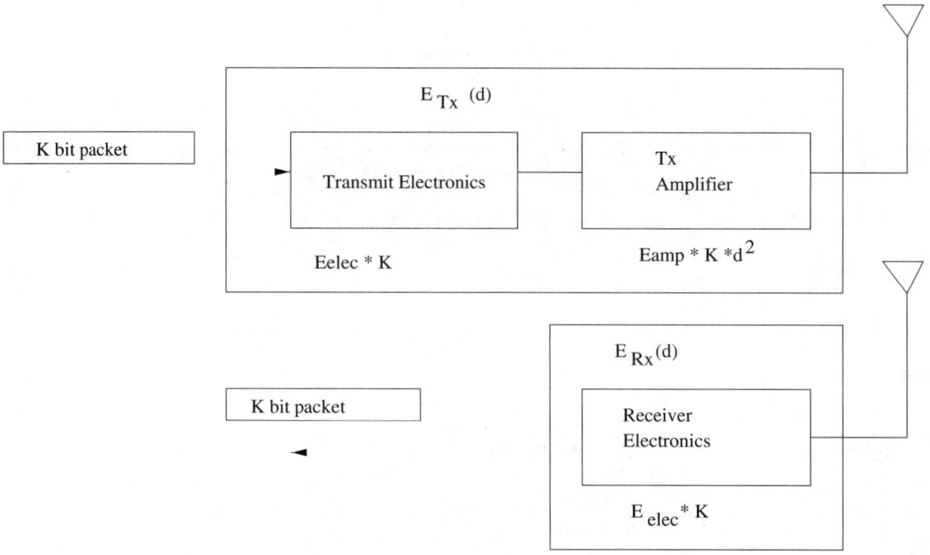

FIGURE 9.9 First-order radio model.

TABLE 9.1 Radio Characteristics

Operation	Energy Dissipated
Transmitter electronics ($E_{Tx-elec}$) Receiver electronics ($E_{Rx-elec}$) ($E_{Tx-elec} = E_{Rx-elec} = E_{elec}$)	50 nJ/b
Transmit amplifier (E_{amp})	100 pJ/b/m^2

Source: W.R. Heinzelman, A. Chandrakasan, and H. Balakrishnan. In *Hawaii Int. Conf. Syst. Sci.*, 2000.

To receive this message, the radio expends:

$$\begin{aligned} E_{Rx}(k) &= E_{Rx-elec}(k) \\ &= E_{elec} * k \end{aligned} \tag{9.2}$$

For simplicity of calculation, assume that the transmission range of each node is equal to each other on one condition: that the value of this transmission range should reach the number of neighbors allowed for each network (maximum number of neighbors). Also, assume that all data packets contain the same number of bits. Thus, a maximum distance $d = 15$ m and number of bits transmitted $k = 512$ bs are assumed. The number of nodes N was chosen to be 36 because it works nicely for two-dimensional and three-dimensional networks with the different topologies considered. This also represents an intermediate value between 16 and 64 node networks that has been used in other studies [7]. For these parameter values, receiving a message is not a low-cost operation; the protocol should thus try to minimize not only the transmit distance but also the number of transmit and receive operations for each message. Next general equations that can be used to estimate the total power used to transmit a message from source to destination will be presented.

9.4.1 Calculation of Power Usage for Each Path

In order to derive the general equations for transmitting a message from a source S to a destination D, two things must be considered for each path: (1) number of transmissions and (2) number of receptions.

Number of transmissions can be measured as the number of hops a packet will travel through a certain path. Number of receptions is the total number of neighbors of each hop taken. Minimizing the number of transmissions and number of receptions will be the mission of any protocol designed. In general, the total power dissipated in the network for one packet to travel from a source to a destination is the sum of total power used for transmission plus the total power used for receiving the packet at each neighbor of each transmitting source.

The next equation presents an estimate for the total power used to transmit a packet over a number of hops from a source S to a destination D;

$$\textit{Total power used} = \textit{total power transmitted} + \textit{total power received} \tag{9.3}$$

Equation 9.3 can be written as:

$$\begin{aligned} \textit{Total power transmitted} &= \textit{number of hops} \times \textit{power transmitted} \\ &= \textit{number of hops} \times E_{Tx}(k,d) \end{aligned} \tag{9.4}$$

$$\begin{aligned} \textit{Total power received} &= \textit{number of hops} \times \textit{number of neighbors} \times \textit{power received} \\ &= \textit{number of hops} \times \textit{number of neighbors} \times E_{rx}(k) \end{aligned} \tag{9.5}$$

Substituting Equation 9.4 and Equation 9.5 in Equation 9.3 yields:

$$\textit{Total power used} = \textit{number of hops} \times (E_{Tx}(k,d) + \textit{number of neighbors} \times (E_{Rx}(k)) \quad (9.6)$$

These equations only estimate the power that will be used for a certain number of hops with a fixed number of neighbors. The idea here is to try to minimize Equation 9.3 by minimizing the total power transmitted; this can be done by minimizing the number of hops by finding the shortest path. Also, Equation 9.3 can be minimized by minimizing the total power received, which can be done by taking the paths that have the least number of neighbors. The next section presents and analyzes the effect of choosing different paths on Equation 9.3.

9.5 Analysis of Power Usage

Various network topologies are studied in this section. First, the routing is considered over the diameter of the network and two possible routes are used along the edge and through the interior. These results show that different paths consume different amounts of power. Next shortest-path routing for the various topologies for a message spanning the diameter of the network is considered. Finally, directional source-aware routing protocol (DSAP) is simulated with and without power-aware routing of arbitrary source–destination pairs and the relative performance of each is shown.

The power dissipated with respect to the network topology will be analyzed with a variable number of neighbors. First, two-dimensional networks with three, four, five, six, seven, and eight neighbors are examined. Then, three-dimensional networks with six neighbors are considered. Two kinds of routing are considered for each of the topologies: (1) edge routing and (2) interior routing.

Edge routing consists of moving messages to the outer edges of the network where there are fewer neighbors. Interior routing keeps the messages in the middle of the network, where there is a consistent number of neighbors for each node. In some cases, longer paths were chosen for some topologies to give a similar number of transmissions. The use of these two methods of routing is only to show the effect of using topologies with different numbers of neighbors. It also shows how useful it is to increase the number of neighbors. Then, shortest-path routing will be studied to see which topology will give the most savings in power. The shortest path will be considered by using the DSAP routing protocol; and also to study the benefit of using a power-aware routing metric by using aware–DSAP will also be studied.

9.5.1 Two-Dimensional Analysis

The degree of routing freedom is the number of alternative paths that a routing protocol can select. Figure 9.2 through Figure 9.7 show that as the number of neighbors increases, the degree of routing freedom increases. For comparison purposes, the source, destination, and number of nodes were fixed to be the same (36 nodes) for all the networks under investigation. An analysis of these networks requires one to classify the routing paths into edge routes and interior routes.

9.5.1.1 Interior Routing

As defined before, interior routing keeps the messages in the middle of the network, where the number of neighbors for each node is consistent. Table 9.2 shows that as the number of neighbors increases, the number of transmissions decreases; however, the number of receptions depends on the topology. This is because, as the number of neighbors increases, the routing protocol has more freedom to choose the shortest path to the destination; by doing so the protocol will dissipate less power to route a packet from source to destination.

9.5.1.2 Edge Routing

Using edge routing is to route the packet using only the edge nodes. This strategy of routing is impossible to use at all times, of course. Here it is used to study the effect of increasing the number of neighbors

TABLE 9.2 Two-Dimensional Interior Routing

Neighbors	T_x	R_x	Energy Used
3	10	27	10.624×10^{-4}
4	10	36	12.928×10^{-4}
5	7	36	11.172×10^{-4}
6	5	27	8.768×10^{-4}
7	5	31	9.792×10^{-4}
8	5	36	10.720×10^{-4}

TABLE 9.3 Two-Dimensional Edge Routing

Neighbors	T_x	R_x	Energy Used
3	14	33	13.645×10^{-4}
4	10	28	10.880×10^{-4}
5	10	37	13.184×10^{-4}
6	10	39	13.696×10^{-4}
7	10	44	14.976×10^{-4}
8	10	46	15.488×10^{-4}

with respect to the edge nodes. As shown in Table 9.3, as the number of neighbors increases, the number of neighbors that receive the packet increases, which will increase the energy used in the network.

9.5.1.3 Edge Routing vs. Interior Routing

From Table 9.2 and Table 9.3, edge routing dissipates more power than interior routing in all cases except for four neighbors. This is because, although the path from the source to the destination in a four-neighbor case is the same, the difference is that taking the edge results in fewer neighbors and interior paths have more neighbors. With either routing strategy, as the number of neighbors increases the power dissipated increases for the same number of transmissions.

9.5.1.4 Fixed Number of Transmissions

This subsection studies the effect of increasing the number of neighbors. In order to do that it is necessary to fix the number of transmissions that a certain path can have and also certain nodes through which a path must pass. These fixed nodes are the nodes that fall on the diagonal of the network, such as nodes (1,1), (2,2), (3,3), (4,4), (5,5), (6,6), (7,7), and (8,8). By using this path, one can control the path and study the effect of increasing the number of neighbors. As shown in Table 9.4, as the number of neighbors increases, the number of receptions increases also. This yields to an increase in the energy used in the network.

TABLE 9.4 Two-Dimensional Fixed Number of Hops

Neighbors	T_x	R_x	Energy Used
3	10	27	10.624×10^{-4}
4	10	36	12.928×10^{-4}
5	10	45	15.232×10^{-4}
6	10	53	17.280×10^{-4}
7	10	61	19.328×10^{-4}
8	10	69	21.376×10^{-4}

TABLE 9.5 Routing Freedom and Power Dissipation Three and Six Neighbors

Neighbors	T_x	R_x	Energy Used
3	10	27	10.624×10^{-4}
6	5	27	8.768×10^{-4}

TABLE 9.6 Routing Freedom and Power Dissipation Four and Eight Neighbors

Neighbors	Tx	Rx	Energy Used
4	10	36	12.928×10^{-4}
8	5	36	10.720×10^{-4}

9.5.1.5 Routing Freedom

Routing freedom means that the routing protocol has the freedom to choose the optimal path. This subsection studies the effect of doubling the number of neighbors, between three and six neighbors and four and eight neighbors, to study the effect of increasing the number of neighbors and the impact it will have on routing freedom.

Table 9.5 considers the power dissipated between the source and destination for a message spanning the diameter of the network for topologies with three and six neighbors as shown in Figure 9.2 and Figure 9.5. As Table 9.5 shows, increasing the number of neighbors decreases the number of transmissions and the total power dissipated in the system. This result can only be attributed to the availability of a shorter path between the source and destination. A similar conclusion can be reached from Table 9.6.

In summary, a trade-off occurs between the number of neighbors and the total power dissipated in the system. However, this trade-off breaks in special cases in which the availability of alternative shortest paths can be used as an advantage for the power budget calculations.

9.5.2 Three-Dimensional Analysis

A three-dimensional network can be constructed from a two-dimensional network with four neighbors just by adding another dimension, which will create a three-dimensional network with six neighbors. The same thing can be done for two-dimensional networks with six neighbors, but implementing such a network with a regular structure is not possible. Figure 9.8 shows a three-dimensional network with six neighbors that has some advantages due to its inherent symmetry.

In a three-dimensional network, the routing paths between any given source and destination without misrouting would always result in the same number of transmissions but a different number of receptions. For example, from source (0,0,0) to destination (2,2,3), the number of transmissions using interior or edge routing is constant and equals seven in Figure 9.8.

From Table 9.7, the following can be concluded:

- Edge routing in the case of the three-dimensional network has lower power dissipation than interior routing does.
- The number of transmissions and receptions as well as the total power dissipated in a three-dimensional network is less than in a two-dimensional network for edge routing as well as interior routing.

For Table 9.8, the number of neighbors was fixed to study the effect of using two different dimensions on the number of transmissions each path will require using edge routing and interior routing. Using interior routing, two dimensions with six neighbors have fewer transmissions than the three dimensions with six neighbors. Also, from the nature of the two-dimensional topology, using edge routing takes

TABLE 9.7 Edge and Interior Routing Power Dissipation

Network	Path	T_x	R_x	Energy Used $\times 10^{-4}$
2D	Interior	10	36	12.928
4 Neighbor	Edge	10	28	10.880
3D	Interior	7	33	11.046
6 Neighbor	Edge	7	25	8.998

TABLE 9.8 Six Neighbors for 2-D and 3-D Routing Power Dissipation

Network	Path	T_x	R_x	Energy Used $\times 10^{-4}$
2D	Interior	5	27	8.768
6 Neighbor	Edge	10	39	13.696
3D	Interior	7	33	11.046
6 Neighbor	Edge	7	25	8.998

longer paths than three dimensions because the three-dimensional topology makes the edges closer than the two-dimensional one. Thus, a trade-off occurs between using edge routing and using interior routing for the two different dimensions.

9.6 Directional Source-Aware Routing Protocol (DSAP)

In order to resolve the problems of power efficiency, a unique identification system has been developed for the networks used. The idea behind this identification system is to identify the location of each node in the network that will help in routing the packets. The system has the following properties:

- Each node has unique ID.
- Each value represents how far the node is from a certain direction.
- Each ID gives how far the node is from the nodes in each direction.
- Each node can compute the direction of other nodes from its ID.

To help in studying the effect of using different numbers of neighbors, a routing scheme based on the identification system has been developed. This identification system is referred to as the *directional value* (DV). To construct the DV, each node in each topology that has been used has a fixed number of neighbors. Each neighbor represents a direction that the node can route through it, as shown in Figure 9.10. How far the node is from the edge of the network in each direction represents the directional value of each node. This number is unique for each node and can be used as the ID number for each node for the purpose of routing.

Each topology was constructed from Figure 9.10 by eliminating the directions that will make that topology. For example, constructing a seven-neighbor topology from an eight-neighbor one is done by eliminating *D*-7 in one node and also eliminating the corresponding direction from the other node. Each direction has a corresponding or an associate direction. *D*-7 has *D*-3, *D*-6 has *D*-2, *D*-5 has *D*-1, *D*-4 has *D*-0, and vice versa.

From this DV, a DSAP [11] was developed. DSAP incorporates the DV and power into routing protocols. For instance, in the four-neighbor case of Figure 9.3, node 31 would have an identifier of (1, 0, 3, 0, 4, 0, 2). This means that there is one node to the edge in direction 0 (left); three in direction 2 (up); four in direction 4 (right); and two in direction 6 (down). Because placement of the nodes is controlled and topology is fixed, this information can be hard-coded into each node with relative ease. However, for a random topology, it is necessary to discover the directional values of each node in the network.

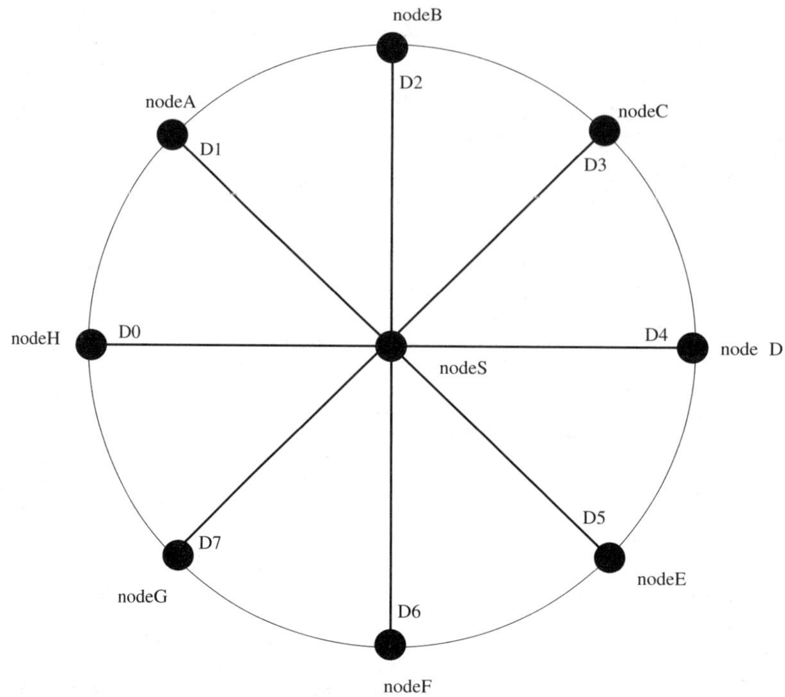

FIGURE 9.10 Directional eight-neighbor node.

In Figure 9.10, node S would have an identifier of $(DV_0,DV_1,DV_2,DV_3,DV_4,DV_5,DV_6,DV_7)$. This means that DV_0 nodes are to the edge in direction D-0; DV_1 in D-1; DV_2 in D-2; and so on. When transmitting a message, the destination node identifier is subtracted from the source node identifier. This yields at most five positive numbers (for a two-dimensional topology with eight neighbors) that describe in which direction the message needs to move. Negative numbers are ignored. The decision to move in any positive direction is determined by the DV of the nodes in question. Taking each of the neighbor's identifiers and subtracting them from the destination node's identifier computes the DV. These eight numbers are added together and the one with the smaller number is chosen. If both nodes have the same DV, then one is randomly picked. This is the basic scheme developed for routing the messages.

For example, in Figure 9.7 consider the source node $S_{1,1}$ with $DV_{1,1}$ = (1, 1, 1, 1, 4, 4, 4, 1) and destination node $D_{4,4}$ with $DV_{4,4}$ = (4, 4, 4, 1, 1, 1, 1, 1). According to the algorithm of DSAP [11], $S - D$ = (−3, −3, −3, 0, 3, 3, 3, 0), which produces D-3, D-4, D-5, D-6, and D-7 as possible positive directions to which the message can be forwarded and then computes the directional value of each positive direction to find which route to take. By doing so, the following values for each direction are obtained: 20, 17, 14, 16, and 20, respectively. By choosing the minimum directional value, the message is forwarded in direction D-5, which is obvious from Figure 9.7. Then the protocol repeats until reaching the final destination, which will have a DV of 0.

This is the basic scheme developed for routing messages. However, the objective is to incorporate energy efficiency as well. This is achieved by considering the maximum available power and minimal directional value when picking which node route to take. Instead of simply picking the node with the lowest directional value, the directional value is divided by the power available at that node. The smaller value of this power-constrained directional value is the path chosen. This allows for a least-transmission path that is also cognizant of power resources, although in some cases a longer path may be chosen if the available power dictates that choice. Salhieh and Schwiebert [10] have presented several power-aware metrics that can be incorporated with DSAP. The idea here is to show that using power-aware methods will extend the life of the network and have a fair load balance between the nodes. The method used here was only to show the effect of using power-aware rather than shortest-path metrics.

9.7 DSAP Analysis

To study the relationship between the number of neighbors and the power dissipated in the network, a controlled environment is used. This has been done to study the effect on the power dissipated in the network when the number of neighbors is increased. The effect of increasing or decreasing the number of neighbors is studied from two viewpoints: (1) power usage in the network; and (2) which topology or number of neighbors will extend the life of the network because extending the life of the network is one of the main objectives of designing WSNs.

In the simulation, two different methods for routing are used: (1) DSAP without the power aware, which is based on the shortest number of hops between a source and a destination; and (2) DSAP with power aware, which incorporates the power available at the next neighbor and tries to balance the load between the neighbors of a source. The simulation has two runs: (1) a fixed run from $S(0, 0)$ to $D(5, 5)$; and (2) a run that each node sends a message to every node in the network. Both of these should help in studying the relationship between the power usage in the system and the number of neighbors.

9.7.1 Two-Dimension Analysis

In Table 9.9, a message is sent from source (0, 0) to destination (5, 5) for 10,000 times. Note that:

- Increasing the number of neighbors, for DSAP in general, results in decreasing the number of transmissions that the network performs because having more neighbors creates shorter paths or alternative routes that are shorter to the destination. This is also reflected in the total power transmitted (TPT) in the network, which is decreased from a sparse topology to a more dense topology.
- Looking at the power used for both protocols, note that DSAP with power aware uses more power, which is reflected throughout Table 9.9. However, looking at Figure 9.11 and Figure 9.12, note that DSAP with power aware has a better power distribution than DSAP without power aware. This means that the life of the network can be extended using the power-aware concept.

Table 9.10 and Table 9.11 concern when the first node dies in the network. Note that:

- In Table 9.10, more than one node died in the network. This is because using DSAP without power aware uses the concept of shortest path, so every message takes the same path and thus these nodes will lose power faster than other nodes.
- In Table 9.11, the first node died at different rounds and even at a higher number of rounds than in Table 9.10 because DSAP with power aware was used in Table 9.11. This gives the routing protocol more alternative paths to use and also balances the load in the network.
- Also notice that in Table 9.11, as the number of neighbors is increased, the number of rounds when the first node dies decreases because more neighbors are hearing the transmission of each source.
- In Table 9.12 through Table 9.14, each node sends a message to every other node in the network. This will be considered as one complete run and is repeated until a fixed round or until the death of the first node. In these tables we ran the simulation for the DSAP without power aware and also for the power-aware protocol.

In Table 9.12:

- As the number of neighbors is increased, the first node dies at a lower number of rounds in both protocols because more nodes will be reached during each transmission, so more nodes will lose power.

TABLE 9.9 Round 10000 from S(0,0) to D(5,5)

Neighbors		DSAP Routing				
		TR	TT	TPA (J)	TPR (J)	TPT (J)
	4	280,000	100,000	25.12	7.16	3.71
	5	370,000	90,000	23.19	9.47	3.34
2D	6	270,000	50,000	27.23	6.91	1.86
	7	310,000	50,000	26.20	7.94	1.86
	8	350,000	50,000	25.18	8.96	1.86
		Aware–DSAP Routing				
Neighbors		TR	TT	TPA (J)	TPR (J)	TPT (J)
	4	314,787	100,000	24.23	8.06	3.71
	5	359,428	87,861	23.54	9.20	3.26
2D	6	301,852	65,926	25.83	7.73	2.45
	7	388,748	73,624	23.32	9.95	2.73
	8	396,424	73,212	23.13	10.15	2.72

Notes:
TR = total number of packets received by the neighbors of a source.
TT = total number of transmissions in the networks.
TPA = total power available for the network.
TPR = total power received by the neighbors of a transmitting source.
TPT = total power used for transmitting these packets.

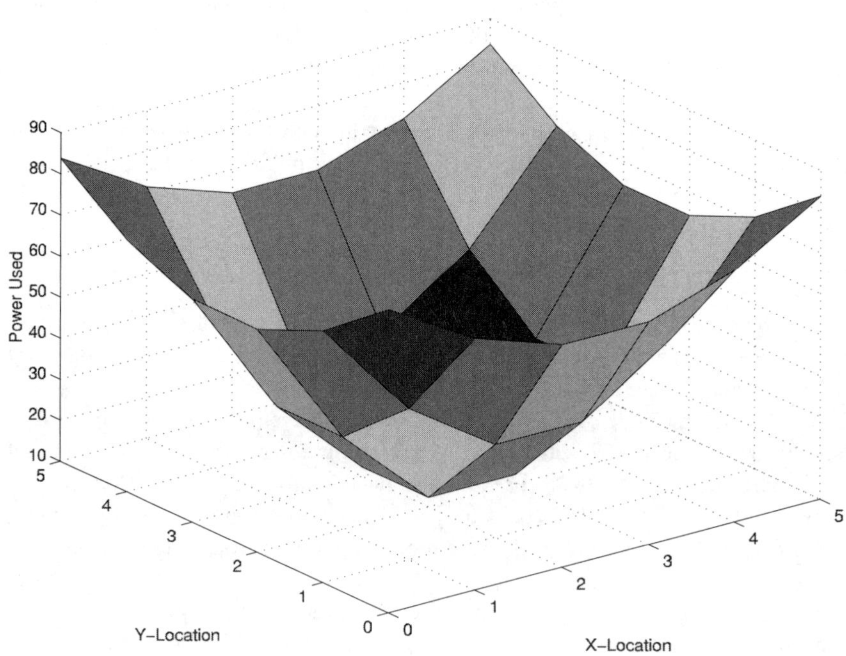

FIGURE 9.11 Remaining power in each node using DSAP.

- The number of rounds in the DSAP with power aware is higher than the DSAP without power aware. This is because alternative paths have been used, resulting in a better load balance than in the DSAP without the power aware.
- Notice that the standard deviation for the DSAP with power aware is less than that of the DSAP without power aware because DSAP with power aware has a better distribution of power usage

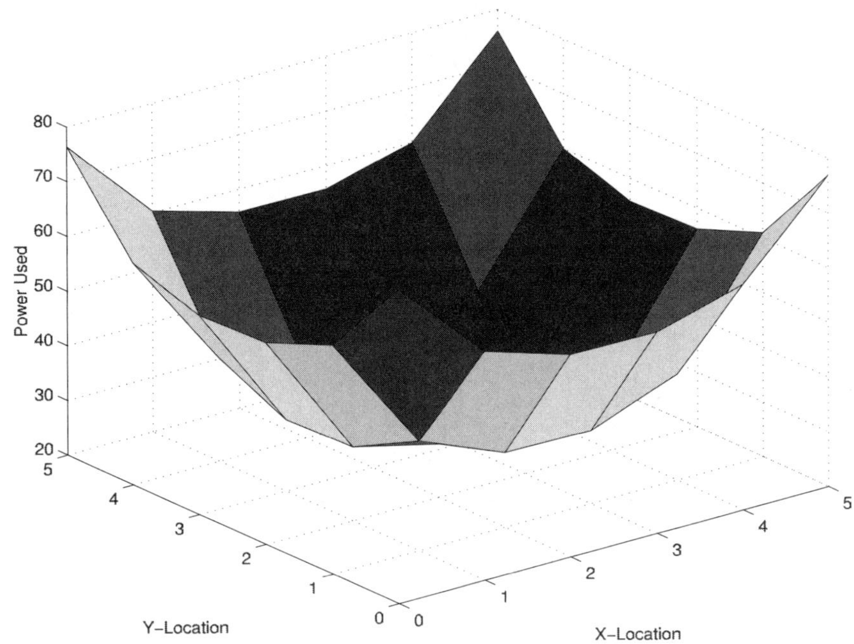

FIGURE 9.12 Remaining power in each node using aware–DSAP.

TABLE 9.10 First Node Dead for DSAP at Round 10191 from S(0,0) to D(5,5)

	Neighbors	Dead Nodes	GeoMean
	4	8	51.89
	5	7	48.20
2D	6	3	64.55
	7	3	62.42
	8	3	60.36

Note: GeoMean = geometric mean.

TABLE 9.11 First Node Dead Aware–DSAP from S(0,0) to D(5,5)

	Neighbors	Round	GeoMean
	4	14,350	49.58
	5	13,563	47.76
2D	6	14,350	52.71
	7	13,060	48.52
	8	11,456	54.82

Note: GeoMean = geometric mean.

than does DSAP without power aware. Also the geometric mean is less in the DSAP with power aware than the DSAP without power aware because DSAP with power aware balances the load among all the nodes.

In Table 9.13 and Table 9.14, the two protocols are compared at round 28,512 to study the geometric mean, the standard deviation, and different power parameters:

TABLE 9.12 First Node Dead for Fixed All Routing

Neighbors		DSAP Routing GeoMean	STDEV	Number of Rounds
	4	39.69	21.33	39,605
	5	39.99	21.82	34,001
2D	6	44.33	22.04	31,715
	7	42.09	21.34	29,485
	8	45.07	22.94	29,120
		Aware–DSAP		
Neighbors		T_x	R_x	Total Power Used
	4	20.75	15.24	56,084
	5	31.04	18.66	30,934
2D	6	27.50	14.31	39,512
	7	28.76	15.71	29,485
	8	24.48	18.17	37,915

Notes:
GeoMean = geometric mean.
STDEV = standard deviation.

TABLE 9.13 Topology at Round 28512 for Fixed All Routing

		DSAP Routing		Aware–DSAP Routing	
Neighbors		GeoMean	STDEV	GeoMean	STDEV
	4	58.79	15.42	61.34	7.81
	5	51.75	18.38	44.66	15.40
2D	6	51.31	19.84	51.96	11.60
	7	44.67	20.59	43.98	13.98
	8	46.74	22.45	47.11	15.61

Notes:
GeoMean = geometric mean.
STDEV = standard deviation.

- In Table 9.13, DSAP aware has a lower standard deviation than the DSAP, but in some cases has a higher geometric mean.
- In Table 9.13, the topology with four neighbors has a lower standard deviation in both protocols.
- In Table 9.14, the number of neighbors increases, the number of transmissions decreases, as noted in Table 9.9.

In general, for the two-dimensional topologies, a trade-off occurs between increasing the number of neighbors and the power dissipated in the networks. As the number of neighbors increases, the protocol will have alternative routes; however, more power will be dissipated in the network. Also, using a power-aware routing protocol will help in extending the life of the network.

9.7.2 Three-Dimension Analysis

In Table 9.15, different runs were done for the three-dimensional topology to try to see how the power dissipated in the network would be affected by using the two different protocols. For the first 1000 rounds, there is only a difference in the number of reception in the network. This is because when the network is used more, the DSAP with power aware tries to find alternative paths with more power. If one looks at 10,000 and 100,000, it is seen that the power used is less in the DSAP with power aware than the DSAP without power aware for the same reasons mentioned before.

TABLE 9.14 Power Values at Round 28512 for Fixed All Routing

Neighbors		DSAP Routing				
		TR	TT	TPA (J)	TPR (J)	TPT (J)
	4	390,720	110,880	21.88	10.0	4.12
	5	478,522	105,292	19.84	12.25	3.91
2D	6	490,776	94,556	19.93	12.56	3.51
	7	570,768	91,718	17.98	14.61	3.4
	8	544,456	78,232	19.16	13.94	2.90
		Aware–DSAP Routing				
Neighbors		TR	TT	TPA (J)	TPR (J)	TPT (J)
	4	376,541	110,880	22.24	9.64	4.12
	5	558,634	127,596	16.96	14.30	4.74
2D	6	507,003	104,465	19.14	12.98	3.88
	7	608,627	103,897	16.56	15.58	3.86
	8	578,045	90,638	17.83	14.79	3.36

Notes:
TR = total number of packets received by the neighbors of a source.
TT = total number of transmissions in the networks.
TPA = total power available for the network.
TPR = total power received by the neighbors of a transmitting source.
TPT = total power used for transmitting these packets.

TABLE 9.15 Power Assessment for 3D Topology

Protocol	DSAP Routing			Aware–DSAP Routing		
Number of rounds	1000	10,000	100,000	1000	10,000	100,000
Total power used (J)	0.416	4.126	41.354	0.4	3.937	39.469
Total transmissions	3051	30,131	302,160	3051	30,131	302,160
Total reception	13,228	131,043	1,312,998	12,573	123,656	1,239,477

9.8 Summary

This chapter has looked at the WSN network topology from a different perspective: a neighborhood point of view. In these topologies, the number of neighboring nodes determines the number of receivers and therefore may result in more overall power usage, even though the number of transmissions decreases. Thus, a fundamental trade-off takes place between decreasing the number of transmissions and increasing the number of receptions. This chapter has presented a variety of topologies and examined this trade-off.

Because the number of neighbors differs with different topologies, one expects different topologies to have different power usage rates. Even simulations of the contention-free case show that different topologies have different levels of power efficiency. The results show that the total power consumption is reduced for topologies with fewer neighbors; although the topologies with more neighbors require fewer hops, the power expended by many nodes to receive these messages increases the power usage. Among the two-dimensional topologies, the best power efficiency is achieved with two dimensions with four neighbors. The three-dimensional topology performs even better, although this topology may not be feasible for some applications.

Many areas remain to be explored within this research topic. This initial set of experiments serves to demonstrate the marked difference between basic and power-aware DSAP routing. These differences are significant enough to warrant further research.

One option would be to rerun the large simulations with each node beginning with a randomly chosen power amount. This would allow for a simulation of a network that has been in use for some time. DSAP can also be extended to include a more efficient power management scheme. Because the message knows in which direction to head, it is not necessary to broadcast to all neighbors. Rather, the nodes in the wrong direction can be put to sleep. This will reduce the power used because it takes more power to transmit the large message than to poll the neighboring nodes.

Contention is also an issue that needs to be addressed in future studies because it is not realistic to have a system that sends only one message at a time. Although previous work has also ignored this issue, it is important to find a solution to give a more accurate comparison of the relative performance of the networks.

References

1. J. Chen, K.M. Sivalingam, and P. Agrawal. Performance comparison of battery power consumption in wireless multiple access protocols. *Wireless Networks*, 5(6):445–460, 1999.
2. W.R. Heinzelman, A. Chandrakasan, and H. Balakrishnan. Energy-efficient communication protocols for wireless microsensor networks. In *33rd Ann. Hawaii Int. Conf. Syst. Sci.*, 2000.
3. W.R. Heinzelman, J. Kulik, and H. Balakrishnan. Adaptive protocols for information dissemination in wireless sensor networks. In *Proc. 5th Annu. ACM/IEEE Int. Conf. Mobile Computing Networking (MobiCom'99)*, 174–185, August 1999.
4. L. Hu. Topology control for multihop packet radio networks. *IEEE Trans. Commun.*, 41(10):1474–1481, October 1993.
5. J. Kulik, W.R. Heinzelman, and H. Balakrishnan. Negotiation-based protocols for disseminating information in wireless sensor networks. In *ACM MOBICOM*, 99.
6. B. Nath and D. Niculescu. Routing on a curve. *ACM SIGCOMM Comp. Commun. Rev.,* 33(1), 155–160, 2003.
7. C. Patel, S.M. Chai, S. Yalamanchili, and D.E. Schimmel. Power/performance trade-offs for direct networks. In *Parallel Computer Routing Commun. Workshop*, 193–206, July 1997.
8. G. Pottie. Wireless sensor networks. In *Inf. Theory Workshop*, 2, 139–140, 1998.
9. R. Ramanathan and R. Rosales–Hain. Topology control of multihop wireless networks using transmit power adjustment. In *INFOCOM*, 404–413, 2000.
10. A. Salhieh and L. Schwiebert. Power-aware metrics for wireless sensor networks. In *14th IASTED Conf. Parallel Distributed Computing Syst. (PDCS 2002) Symp.*, 326–331, November 2002.
11. A. Salhieh, J. Weinmann, M. Kochhal, and L. Schwiebert. Power-efficient topologies for wireless sensor networks. In *Int. Conf. Parallel Process.*, 156–163, Sep. 2001.
12. Z. Tang and J.J. Garcia–Luna–Aceves. A protocol for topology-dependent transmission scheduling in wireless networks. In *IEEE Wireless Commun. Networking Conf.*, 3, 1333–1337, 1999.
13. A. Wang, W.R. Heinzelman, and A. Chandrakasan. Energy-scalable protocols for battery-operated microsensor networks. In *IEEE Workshop Signal Process. Syst.*, 483–492, Oct. 1999.

10 Overview of Communication Protocols for Sensor Networks

Weilian Su
Georgia Institute of Technology

Erdal Cayirci
Istanbul Technical University

Özgür B. Akan
Georgia Institute of Technology

10.1 Introduction 10-1
10.2 Applications/Application Layer Protocols 10-2
Sensor Network Applications • Application Layer Protocols
10.3 Localization Protocols 10-4
10.4 Time Synchronization Protocols 10-5
10.5 Transport Layer Protocols 10-7
Event-to-Sink Transport • Sink-to-Sensors Transport
10.6 Network Layer Protocols 10-9
10.7 Data Link Layer Protocols 10-11
Medium Access Control • Error Control
10.8 Conclusion 10-14

10.1 Introduction

As the technology for wireless communications advances and the cost of manufacturing a sensor node continues to decrease, a low-cost but yet powerful sensor network may be deployed for various applications that can be envisioned for daily life. Although each sensor node may seem to be much less capable than a traditional stationary sensor, a collective effort of the sensor nodes may provide sensing capabilities in space and time that surpass the stationary sensor.

The communication protocols for sensor networks may leverage the capabilities of collective efforts to provide users with specialized applications. These protocols may fuse, extract, or aggregate data from the sensor field. In addition, they may self-organize the sensor nodes into clusters to complete a task or overcome certain obstacles, e.g., hills. In essence, sensor networks may provide end users with intelligence and details that traditional stationary sensors may not be able to do.

Although the sensor nodes communicate through the wireless medium, protocols and algorithms proposed for traditional wireless ad hoc networks may not be well suited for sensor networks. As previously explained, sensor networks are application specific, and the sensor nodes work collaboratively together. In addition, the sensor nodes are very energy constrained compared to traditional wireless ad hoc devices. The differences between sensor networks and ad hoc networks [29] are:

- The number of sensor nodes in a sensor network can be several orders of magnitude higher than the nodes in an ad hoc network.

- Sensor nodes are densely deployed.
- Sensor nodes are prone to failures.
- The topology of a sensor network changes very frequently.
- Sensor nodes mainly use a broadcast communication paradigm whereas most ad hoc networks are based on point-to-point communications.
- Sensor nodes are limited in power, computational capacities, and memory.
- Sensor nodes may not have global identification (ID) because of the large amount of overhead and large number of sensor nodes.
- Sensor networks are deployed with a specific sensing application in mind; ad hoc networks are mostly constructed for communication purposes.

With these differences, the design of communication protocols for sensor networks requires specific attention. Some of the potential applications as well as some application layer protocols for sensor networks are presented in Section 10.2. Next, because many of the communication protocols require the knowledge of location and time in order to function properly, localization and time synchronization protocols are described in Section 10.3 and Section 10.4. Furthermore, protocols and challenges for the transport, network, and data-link layers are consecutively explained in Section 10.5 through Section 10.7, respectively.

10.2 Applications/Application Layer Protocols

Sensor nodes can be used for continuous sensing, event detection, event identification, location sensing, and local control of actuators. The concept of microsensing and wireless connection of these nodes promise many new application areas, e.g., military, environment, health, home, commercial, space exploration, chemical processing, and disaster relief, etc. Some of these application areas are described in the next subsection. In addition, Subsection 10.2.2 introduces some application layer protocols used to realize these applications.

10.2.1 Sensor Network Applications

The number of potential applications for sensor networks is huge. Actuators may also be included in the sensor networks, thus making the number of applications that can be developed much higher. In this section, some example applications are given to provide the reader with a better insight about the potentials of sensor networks.

Military applications. Sensor networks can be an integral part of military command, control, communications, computers, intelligence, surveillance, reconnaissance and tracking (C4ISRT) systems. The rapid deployment, self-organization, and fault tolerance characteristics of sensor networks make them a very promising sensing technique for military C4ISRT. Because sensor networks are based on dense deployment of disposable and low-cost sensor nodes, destruction of some nodes by hostile actions does not affect a military operation as much as the destruction of a traditional sensor does. Military applications include: monitoring friendly forces, equipment, and ammunition; battlefield surveillance; reconnaissance of opposing forces and terrain; targeting; battle damage assessment; and nuclear, biological, and chemical attack detection and reconnaissance.

Environmental applications. Some environmental applications of sensor networks include tracking the movements of species, i.e., habitat monitoring; monitoring environmental conditions that affect crops and livestock; irrigation; macroinstruments for large-scale Earth monitoring and planetary exploration; and chemical/biological detection [1, 3, 4, 6, 15, 17, 19, 20, 39, 45].

Commercial applications. The sensor networks are also applied in many commercial applications, including building virtual keyboards; managing inventory control; monitoring product quality; constructing smart office spaces; and environmental control in office buildings [1, 6, 11, 12, 20, 31, 33, 34, 38, 45].

10.2.2 Application Layer Protocols

Although many application areas for sensor networks are defined and proposed, potential application layer protocols for sensor networks remain largely unexplored. Three possible application layer protocols are introduced in this section: sensor management protocol; task assignment and data advertisement protocol; and sensor query and data dissemination protocol. These protocols may require protocols at other stack layers (explained in the remaining sections of this chapter).

10.2.2.1 Sensor Management Protocol (SMP)

Designing an application layer management protocol has several advantages. Sensor networks have many different application areas; accessing them through networks such as the Internet is the aim in some current projects [31]. An application layer management protocol makes the hardware and software of the lower layers transparent to the sensor network management applications.

System administrators interact with sensor networks by using sensor management protocol (SMP). Unlike many other networks, sensor networks consist of nodes that do not have global ID, and they are usually infrastructureless. Therefore, SMP needs to access the nodes by using attribute-based naming and location-based addressing, which are explained in detail in Section 10.6. SMP is a management protocol that provides software operations needed to perform the following administrative tasks:

- Introducing rules related to data aggregation, attribute-based naming, and clustering to the sensor nodes
- Exchanging data related to location-finding algorithms
- Time synchronization of the sensor nodes
- Moving sensor nodes
- Turning sensor nodes on and off
- Querying the sensor network configuration and the status of nodes, and reconfiguring the sensor network
- Authentication, key distribution, and security in data communications

Descriptions of some of these tasks are given in references 8, 11, 30, 36, and 37.

10.2.2.2 Task Assignment and Data Advertisement Protocol (TADAP)

Another important operation in the sensor networks is interest dissemination. Users send their interest to a sensor node, a subset of the nodes, or the whole network. This interest may be about a certain attribute of the phenomenon or a triggering event. Another approach is the advertisement of available data in which the sensor nodes advertise the available data to the users and the users query the data in which they are interested. An application layer protocol that provides the user software with efficient interfaces for interest dissemination is useful for lower layer operations, such as routing.

10.2.2.3 Sensor Query and Data Dissemination Protocol (SQDDP)

The sensor query and data dissemination protocol (SQDDP) provides user applications with interfaces to issue queries, respond to queries, and collect incoming replies. These queries are generally not issued to particular nodes; instead, attribute-based or location-based naming is preferred. For instance, "the locations of the nodes that sense temperature higher than 70°F" is an attribute-based query. Similarly, "temperatures read by the nodes in Region A" is an example of location-based naming.

Similarly, sensor query and tasking language (SQTL) [37] is proposed as an application that provides even a larger set of services. SQTL supports three types of events, which are defined by keywords *receive*, *every*, and *expire*. The *receive* keyword defines events generated by a sensor node when it receives a message; *every* keyword defines events occurring periodically due to a timer time-out; and *expire* keyword defines events occurring when a timer is expired. If a sensor node receives a message intended for it that contains a script, it then executes the script. Although SQTL is proposed, different types of SQDDP can be developed for various applications. The use of SQDDPs may be unique to each application.

SQDDP provides interfaces to issue queries, responds to queries, and collects incoming replies. Other types of protocols are also essential to sensor network applications: the localization and time synchronization protocols. The localization protocol enables sensor nodes to determine their locations; the time synchronization protocol provides sensor nodes with a common view of time throughout the sensor network. Because many communication protocols require knowledge of location and time, it is important to describe the localization and time synchronization techniques in detail in the following sections before transport, network, and data link protocols are discussed later.

10.3 Localization Protocols

Because sensor nodes may be randomly deployed in any area, they must be aware of their locations in order to provide meaningful data to the users. In addition, location information may be required by the network and data-link layer protocols described in Section 10.6 and Section 10.7, respectively. In order to meet design challenges, a localization protocol must be:

- Robust to node failures
- Less sensitive to measurement noise
- Low error in location estimation
- Flexible in any terrain

Currently, two types of localization techniques address these challenges: (1) beacon based and (2) relative location based. Both techniques may use range and angle estimations for sensor node localization via received signal strength (RSS) [23, 42]; time of arrival (TOA) [13, 41]; time difference of arrival (TDOA); and angle of arrival (AOA).

Current localization methods [27, 36] are based on beacons with position known. The ad hoc localization system (AHLoS) [36] requires few nodes to have known location through GPS or through manual configuration. This allows nodes to discover their location through a two-phase process: ranging and estimation. During the ranging phase, each node estimates the range of its neighbors. The estimation phase then allows neighbors that do not have location to use the range estimated in the ranging phase and the known location of the beacons to estimate their locations.

Also, some methods [5, 6] assume beacon signals at known locations. This assumption may be fine for some applications, but sensor nodes may be deployed in regions in which known location is not possible. As a result, Moses and colleagues are investigating self-localization using sources at unknown locations [27]. Although these authors relax the assumption that beacons require fixed locations, the beacons still need a number of signal sources. These signal sources are deployed in the same region as the sensor nodes and used as references by the neighbor nodes to estimate the unknown locations and orientations from the signal sources.

The work of Moses et al. [27] and Savvides et al. [36] is based on signal sources. Other work [7] estimates locations of the sensor nodes by viewing the location estimation problem as a convex optimization problem because a proximity constraint exists between two nodes, i.e., the range of broadcast. In addition to these localization methods, Patwari and coworkers [28] provide the Cramer–Rao bound of sensor location accuracy based on fixed base stations capable of peer-to-peer time of arrival or received signal strength measurements.

Although beacon-based localization protocols are sufficient for certain sensor network applications, some sensor networks may be deployed in areas unreachable by beacons or GPS; they may be frequently jammed by environmental or manually induced noise. In addition, low-end sensor nodes may exhibit nonlinear device behavior and non-Gaussian measurement noise. To overcome these challenges, the location information is relayed hop by hop from the source to the sink. In order to obtain precise relative location information, the sensor nodes must collaboratively work together to assist each other. Furthermore, energy may be additionally conserved by enabling sensor nodes to track the locations of their neighbor nodes.

This relative localization technique is further explored by the perceptive localization framework (PLF) [43]. In this framework, a node is able to detect and track the location of the neighboring node by using a collaborative estimation technique and a particle filter applied to an array of sensors. To increase the accuracy of the location estimation, the sink may request all the nodes along the path to the sources to increase the number of samples (particles) for particle filtering. This process of local interaction does not require any beacon in place. In addition, a central processing unit is not required in order to determine the locations of the sources.

Whether the beacon- or relative location-based localization protocol is used, the location information is required by the protocols in the transport, network, and data-link layers. Each type of localization protocols offers different capabilities. Future sensor network applications may utilize a combination of localization techniques.

10.4 Time Synchronization Protocols

Instead of time synchronization between the sender and receiver during an application, such as in the Internet, the sensor nodes in the sensor field must maintain a similar time within a certain tolerance throughout the lifetime of the network. Combining with the criteria that sensor nodes must be energy efficient, low cost, and small in a multihop environment as described in Section 10.1, this requirement offers a challenging problem. In addition, the sensor nodes may be left unattended for a long period of time, e.g., in deep space or on an ocean floor. For short-distance multihop broadcast, data processing time and the variation of data processing time may contribute the most in time fluctuations and differences in path delays. Also, the time difference between two sensor nodes is significant over time due to the wandering effect of the local clocks.

Small and low-end sensor nodes may exhibit device behaviors much worse than those of large systems such as personal computers (PCs). Some of the factors influencing time synchronization in large systems also apply to sensor networks [21]:

- *Temperature.* Because sensor nodes are deployed in various places, the temperature variation throughout the day may cause the clock to speed up or slow down. For a typical PC, the clock drifts few parts per million during the day [25]. For low-end sensor nodes, the drifting may be even worse.
- *Phase noise.* Some of the causes of phase noise are due to access fluctuation at the hardware interface, response variation of the operating system to interrupts, and jitter in the network delay. The latter may be due to medium access and queueing delays.
- *Frequency noise.* The frequency noise is due to the instability of the clock crystal. A low-end crystal may experience large frequency fluctuation because the frequency spectrum of the crystal has large sidebands on adjacent frequencies.
- *Asymmetric delay.* Because sensor nodes communicate with each other through the wireless medium, the delay of the path from one node to another may be different from that of the return path. As a result, an asymmetric delay may cause an offset to the clock that cannot be detected by a variance type method [21]. If the asymmetric delay is static, the time offset between any two nodes is also static. The asymmetric delay is bounded by one half the round trip time between the two nodes [21].
- *Clock glitches.* Clock glitches are sudden jumps in time that may be caused by hardware or software anomalies such as frequency and time steps.

Table 10.1 shows three types of timing techniques, each of which must address the challenges mentioned earlier. In addition, the timing techniques must be energy aware because the batteries of the sensor nodes are limited. Also, they must address the mapping between the sensor network time and the Internet time, e.g., universal coordinated time. Next, examples of these types of timing techniques are described, namely, the network time protocol (NTP) [24]; the reference-broadcast synchronization (RBS) [9]; and the time-diffusion synchronization protocol (TDP) [44].

TABLE 10.1 Three Types of Timing Techniques

Type	Description
(1) Relies on fixed time servers to synchronize the network	The nodes are synchronized to time servers that are readily available. These time servers are expected to be robust and highly precise.
(2) Translates time throughout the network	The time is translated hop-by-hop from the source to the sink. In essence, it is a time translation service.
(3) Self-organizes to synchronize the network	The protocol does not depend on specialized time servers. It automatically organizes and determines the master nodes as the temporary time-servers.

In the Internet, the NTP is used to discipline the frequency of each node's oscillator. It may be useful to use NTP to discipline the oscillators of the sensor nodes, but connection to the time servers may not be possible because of frequent sensor node failures. In addition, disciplining all the sensor nodes in the sensor field may be a problem because of interference from the environment and large variation of delay between different parts of the sensor field. The interference can temporarily disjoint the sensor field into multiple smaller fields, causing undisciplined clocks among these smaller fields. The NTP protocol may be considered type 1 of the timing techniques; in addition, it must be refined to address timing challenges in the sensor networks.

The RBS, type 2 of the timing techniques, provides instantaneous time synchronization among a set of receivers within the reference broadcast of the transmitter. The transmitter broadcasts m reference packets. Each of the receivers within the broadcast range records the time of arrival of the reference packets. Afterwards, the receivers communicate with each other to determine the offsets. To provide multihop synchronization, it is proposed to use nodes receiving two or more reference broadcasts from different transmitters as translation nodes. These translation nodes are used to translate the time between different broadcast domains. As shown in Figure 10.1, nodes A, B, and C are the transmitter, receiver, and translation nodes, respectively.

Another emerging timing technique is the TDP, which is used to maintain the time throughout the network within a certain tolerance. The tolerance level can be adjusted based on the purpose of the sensor networks. The TDP automatically self-configures by electing master nodes to synchronize the sensor network. In addition, the election process is sensitive to energy requirement as well as the quality of the

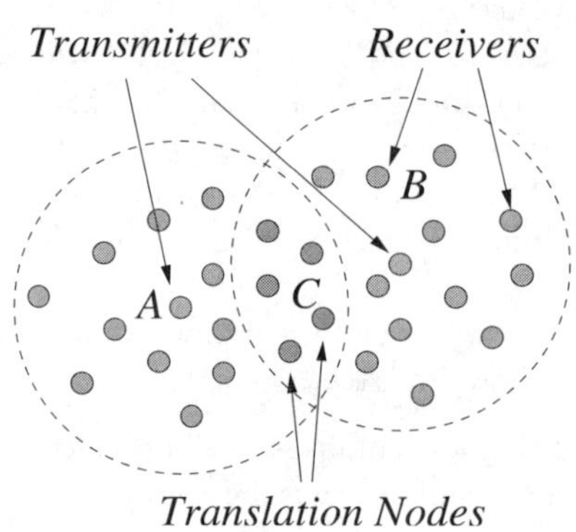

FIGURE 10.1 The RBS.

clocks. The sensor network may be deployed in unattended areas, and the TDP still synchronizes the unattended network to a common time. It is considered type 3 of the timing techniques.

In summary, these timing techniques may be used for different types of applications as discussed in Section 10.2; each has its benefits. A time-sensitive application must choose not only the type of timing techniques but also the type of transport, network, and data-link schemes described in the following sections. This is because different protocols provide different features and services to the time-sensitive application.

10.5 Transport Layer Protocols

The collaborative nature of the sensor network paradigm brings several advantages over traditional sensing, including greater accuracy, larger coverage area, and extraction of localized features. The realization of these potential gains, however, directly depends on efficient, reliable communication between the sensor network entities, i.e., the sensor nodes and the sink. To accomplish this, a reliable transport mechanism is imperative.

In general, the main objectives of the transport layer are (1) to bridge application and network layers by application multiplexing and demultiplexing; (2) to provide data delivery service between the source and the sink with an error control mechanism tailored according to the specific reliability requirement of the application layer; and (3) to regulate the amount of traffic injected into the network via flow and congestion control mechanisms. Nevertheless, the required transport layer functionalities to achieve these objectives in the sensor networks are subject to significant modifications in order to accommodate unique characteristics of the sensor network paradigm. Energy, processing, and hardware limitations of the sensor nodes bring further constraints on the transport layer protocol design. For example, conventional end-to-end, retransmission-based error control mechanisms and window-based, additive-increase, multiplicative-decrease congestion control mechanisms adopted by the vastly used transport control protocol (TCP) may not be feasible for the sensor network domain and thus may lead to waste of scarce resources.

On the other hand, unlike other conventional networking paradigms, the sensor networks are deployed with a specific sensing application objective, such as event detection, event identification, location sensing, and local control of actuators, for a wide range of applications (e.g., military, environment, health, space exploration, and disaster relief). The specific objective of the sensor network also influences the design requirements of the transport layer protocols. For example, the sensor networks deployed for different applications may require different reliability levels as well as different congestion control approaches. Consequently, development of transport layer protocols is a challenge because the limitations of the sensor nodes and the specific application requirements primarily determine design principles of transport layer protocols.

Due to the application-oriented and collaborative nature of the sensor networks, the main data flow takes place in the forward path, where the source nodes transmit their data to the sink. The reverse path, on the other hand, carries the data originated from the sink, such as programming/retasking binaries, queries, and commands to the source nodes. Therefore, different functionalities are required to handle the transport needs of the forward and reverse paths. Transport layer issues pertaining to these distinct cases are investigated separately in the following subsections.

10.5.1 Event-to-Sink Transport

Under the premise that data flows from source to sink are generally loss tolerant, Wan and coworkers questioned the need for a transport layer for data delivery in the sensor networks [32]. Although the need for end-to-end reliability may not exist because of the sheer amount of correlated data flows, an event in the sensor field needs to be tracked with a certain amount of accuracy at the sink. Therefore, unlike traditional communication networks, the sensor network paradigm necessitates an event-to-sink reliability notion at the transport layer [35]. This involves a reliable communication of the event features

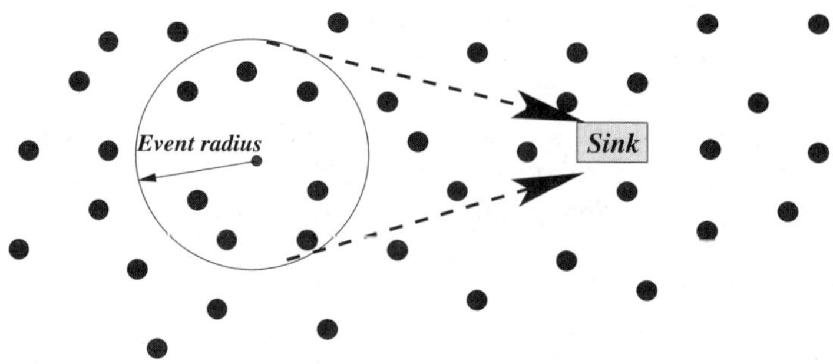

FIGURE 10.2 Typical sensor network topology with event and sink. (The sink is only interested in collective information of sensor nodes within the even radius and not in their individual data.)

to the sink rather than conventional packet-based reliable delivery of the individual sensing reports/packets generated by each sensor node in the field. Figure 10.2 illustrates an event-to-sink reliable transport notion based on collective identification of data flows from the event to the sink.

In order to provide reliable event detection at the sink, possible congestion in the forward path should also be addressed by the transport layer. Once the event is sensed by a number of sensor nodes within the coverage of the phenomenon, i.e., event radius, a significant amount of traffic is triggered by these sensor nodes; this may easily lead to congestion in the forward path. The need for transport layer congestion control to assure reliable event detection at the sink is revealed by the results of Tilak and colleagues [18], who have shown that exceeding network capacity can be detrimental to the observed goodput at the sink. Moreover, although the event-to-sink reliability may be attained even in the presence of packet loss due to network congestion (thanks to the correlated data flows), a suitable congestion control mechanism can also help conserve energy while maintaining desired accuracy levels at the sink.

On the other hand, although the transport layer solutions in conventional wireless networks are relevant, they are simply inapplicable for event-to-sink reliable transport in the sensor networks. These solutions mainly focus on reliable data transport following end-to-end TCP semantics and are proposed to address challenges posed by wireless link errors and mobility [2]. The primary reason for their inapplicability is their notion of end-to-end reliability, which is based on acknowledgments and end-to-end retransmissions. Because of inherent correlation in the data flows generated by the sensor nodes, however, these mechanisms for strict end-to-end reliability are superfluous and drain significant amounts of energy.

In contrast to the transport layer protocols for conventional end-to-end reliability, the event-to-sink reliable transport (ESRT) protocol [35] is based on the event-to-sink reliability notion and provides reliable event detection without any intermediate caching requirements. ESRT is a novel transport solution developed to achieve reliable event detection in the sensor networks with minimum energy expenditure. It includes a congestion control component that serves the dual purpose of achieving reliability and conserving energy. ESRT also does not require individual sensor identification, i.e., an event ID suffices. Importantly, the algorithms of ESRT mainly run on the sink, with minimal functionality required at resource-constrained sensor nodes.

10.5.2 Sink-to-Sensors Transport

Although data flows in the forward path carry correlated sensed/detected event features, the flows in the reverse path mainly contain data transmitted by the sink for an operational or application-specific purpose. This may include operating system binaries; programming/retasking configuration files; and application-specific queries and commands. Dissemination of this type of data mostly requires 100% reliable delivery. Therefore, the event-to-sink reliability approach introduced before would not suffice to address the tighter reliability requirements of flows in the reverse paths.

This strict reliability requirement for the sink-to-sensors transport of operational binaries and application-specific queries and commands involves a certain level of retransmission as well as acknowledgment mechanisms. However, these mechanisms should be incorporated into the transport layer protocols cautiously in order not to compromise scarce sensor network resources totally. In this respect, local retransmissions and negative acknowledgment approaches would be preferable over end-to-end retransmissions and acknowledgments to maintain minimum energy expenditure.

On the other hand, the sink is involved more in the sink-to-sensor data transport on the reverse path, so a sink with plentiful energy and communication resources can broadcast data with its powerful antenna. This helps to reduce the amount of traffic forwarded in the multihop sensor network infrastructure and thus helps sensor nodes conserve energy. Therefore, data flows in the reverse path may experience less congestion compared to the forward path, which is totally based on multihop communication. This calls for less aggressive congestion control mechanisms for the reverse path compared to the forward path in the sensor networks.

Wan and colleagues [32] propose the pump slowly, fetch quickly (PSFQ) mechanism for reliable retasking/reprogramming in the sensor networks. PSFQ is based on slowly injecting packets into the network but performing aggressive hop-by-hop recovery in case of packet loss. The pump operation in PSFQ simply performs controlled flooding and requires each intermediate node to create and maintain a data cache to be used for local loss recovery and in-sequence data delivery. Although this is an important transport layer solution for the sensor networks, PSFQ does not address packet loss due to congestion.

In summary, the transport layer mechanisms that can address the unique challenges posed by the sensor network paradigm are essential to realize the potential gains of the collective effort of sensor nodes. As discussed in the preceding two subsections, promising solutions exist for event-to-sink and sink-to-sensors reliable transports. These solutions and those currently under development, however, need to be exhaustively evaluated under real sensor network deployment scenarios to reveal their shortcomings. Therefore, necessary modifications may be required to provide a complete transport layer solution for the sensor networks.

10.6 Network Layer Protocols

Sensor nodes may be scattered densely in an area to observe a phenomenon. As a result, they may be very close to each other. In such a scenario, multihop communication may be a good choice for sensor networks with strict requirements on power consumption and transmission power levels. As compared to long distance wireless communication, multihop communication may be an effective way to overcome some of the signal propagation and degradation effects. In addition, the sensor nodes consume much less energy when transmitting a message because the distances between sensor nodes are shorter.

As discussed in Section 10.1, ad hoc routing techniques already proposed in the literature [29] do not usually fit requirements of the sensor networks. As a result, the network layer of the sensor networks is usually designed according to the following principles:

- Energy efficiency is always an important consideration.
- Sensor networks are mostly data centric.
- An ideal sensor network has attribute-based addressing and location awareness.
- Data aggregation is useful only when it does not hinder the collaborative effort of the sensor nodes.
- The routing protocol is easily integrated with other networks, e.g., Internet.

These design principles serve as a guideline when designing a routing protocol for sensor networks. Each of them is further explained to emphasize its importance. As described in the preceding section, a transport layer protocol must be energy efficient. This requirement also applies to a routing protocol because the network lifetime depends on the nodes' energy consumption when relaying messages. As a result, energy efficiency plays an important role in various protocol stack layers in addition to the network layer.

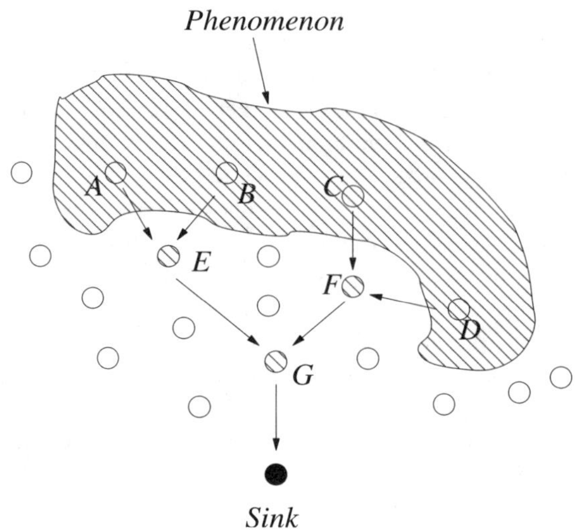

FIGURE 10.3 Data aggregation.

In sensor networks, information or data may be described by using attributes. In order to integrate tightly with the information or data, a routing protocol may be designed according to data-centric techniques. A data-centric routing protocol requires attribute-based naming [8, 10, 26, 37], which is used to carry out queries by using the attributes of the phenomenon. In essence, the users are more interested in the data gathered by the sensor networks in the phenomenon rather than by an individual node. They query the sensor networks by using attributes of the phenomenon that they want to observe. For example, the users may send out a query such as, "find the locations of areas where the temperature is over 70°F."

Furthermore, a data-centric routing protocol should also utilize the design principle of data aggregation — a technique used to solve the implosion and overlap problems in data-centric routing [15]. As shown in Figure 10.3, the sink queries the sensor network to observe the ambient condition of the phenomenon. The sensor network used to gather the information can be perceived as a reverse multicast tree, where the nodes within the area of the phenomenon send the collected data toward the sink. Data coming from multiple sensor nodes are aggregated as if they are about the same attribute of the phenomenon when they reach the same routing node on the way back to the sink. For example, sensor node *E* aggregates the data from sensor nodes *A* and *B* while sensor node *F* aggregates the data from sensor nodes *C* and *D* in Figure 10.3.

Data aggregation can be perceived as a set of automated methods of combining data from many sensor nodes into a set of meaningful information [16]. In this respect, data aggregation is known as data fusion [15]. Also, care must be taken when aggregating data because the specifics of the data, e.g., the locations of reporting sensor nodes, should not be left out. Such specifics may be needed by certain applications.

One of the design principles for the network layer is to allow easy integration with other networks such as the satellite network and the Internet. As shown in Figure 10.4, the sinks are the basis of a communication backbone that serves as a gateway to other networks. The users may query the sensor networks through the Internet or the satellite network, depending on the purpose of the query or the type of application the users are running.

A brief summary of the state of the art in the networking area is shown in Table 10.2. The schemes listed in the table utilize some of the design principles previously discussed. For example, the SMECN [22] creates an energy-efficient subgraph of the sensor networks. It tries to minimize the energy consumption while maintaining connectivity of the nodes in the network. In addition, the directed diffusion protocol [17] is a data-centric dissemination protocol in which the queries and collected data use attribute-based naming schemes.

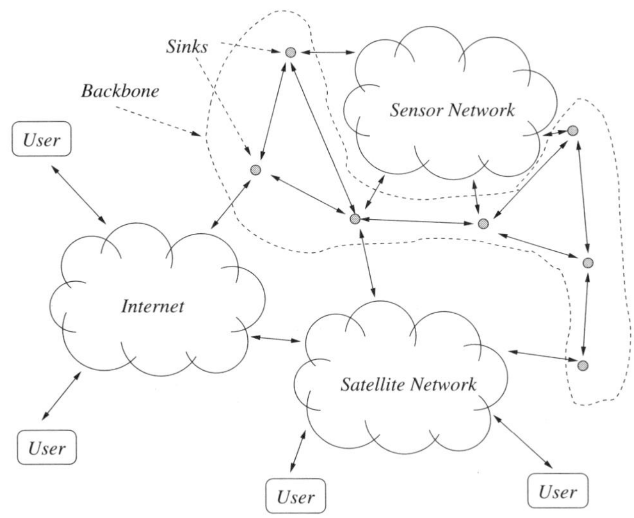

FIGURE 10.4 Internetworking between sensor nodes and user through Internet or satellite network.

TABLE 10.2 Overview of Network Layer Schemes

Network Layer Scheme	Description
SMECN [22]	Creates a sub graph of the sensor network that contains the minimum-energy path.
LEACH [16]	Forms clusters to minimize energy dissipation.
SAR [40]	Creates multiple trees where the root of each tree is one hop neighbor from the sink; select a tree for data to be routed back to the sink according to the energy resources and additive QoS Metric.
Flooding	Broadcasts data to all neighbor nodes regardless if they receive it before or not.
Gossiping [14]	Sends data to one randomly selected neighbor.
SPIN [15]	Sends data to sensor nodes only if they are interested; has three types of messages, i.e., ADV, REQ, and DATA.
Directed Diffusion [17]	Sets up gradients for data to flow from source to sink during interest dissemination.

Because different applications may require different types of network layer protocols, more advanced data-centric routing protocols are needed. In essence, application-specific requirements are part of the driving forces urging for new transport layer protocols, as described in the previous section. In addition, they push for the new data-link schemes described in the following section.

10.7 Data Link Layer Protocols

Although the transport layer mechanisms discussed in Section 10.5 are essential to achieving higher level error and congestion control, it is still imperative to have data-link layer functionalities in the sensor networks. In general, the data link layer is primarily responsible for multiplexing data streams, data frame detection, medium access, and error control; it ensures reliable point-to-point and point-to-multipoint connections in a communication network. Nevertheless, the collaborative and application-oriented nature of the sensor networks and the physical constraints of the sensor nodes, such as energy and processing limitations, determine the way in which these responsibilities are fulfilled. In the following two subsections, data-link layer issues are explored within the discussion of medium access and error control strategies in the sensor networks.

10.7.1 Medium Access Control

The medium access control (MAC) layer protocols in a multihop self-organizing sensor network must achieve two objectives:

- Establish data communication links for creating a basic network infrastructure needed for multihop wireless communication in a densely scattered sensor field.
- Regulate access to shared media so that communication resources are fairly and efficiently shared among the sensor nodes.

Due to the unique resource constraints and application requirements of sensor networks, however, the MAC protocols for the conventional wireless networks are inapplicable to the sensor network paradigm. For example, the primary goal of a MAC protocol in an infrastructure-based cellular system is to provide high QoS and bandwidth efficiency, mainly with dedicated resource assignment strategy. Such an access scheme is impractical for sensor networks because there is no central controlling agent like the base station. Moreover, power efficiency directly influences network lifetime in a sensor network and thus is of prime importance.

Although Bluetooth and the *mobile ad hoc network* (MANET) show similarities to the sensor networks in terms of communication infrastructure, both consist of nodes with portable battery-powered devices that can be replaced by the user. Therefore, unlike in the sensor networks, power consumption is only of secondary importance in these systems. Therefore, none of the existing Bluetooth or MANET MAC protocols can be directly used in the sensor networks because of network lifetime concerns.

It is evident that the MAC protocol for sensor networks must have built-in power conservation, mobility management, and failure recovery strategies. Thus far, *fixed allocation* and *random access* versions of medium access have been proposed [40, 46]. *Demand-based* MAC schemes may be unsuitable for sensor networks due to their large messaging overhead and link setup delay. Furthermore, contention-based channel access is deemed unsuitable because of the requirement to monitor the channel at all times — an energy-draining task.

The applicability of the fundamental MAC schemes in the sensor networks is discussed along with some proposed MAC solutions using that access method as follows:

- *TDMA-based medium access.* Time-division multiple-access (TDMA) access schemes inherently conserve more energy compared to contention-based schemes because the duty cycle of the radio is reduced and no contention-introduced overhead and collisions are present. Pottie and Kaiser [31] have reasoned that a MAC scheme for energy-constrained sensor networks should include a variant of TDMA because radios must be turned off during idling for precious power savings. The self-organizing medium access control for sensor networks (SMACS) [40] is such a time slot-based scheme; each sensor node maintains a TDMA-like super frame in which the node schedules different time slots to communicate with its known neighbors. SMACS achieves power conservation by using a random wake-up schedule during the connection phase and by turning the radio off during idle time slots. However, although a TDMA-based access scheme minimizes the transmit on time, it is not always preferred because of associated time synchronization costs.
- *Hybrid TDMA/FDMA-based medium access.* A pure TDMA-based access scheme dedicates the entire channel to a single sensor node; however, a pure frequency-division multiple access (FDMA) scheme allocates minimum signal bandwidth per node. Such contrast brings the trade-off between the access capacity and the energy consumption. An analytical formula is derived in Shih et al. [38] to find the optimum number of channels, which gives the minimum system power consumption. This determines the hybrid TDMA/FDMA scheme to be used. The optimum number of channels depends on the ratio of the power consumption of the transmitter to that of the receiver.

If the transmitter consumes more power, a TDMA scheme is favored, while the scheme leans toward FDMA when the receiver consumes greater power [38].

- *CSMA-based medium access.* Based on carrier sensing and backoff mechanism, traditional carrier-sense multiple access (CSMA)-based schemes are inappropriate because they make the fundamental assumption of stochastically distributed traffic and tend to support independent point-to-point flows. On the other hand, the MAC protocol for sensor networks must be able to support variable, but highly correlated and dominantly periodic traffic. Any CSMA-based medium access scheme has two important components: the listening mechanism and the backoff scheme. Woo and Culler [46] present a CSMA-based MAC scheme for sensor networks and observe from the simulations that the constant listen periods are energy efficient and the introduction of random delay provides robustness against repeated collisions.

10.7.2 Error Control

In addition to medium access control, error control of the transmitted data in the sensor networks is another extremely important function of the data-link layer. Error control is critical, especially in some sensor network applications such as mobile tracking and machine monitoring. In general, the error control mechanisms in communication networks can be categorized into two main approaches: forward error correction (FEC) and automatic repeat request (ARQ).

ARQ-based error control mainly depends on retransmission for the recovery of lost data packets/frames. It is clear that this ARQ-based error control mechanism incurs significant additional retransmission cost and overhead. Although ARQ-based error control schemes are utilized at the data-link layer for the other wireless networks, the usefulness of ARQ in sensor network applications is limited due to the scarcity of the energy and processing resources of the sensor nodes. On the other hand, FEC schemes have inherent decoding complexity that require relatively considerable processing resources in the sensor nodes. In this respect, simple error control codes with low-complexity encoding and decoding might present the best solutions for error control in the sensor networks.

On the other hand, for the design of efficient FEC schemes, it is important to have good knowledge of channel characteristics and implementation techniques. Channel bit error rate (BER) is a good indicator of link reliability. In fact, a good choice of the error correcting code can result in several orders of magnitude reduction in BER and an overall gain. The coding gain is generally expressed in terms of the additional transmit power needed to obtain the same BER without coding.

Therefore, the link reliability can be achieved by increasing the output transmit power or the use of suitable FEC scheme. Due to energy constraints of the sensor nodes, increasing the transmit power is not a feasible option. Therefore, using FEC is still the most efficient solution, given the constraints of the sensor nodes. Although the FEC can achieve significant reduction in the BER for any given value of the transmit power, the additional processing power consumed during encoding and decoding must be considered when designing an FEC scheme. If this additional power is greater than the coding gain, the whole process is not energy efficient and thus the system is better without coding. On the other hand, the FEC is a valuable asset to the sensor networks if the additional processing power is less than the transmission power savings. Thus, the trade-off between this additional processing power and the associated coding gain should be optimized in order to have powerful, energy-efficient, and low-complexity FEC schemes for error control in the sensor networks.

As researchers continue to investigate new FEC schemes for sensor networks, designers must bear in mind that the new schemes may be application specific. The data-link layer remains a challenging area in which to work because sensor nodes are inherently low end. Combining the low-end characteristic of the sensor nodes with harsh deployed terrains calls for new medium-access as well as error-control schemes.

10.8 Conclusion

An overview of the communication protocols for sensor networks is given in this chapter. Challenges and design guidelines for localization, time synchronization, application layer, transport layer, network layer, and data-link layer protocols are explored. As technology advances in the sensor network area, sensor network technologies may become an integral part of our lives.

Acknowledgment

The authors thank Dr. Ian F. Akyildiz for his encouragement and support.

References

1. Agre, J. and Clare, L., An integrated architecture for cooperative sensing networks, *IEEE Computer Mag.*, 106–108, May 2000.
2. Balakrishnan, H. et al., A comparison of mechanisms for improving TCP performance over wireless links, *IEEE/ACM Trans. Networking*, 5(6), 756–769, December 1997.
3. Bhardwaj, M., Garnett, T., and Chandrakasan, A.P., Upper bounds on the lifetime of sensor networks, *IEEE Int. Conf. Commun. '01,* Helsinki, Finland, June 2001.
4. Bonnet, P., Gehrke J., and Seshadri, P., Querying the physical world, *IEEE Personal Commun.,* 10–15, October 2000.
5. Bulusu, N., Heidemann, J., and Estrin, D., GPS-less low-cost outdoor localization for very small devices, *IEEE Personal Commun.,* 7, 28–34, October 2000.
6. Bulusu, N. et al., Scalable coordination for wireless sensor networks: self-configuring localization systems, *Int. Symp. Commun. Theory Applications (ISCTA 2001),* Ambleside, U.K., July 2001.
7. Doherty, L., Pister, K.S.J., and Ghaoui, L.E., Convex position estimation in wireless sensor networks, *INFOCOM'01*, Anchorage, AK, April 2001.
8. Elson, J. and Estrin, D., Random, ephemeral transaction identifiers in dynamic sensor networks, *Proc. 21st Int. Conf. Distributed Computing Syst.,* 459–468, Phoenix, AZ, April 2001.
9. Elson, J., Girod, L., and Estrin, D., Fine-grained network time synchronization using reference broadcasts, *Proc. 5th Symp. Operating Syst. Design Implementation (OSDI 2002),* Boston, MA, December 2002.
10. Estrin, D. et al., Instrumenting the world with wireless sensor networks, *Int. Conf. Acoustics, Speech, Signal Process. (ICASSP 2001),* Salt Lake City, UT, May 2001.
11. Estrin, D. et al., Next century challenges: scalable coordination in sensor networks, *ACM Mobicom '99,* 263–270, Seattle, WA, August 1999.
12. Estrin, D., Govindan R., and Heidemann J., Embedding the Internet, *Commun. ACM,* 43, 38–41, May 2000.
13. Fischer, S. et al., System performance evaluation of mobile positioning methods, *Proc. IEEE Vehicular Technol. Conf.,* Houston, TX, May 1999.
14. Hedetniemi, S., Hedetniemi, S., and Liestman, A., A survey of gossiping and broadcasting in communication networks, *Networks,* 18(4), 319–349, 1988.
15. Heinzelman, W.R., Kulik, J., and Balakrishnan, H., Adaptive protocols for information dissemination in wireless sensor networks, *ACM Mobicom '99,* 174–185, Seattle, WA, August 1999.
16. Heinzelman, W.R., Chandrakasan, A., and Balakrishnan, H., Energy-efficient communication protocol for wireless microsensor networks, *IEEE Proc. Hawaii Int. Conf. Syst. Sci.,* 1–10, Maui, HI, January 2000.
17. Intanagonwiwat, C., Govindan, R., and Estrin, D., Directed diffusion: a scalable and robust communication paradigm for sensor networks, *ACM Mobicom '00,* 56–67, Boston, MA, August 2000.
18. Tilak, S., Abu–Ghazaleh, N.B., and Heinzelman, W., Infrastructure trade-offs for sensor networks, *Proc. WSNA 2002,* Atlanta, GA, September 2002.

19. Jaikaeo, C., Srisathapornphat, C., and Shen, C., Diagnosis of sensor networks, *IEEE Int. Conf. Commun. '01,* Helsinki, Finland, June 2001.
20. Kahn, J.M., Katz, R.H., and Pister, K.S.J., Next century challenges: mobile networking for Smart Dust, *ACM Mobicom '99,* 271–278, Seattle, WA, August 1999.
21. Levine, J., Time synchronization over the Internet using an adaptive frequency-locked loop, *IEEE Trans. Ultrasonics, Ferroelectrics, Frequency Control,* 46(4), 888–896, July 1999.
22. Li, L. and Halpern, J.Y., Minimum-energy mobile wireless networks revisited, *IEEE Int. Conf. Commun. ICC '01,* Helsinki, Finland, June 2001.
23. Mark, B. and Zaidi, Z., Robust mobility tracking for cellular networks, *Proc. IEEE Int. Commun. Conf.,* New York, 2002.
24. Mills, D.L., Internet time synchronization: the network time protocol, *Global States and Time in Distributed Systems, IEEE,* Computer Society Press, 1994.
25. Mills, D.L., Adaptive hybrid clock discipline algorithm for the network time protocol, *IEEE/ACM Trans. Networking,* 6(5), 505–514, October 1998.
26. Mirkovic, J. et al., A self-organizing approach to data forwarding in large-scale sensor networks, *IEEE Int. Conf. Commun. ICC '01,* Helsinki, Finland, June 2001.
27. Moses, R., Krishnamurthy, D., and Patterson, R., A self-localization method for wireless sensor networks, *Eurasip J. Appl. Signal Process.,* 4, 348–358, 2003.
28. Patwari, N. et al., Relative location estimation in wireless sensor networks, *IEEE Trans. Signal Process.,* August 2003.
29. Perkins, C., *Ad Hoc Networks,* Addison-Wesley, Reading, MA, 2000.
30. Perrig, A. et al., SPINS: security protocols for sensor networks, *Proc. ACM MobiCom '01,* 189–199, Rome, July 2001.
31. Pottie, G.J. and Kaiser, W.J., Wireless integrated network sensors, *Commun. ACM,* 43(5), 551–558, May 2000.
32. Wan, C.Y., Campbell, A.T., and Krishnamurthy, L., PSFQ: a reliable transport protocol for wireless sensor networks, *Proc. WSNA 2002,* Atlanta, GA, September 2002.
33. Rabaey, J. et al., PicoRadio: ad hoc wireless networking of ubiquitous low-energy sensor/monitor nodes, *Proc. IEEE Computer Soc. Annu. Workshop VLSI (WVLSI '00),* 9–12, Orlando, FL, April 2000.
34. Rabaey, J.M. et al., PicoRadio supports ad hoc ultra-low power wireless networking, *IEEE Computer Mag.,* 33, 42–48, July 2000.
35. Sankarasubramaniam, Y., Akan, O.B., and Akyildiz, I.F., ESRT: event-to-sink reliable transport for wireless sensor networks, *Proc. ACM MOBIHOC 2003,* 177–188, Annapolis, MD, June 2003.
36. Savvides, A., Han, C., and Srivastava, M., Dynamic fine-grained localization in ad hoc networks of sensors, *Proc. ACM MobiCom '01,* 166–179, Rome, July 2001.
37. Shen, C., Srisathapornphat, C., and Jaikaeo, C., Sensor information networking architecture and applications, *IEEE Personal Commun.,* 52–59, August 2001.
38. Shih, E. et al., Physical layer driven protocol and algorithm design for energy-efficient wireless sensor networks, *ACM Mobicom '01,* 272–286, Rome, July 2001.
39. Slijepcevic, S. and Potkonjak, M., Power efficient organization of wireless sensor networks, *IEEE Int. Conf. Commun. '01,* Helsinki, Finland, June 2001.
40. Sohrabi, K. et al., Protocols for self-organization of a wireless sensor network, *IEEE Personal Commun.,* 16–27, October 2000.
41. Spirito, M.A. and Mattioli, A.G., On the hyperbolic positioning of GSM mobile stations, *Proc. Int. Symp. Signals, Syst. Electron.,* September 1998.
42. Spirito, M.A., Further results on GSM mobile station location, *IEEE Electron. Lett.,* 35(22), 1999.
43. Su, W. and Akyildiz, I.F., Perceptive localization framework for sensor networks, *Georgia Tech Technical Report,* 2003.
44. Su, W. and Akyildiz, I.F., Time-diffusion synchronization protocol for sensor networks, *Georgia Tech Technical Report,* 2003.

45. Warneke, B., Liebowitz, B., and Pister, K.S.J., Smart Dust: communicating with a cubic-millimeter computer, *IEEE Computer Mag.*, 2–9, January 2001.
46. Woo, A. and Culler, D., A transmission control scheme for media access in sensor networks, *ACM Mobicom '01,* 221–235, Rome, July 2001.

11

Positioning and Location Tracking in Wireless Sensor Networks

Yu-Chee Tseng
National Chiao-Tung University

Chi-Fu Huang
National Chiao-Tung University

Sheng-Po Kuo
National Chiao-Tung University

11.1 Introduction 11-1
11.2 Fundamentals 11-2
ToA, TDoA, and AoA • Positioning by Signal Strength
11.3 Positioning and Location Tracking Algorithms 11-4
Trilateration • Multilateration • Pattern Matching • Location Tracking • Network-Based Tracking
11.4 Experimental Location Systems 11-10
Active Badge and Bat • Cricket • RADAR and Nibble • CSIE/NCTU Indoor Tour Guide
11.5 Conclusions 11-12

11.1 Introduction

Locations of devices or objects are important information in many applications. This is particularly true for wireless sensor networks, which usually need to determine devices' context. For outdoor environments, the most well-known positioning system is the global positioning system (GPS) [5]. This positioning system uses 24 satellites set up by the U.S. Department of Defense to enable global three-dimensional positioning services; it has two levels of accuracy: stand positioning service (SPS) and precise positioning service (PPS). The accuracy provided by GPS is around 20 to 50 m.

In addition to the GPS system, positioning can also be done using some wireless networking infrastructures. Taking the PCS cellular networks as an example, the E911 emergency service requires determining the location of a phone call via the base stations of the cellular system. Several location estimation models, such as angle of arrival (AoA); time of arrival (ToA); received signal strength (RSS); phase of arrival (PoA); and assisted global positioning system (A-GPS), are widely used in cellular networks and wireless sensor networks.

Much work has been dedicated recently to positioning and location tracking in the area of wireless ad hoc and sensor networks. The purpose of this chapter is to review the recent progress in this direction. GPS is not suitable for wireless sensor networks for several reasons:

- It is not available in an indoor environment because satellite signals cannot penetrate buildings.
- For more fine-grained applications, higher accuracy is usually necessary in the positioning result.
- Sensor networks have their own battery constraint, which requires special design.

Location information can be used to improve the performance of wireless networks and provide new types of services. For example, it can facilitate routing in a wireless ad hoc network to reduce routing overhead. This is known as geographic routing [7, 9]. Through location-aware network protocols, the number of control packets can be reduced. Service providers can also use location information to provide some novel location-aware or follow-me services. The navigation system based on GPS is an example. A user can tell the system his destination and the system will guide him there. Phone systems in an enterprise can exploit locations of people to provide follow-me services. Other types of location-based services include *geocast* [6, 8], by which a user can request to send a message to a specific area, and *temporal geocast*, by which a user can request to send a message to a specific area at specific time. In contrast to traditional multicast, such messages are not targeted at a fixed group of members, but rather at members located in a specific physical area.

Section 11.2 introduces some fundamental distance estimation models; Section 11.3 discusses some positioning and location tracking algorithms. In Section 11.4, some experimental systems are reviewed and Section 11.5 gives a summary.

11.2 Fundamentals

To position an object or a device, the basic step is to use a reference point to determine the distance and angle between the device and the reference point. This has been exploited in the radar systems widely used in military applications. This section describes several such basic approaches. The next subsection discusses how to use multiple reference points jointly to estimate the location of a device.

11.2.1 ToA, TDoA, and AoA

In the ToA (time of arrival) approach, signal traveling time is used to estimate the distance between a device and the reference point. Such systems typically use signals that move at a slower speed, such as ultrasound, to measure the time of signal arrival. Figure 11.1(a) illustrates this idea. An ultrasound signal is sent from the transmitter to the receiver; in return, the receiver sends a signal back to the transmitter. After this two-way handshake, the transmitter can infer the distance from the round-trip delay of the signals:

$$\frac{((T_3 - T_0) - (T_2 - T_1)) * V}{2},$$

where V is the velocity of the ultrasound signals. The error of such measurement may come from the processing time of signals (such as computing latency and the unknown delay $T_2 - T_1$ at the receiver's side).

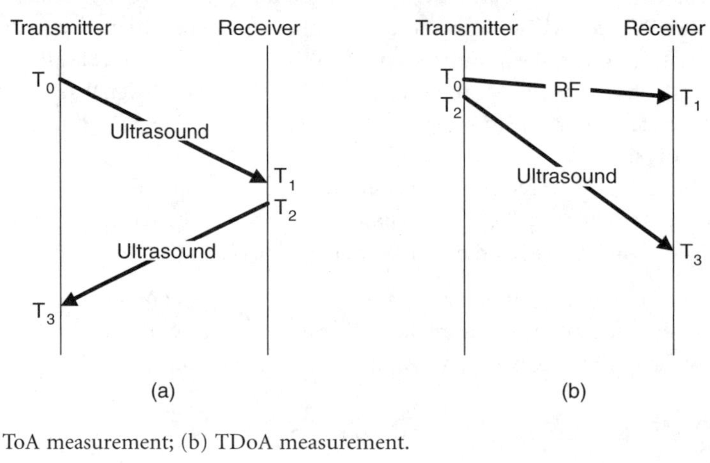

FIGURE 11.1 (a) ToA measurement; (b) TDoA measurement.

Another distance estimation technique is the time difference of arrival (TDoA). Although similar to the ToA scheme, this method uses two signals that travel at different speeds, such as the radio frequency (RF) and ultrasound. Figure 11.1(b) shows how TDoA works; transmission in one direction is sufficient. At T_0 the transmitter sends an RF signal, followed by an ultrasound signal at time T_2. The receiver can then determine its distance to the transmitter by

$$((T_3 - T_1) - (T_2 - T_0)) * (\frac{V_{RF} * V_{US}}{V_{RF} - V_{US}}),$$

where V_{RF} and V_{US} are the traveling speeds of RF and ultrasound signals, respectively. For TDoA, in addition to errors caused by processing time, the receiver also must know the precise value of $(T_2 - T_0)$ to determine the distance.

The AoA approach is another commonly used method for positioning [10, 13]. Such approaches require an antenna array or an array of ultrasound receivers, which can determine the angle and orientation of received signals.

11.2.2 Positioning by Signal Strength

Besides using the signal traveling time, another distance estimation technique is to use the property of signal degradation while traveling in a space to determine the mutual distance. Because signals traveling in a space typically reduce in strength with respect to the distance that they travel, the *received signal strength (RSS)* can be measured at the receiver's side. A mathematical propagation model can be derived to estimate the distance d between a transmitter and a receiver [14] as follows

$$PL(d) = PL(d_0) + 10n \log\left(\frac{d}{d_0}\right) \propto \left(\frac{d}{d_0}\right)^n,$$

where $PL()$ is the path loss function with respect to distance measured in decibels; n is a loss exponent that indicates the rate at which loss increases with distance; and d_0 is the reference distance determined from a measurement close to the transmitter. The path loss exponent n usually ranges from 2 to 4.

Using path loss may incur significant errors. For example, Figure 11.2 shows an experimental result based on IEEE 802.11b. As can be seen, a trend for the relation between distance and signal strength does exist; however, the curve is unstable in small ranges. The true signal strength model is complex and many uncontrollable environmental factors (such as shadows and terrain) are present.

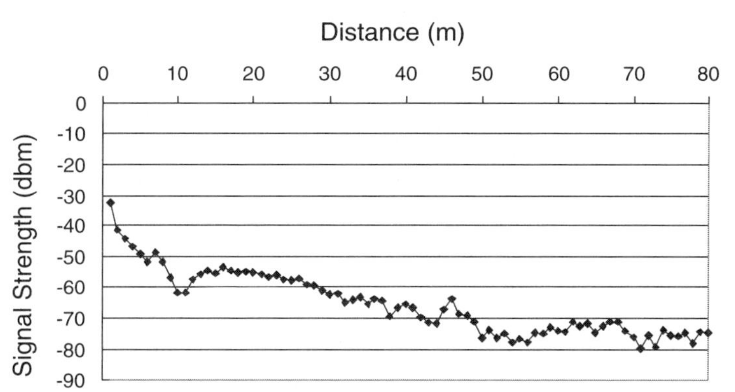

FIGURE 11.2 Signal strength vs. distance in IEEE 802.11b.

To solve the preceding problem, it is necessary to model the error for signal attenuation. One possibility is to include a random variable in the preceding path loss function as follows

$$PL(d) = PL(d_0) + 10n\log\left(\frac{d}{d_0}\right) + X_\rho,$$

where X_ρ is a zero-mean Gaussian random variable with a standard deviation ρ. Due to the existence of such errors, errors will occur as well when positioning a device based on signal strength. Assuming the similar error model in measuring distances, Slijepcevic and colleagues [16] further analyzed the location errors in a wireless sensor network and proved that the distribution of location error can be approximated by a family of Weibull distributions.

11.3 Positioning and Location Tracking Algorithms

The previous section discussed how to estimate the distance between two devices. If an object knows its distances to multiple devices at known locations, one may estimate its location. Several such methods are discussed here.

11.3.1 Trilateration

Trilateration is a well-known technique in which the positioning system has a number of *beacons* at known locations. These beacons can transmit signals so that other devices can determine their distances to these beacons based on received signals. If a device can hear at least three beacons, its location can be estimated. Figure 11.3(a) shows how trilateration works; A, B, and C are beacons with known locations. From A's signal, one can determine that the object should be located at the circle centered at A. Similarly, from B's and C's signals, it can be determined that the object should be located at the circles centered at B and C, respectively. Thus, the intersection of the three circles is the estimated location of the device.

The preceding discussion has assumed an ideal situation; however, as mentioned earlier, distance estimation always contains errors that will, in turn, lead to location errors. Figure 11.3(b) illustrates an example in practice. The three circles do not intersect in a common point. In this case, the maximum likelihood method may be used to estimate the device's location. Let the three beacons A, B, and C be located at (x_A, y_A), (x_B, y_B), and (x_C, y_C), respectively. For any point (x, y) on the plane, a difference function is computed:

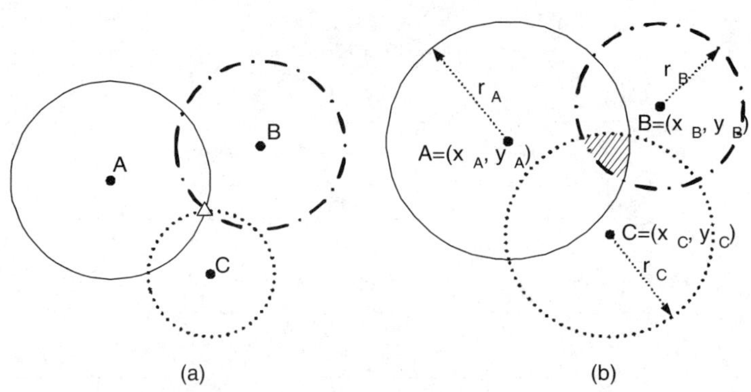

FIGURE 11.3 Trilateration method: (a) ideal situation; (b) real situation with errors.

$$\sigma_{x,y} = \left|\sqrt{(x-x_A)^2+(y-y_A)^2}-r_A\right| + \left|\sqrt{(x-x_B)^2+(y-y_B)^2}-r_B\right| + \left|\sqrt{(x-x_C)^2+(y-y_C)^2}-r_C\right|$$

where r_A, r_B, and r_C are the estimated distances to A, B, and C, respectively. The location of the object can then be predicted as the point (x, y) among all points such that $\sigma_{x,y}$ is minimized.

In addition to using the ToA approach for positioning, the AoA approach can be used. For example, in Figure 11.4, the unknown node *D* measures the angle of, *ADB*, *BDC*, and *ADC* by the received signals from beacons *A*, *B*, and *C*. From this information, *D*'s location can be derived [10].

11.3.2 Multilateration

The trilateration method has its limitation in that at least three beacons are needed to determine a device's location. In a sensor network, in which nodes are randomly deployed, this may not be true. Several multilateration methods are proposed to relieve this limitation.

The AHLoS (Ad Hoc Localization System) [1] is a distributed system for location discovery. In the network, some beacons have known locations and some devices have unknown locations. The AHLoS enables nodes to discover their locations by using a set of distributed iterative algorithms. The basic one is *atomic multilateration*, which can estimate the location of a device of unknown location if at least three beacons are within its sensing range. Figure 11.5 shows an example in which, initially, beacon nodes contain only nodes marked as having a GPS. Device nodes 1, 2, 3, and 4 are at unknown locations. In the first iteration, as Figure 11.5(a) shows, the locations of nodes 1, 2, and 3 will be determined.

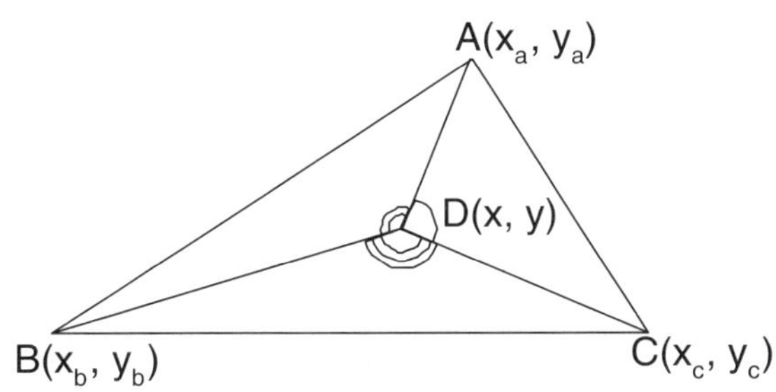

FIGURE 11.4 Angle measurement from three beacons, *A*, *B*, and *C*.

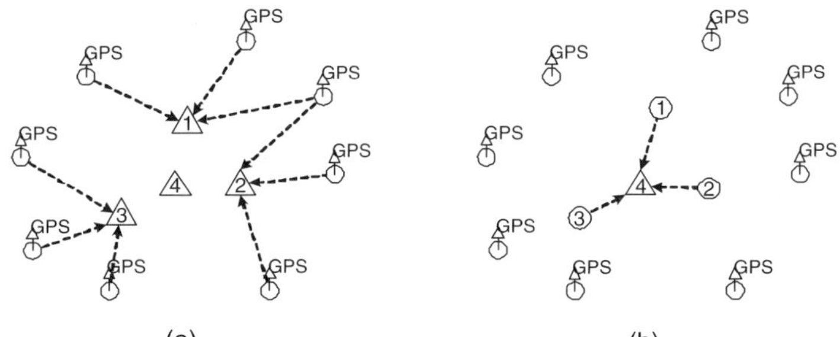

FIGURE 11.5 (a) Atomic multilateration; (b) iterative multilateration in AHLoS.

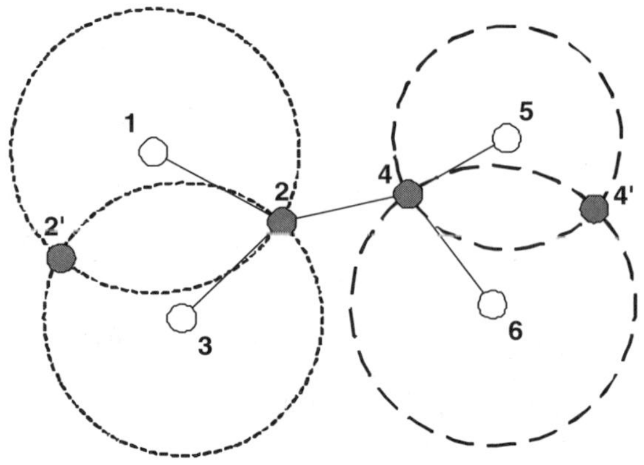

FIGURE 11.6 Collaborative multilateration in AHLoS.

The atomic multilateration is further extended to an *iterative multilateration* method. Specifically, once the location of a device is estimated, its role is changed to a beacon node so as to help determine other devices' locations. This is repeated until all hosts' locations are determined (if possible). As Figure 11.5(b) shows, in the second iteration, the location of node 4 can be determined with the help of nodes 1, 2, and 3, which are now serving as beacons.

The iterative multilateration still has its limitation. For example, as Figure 11.6 shows, it is impossible to determine node 2's and node 4's locations even if the locations of nodes 1, 3, 5, and 6 are known. The *collaborative multilateration* method may relieve this problem because it allows one to predict multiple potential locations of a node if it can hear fewer than three beacons. For example, in Figure 11.6, from beacon nodes 1 and 3, two potential locations of node 2 may be guessed (the other potential location is marked by 2′). Similarly, from beacon nodes 5 and 6, one may guess two potential locations of node 4 (the other potential location is marked by 4′). Collaborative multilateration allows estimation of the distance between nodes 2 and 4. With this information, the locations of nodes 2 and 4 can be estimated, as the figure shows.

11.3.3 Pattern Matching

Another type of location discovery is by pattern matching. Instead of estimating the distance between a beacon and a device, this approach tries to compare the received signal pattern against the training patterns in the database. Thus, this method is also known as the *fingerprinting* approach. The basic idea is that signal strength received at a fixed location is not necessarily a constant. It typically moves up and down, so it would be better to model signal strength by a random variable. This is especially true for indoor environments.

The main idea is to compare the received signals against those in the database and determine the likelihood that the device is currently located in a position. A typical solution has two phases (refer to Figure 11.7):

- *Off-line phase.* The purpose of this phase is to collect signals from all base stations at each training location. The number of training locations is decided first. Then, the received signal strengths are recorded (for a base station that is too far away, the signal strength is indicated as zero). Each entry in the database has the format: (x, y, $\langle ss_1, ss_2, \ldots, ss_n\rangle$), where (x, y) is the coordinate of the training location, and ss_i, $i = 1 \ldots n$, is the signal strength received at the training location from the ith base station. These entries are stored in the database. Note that for higher accuracy, one may establish multiple entries in the database for the same training location. From the database, some positioning rules, which form the positioning model, will then be established.
- *Real-time phase.* With a well-trained positioning model, one can estimate a device's location given the signal strengths collected by the device from all possible base stations. The positioning model may determine a number of locations, each associated with a probability. However, the typical solution is to output only the location with the highest likelihood.

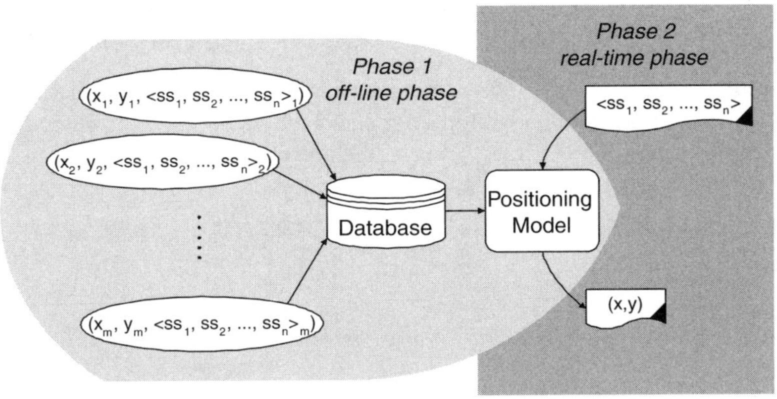

FIGURE 11.7 Pattern matching approach.

There are several similarity searching methods in the matching process; two approaches are introduced next.

11.3.3.1 Nearest Neighbor Algorithms

The simplest approach is the *nearest neighbor in signal space (NNSS)* approach [3, 11]. In the first phase, only the average signal strength of each base station at each training location is recorded. Then, in the second phase, the NNSS algorithm computes the *Euclidean distance* in signal space between the received signal and each record in the database. Euclidean distance means the square root of the summation of square of the difference between each received signal strength and the corresponding average signal strength from the access point under consideration. The training location with the minimum Euclidean distance is then chosen as the estimated location of the device. Because this algorithm only picks existing locations in the database, to improve its accuracy, it is suggested that the training set be dense enough.

One variant of the basic NNSS algorithm is *NNSS-AVG*. To take the uncertainty of a device's location into consideration, this method tries to pick a small number of training locations that closely match the received signal strengths (such as those with smaller Euclidean distances). Then, it infers the location of the device to be a function of the coordinates of the selected training locations. For example, one may take the average of the x and y coordinates of the selected training locations as the estimated result.

11.3.3.2 Probability-Based Algorithms

The probability-based positioning approach regards signal strength as a probability distribution [15]. In NNSS, because the received signal strengths are averaged out, the probability distribution would disappear. So the probability-based approach will try to maintain more complete information of signal strength distribution. The prediction result is typically more accurate.

The core of the probability-based model is the Bayes rule:

$$p(l \mid o) = \frac{p(o \mid l)\,p(l)}{p(o)} = \frac{p(o \mid l)\,p(l)}{\sum_{l' \in L} p(o \mid l')\,p(l')},$$

where $p(l \mid o)$ is the probability that the device is at location l given an observed signal strength pattern o. The prior probability that a device is resident at l is $p(l)$, which may be inferred from history or experience. For example, people may have a higher probability to appear in a hallway or lobby. If this is not available, $p(l)$ may be assumed to be a uniform distribution. L is the set of all training locations. The denominator $p(o)$ does not depend on the location variable l, so it can be treated as a normalized constant whenever only relative probabilities are required.

The term $p(o\,|\,l)$ is called the likelihood function; this represents the core of the positioning model and can be computed in the off-line phase. There are two ways to implement the likelihood function [15]:

- *Kernel method.* For each observation o_i in the training data, it is assumed that the signal strength exhibits a Gaussian distribution with mean o_i and standard deviation σ, where σ is an adjustable parameter in the model. Specifically, given o_i, the probability to observe o is

$$K(o;o_i) = \frac{1}{\sqrt{2\pi}\,\sigma} exp\left(\frac{(o-o_i)^2}{2\sigma^2}\right).$$

  Based on the kernel function, the probability $p(o\,|\,l)$ can be defined as

$$p(o\,|\,l) = \frac{1}{n_1} \sum_{l_i \in L, l_i = l} K(o;o_i),$$

  where n_1 is the number of training vectors in L obtained at location l. Intuitively, the probability function is a mixture of n_1 equally weighted density functions. Also note that the preceding formulas are derived assuming that only one base station exists. With multiple base stations, the probability function will be multivariated, and the probability will become the multiplication of multiple independent probabilities, each for one base station.
- *Histogram method.* Another method to estimate the density functions is to use histogram, which is related to *discretization* of continuous values to discrete ones. A number of *bins* can be defined as a set of nonoverlapping intervals that cover the whole random variables. The advantage of this method is in its ease in implementation and low computational cost. Another reason is that its discrete property can smooth out the instability of signal strengths.

The probability-based methods can adapt to different environments. To further reduce the computational overhead, Youssef and colleagues [19] proposed a method by clustering training data in the database.

11.3.4 Location Tracking

Location tracking means that a device's location can be derived based on some history traces. Because the trace of a device may indicate where it may move in the next step, this information can be used to improve the accuracy of positioning results. For example, one possibility is to consider the relative distances between consecutive moves of a device in a short period of time. These distances are typically not long. Using this information can reduce errors in tracking results.

In Bahl et al. [3], a Viterbi-like tracking algorithm is proposed for location tracking. The Viterbe algorithm is typically used in communications theory for recognizing the most likely message that is transmitted over a noisy channel. In location tracking, because various environmental factors may interfere with signals, the Viterbi algorithm is also suitable for selecting the most likely location of a device. The idea behind the Viterbi-like tracking algorithm is to take the continuity of a user's track in the past into consideration so as to come up with a better guess of the user's current location.

Figure 11.8 shows the details of the Viterbi-like tracking algorithm. Each time the mobile device receives signals from the access points, it computes a set of k most likely locations. This may be obtained from the NNSS-AVG algorithm described earlier. After receiving continuous h samples, the Viterbi-like algorithm can generate an $h * k$ map, which is an h-stage graph in which each stage contains k possible locations of the device at that stage. The possible locations are modeled by vertices. Edges are established between continuous stages and a weight is assigned to each edge equal to the Euclidean distance of the two incident vertices.

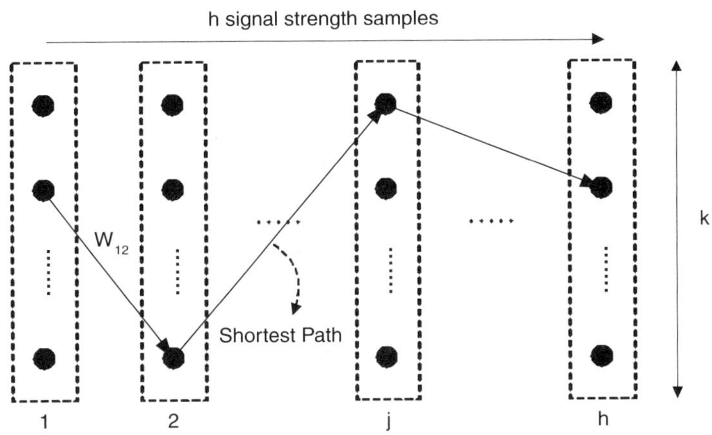

FIGURE 11.8 Viterbi-like location tracking algorithm.

Under the assumption that a user may not move too far away from his current location in a short period of time, the Viterbi-like tracking algorithm computes a shortest path in the $k * h$ map. This shortest path can be viewed as the most likely trajectory of the mobile user. Then the user's current location can be guessed to be the head of this shortest path. Note that, for this reason, the Viterbi-like algorithm may have $h - 1$ periods of delay.

The variances of environments may also complicate the problem. The radio channel condition in working hours may significantly differ from that during off hours. The positioning model may need to adapt to such factors. Recalibration is sometimes inevitable, but laborious. An environmental profile may need to be established to conquer this problem.

11.3.5 Network-Based Tracking

Special concerns— power saving, bandwidth conservation, and fault tolerance — arise when a solution is designed for a wireless sensor network. At the network level, location tracking may be done via the cooperation of sensors. Tseng and colleagues [17] addressed these issues using an agent-based paradigm. Once a new object is detected by the network, a mobile agent will be initiated to track the roaming path of the object. The agent is mobile because it will choose the sensor closest to the object to stay. The agent may invite some nearby slave sensors to cooperatively position the object and inhibit other irrelevant (i.e., farther) sensors from tracking the object. More precisely, only three agents will be used for the tracking purpose at any time and they will move as the object moves. The trilateration method is used for positioning. As a result, the communication and sensing overheads are greatly reduced. Because data transmission may consume a lot of energy, this agent-based approach tries to merge the positioning results locally before sending them to the data center. These authors also address how to conduct data fusion.

Figure 11.9 shows an example. The sensor network is deployed in a regular manner and it is assumed that each sensor's sensing distance equals the distance between two neighboring sensors. Initially, each sensor is in the *idle* state, searching for new objects. Once detecting a target, a sensor will transit to the *election* state, trying to serve as the *master agent*. The nearest sensor will win. The master agent will then dispatch two neighboring sensors as the *slave agents*; master and slave agents will cooperate to position the object. In the figure, the object is first tracked by sensors $\{S_0, S_1, S_2\}$ when resident in A_0, then by $\{S_0, S_1, S_6\}$ when in A_1, by $\{S_0, S_5, S_6\}$ when in A_2, etc.

The master agent is responsible for collecting all sensing data and performing the trilateration algorithm. It also conducts data fusion by keeping the tracking results while it moves around. At a proper time, the master agent will forward the tracking result to the data center. Two strategies are proposed for this purpose: *threshold-based (TB)* strategy, which will forward the result when the amount of data

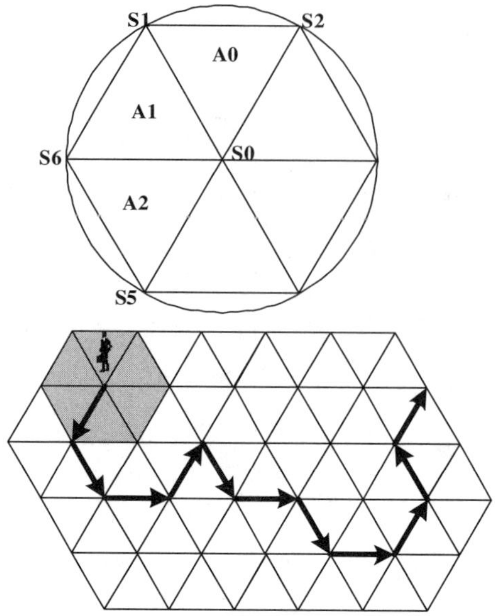

FIGURE 11.9 Roaming path of an object (dashed line) and the migration path of the corresponding master agent (arrow). Sensors that ever host a slave agent are marked by black. (From Y.-C. Tseng et al., *Int. Workshop Inf. Process. Sensor Networks (IPSN)*, 2634, 625–641, 2003. Also to be published in *The Computer Journal.* With permission.)

reaches a predefined threshold value *T*, and *distance-based (DB)* strategy, which will make a decision based on the routing distance from the agent's current location to the data center and the direction in which the agent is moving.

11.4 Experimental Location Systems

In this section, several location systems are introduced. Although they may not be specially designed for wireless sensor networks, these design concepts and experiences will benefit future implementations of positioning systems in wireless sensor networks.

11.4.1 Active Badge and Bat

The *Active Badge* system [18] is a cell-based location system in which objects are each attached with a badge that periodically emits infrared signals with a unique ID. Infrared receivers mounted at known positions collect these signals and relay them over a wired network. As a result, the system knows in which infrared cell a badge currently stays. The disadvantage of this badge system is that it is hard to deploy in a large-scale environment and that infrared is sensitive to external light, such as sunlight.

A successor of the Active Badge system is the *Bat* system [2], which consists of a collection of wireless transmitters, a matrix of receiver elements, and a central RF base station. The wireless transmitters, called bats, can be carried by a tagged object and/or attached to equipment. The sensor system measures the time of flight of the ultrasonic pulses emitted from a bat to receivers installed in known and fixed positions. It uses the time difference to estimate the position of each bat by trilateration.

The RF base station coordinates the activity of bats by periodically broadcasting messages to them. Upon hearing a message, a bat sends out an ultrasonic pulse. A receiver that receives the initial RF signal from the base station determines the time interval between receipt of the RF signal and receipt of the corresponding ultrasonic signal. It then estimates its distance from the bat. These distances are sent to the computer, which performs data analysis. By collecting enough distance readings, it can determine the location of the bat within 3 cm of error in a three-dimensional space at 95% accuracy. This accuracy is quite enough for most location-aware services; however, the deployment cost is high.

11.4.2 Cricket

Cricket is a system that can provide location-dependent applications [12]. Rather than explicitly tracking users' locations, Cricket helps devices learn their locations and lets them decide whether to advertise them, for preservation of privacy. Cricket does not rely on any centralized management or control and no explicit coordination occurs between beacons. To obtain information about a space, every object is attached to a *listener*, a small device that listens to messages from beacons mounted on ceilings and walls.

Similar to the Bat system, Cricket uses a combination of an RF signal and ultrasound to evaluate the distances between beacons and listeners (i.e., TDoA). A beacon sends the space information over an RF and an ultrasonic pulse at the same time. When the listener hears the RF signal, it uses the first few bits as training information and then turns on its ultrasonic receiver. It then listens for the ultrasonic pulse, which will usually arrive in a short time. The listener uses the time difference between the receipt of the first bit of RF information and the ultrasonic signal to determine its distance from the beacon.

11.4.3 RADAR and Nibble

The *RADAR* location system [11] tries to take advantage of the already existing RF data network formed by IEEE 802.11 access points. IEEE 802.11 networks are now becoming more prevalent in many office and public areas, so no extra hardware cost is incurred. In addition, users can enjoy data communications. RADAR uses the nearest neighbor technology of pattern matching discussed in Section 11.3 to infer objects' locations.

The *Nibble* [4] also adopts the IEEE 802.11 infrastructure for positioning purposes. Nibble uses the probability-based approach in Subsection 11.3.3.2. It relies on a fusion service to infer the location of an object from measured signal strengths. Data are characterized probabilistically and input into the fusion service. The output of the fusion service is a probability distribution over a random variable that represents some context.

11.4.4 CSIE/NCTU Indoor Tour Guide

The authors have also developed a prototype indoor tour guide system at the Department of Computer Science and Information Engineering, National Chiao Tung University (CSIE/NCTU), Taiwan. The hardware platforms of this project include several Compaq iPAQ PDAs and laptops. Each mobile station is equipped with a Lucent Orinoco Gold wireless card. Signal strengths are used for indoor positioning. The probability-based pattern-matching algorithm in Subsection 11.3.3.2 is used. Figure 11.10 shows the system architecture. The concept of logical areas is used to identify offices, rooms, lobbies, etc. The manager is the control center responsible for monitoring each user's movements, configuring the system, and planning logical areas and events. The location server takes care of the location discovery job and the service server is in charge of message delivery. The database can record users' profiles; the gateway can conduct location-based access control to the Internet.

One of the innovations in this project is that an event-driven messaging system has been designed. A short message can be delivered to a user when he enters or leaves a logical area. The event-driven message can also be triggered by a combination of time, location, and property of location (such as who is in the location and when the location is reserved for meetings). A user can set up a message and a corresponding event to trigger the delivery of the message. The manager will check the event list periodically and initiate messages, when necessary, with the service server. Messages can be unicast or broadcast. The expectation is that streaming multimedia can be delivered in the next stage. The system can also be applied to support a smart library. Another innovation is to provide location-based access control. In certain rooms, such as classrooms and meeting rooms, users may be prohibited from accessing certain sensitive Web pages. These rules can be organized through the manager and set up at the gateway.

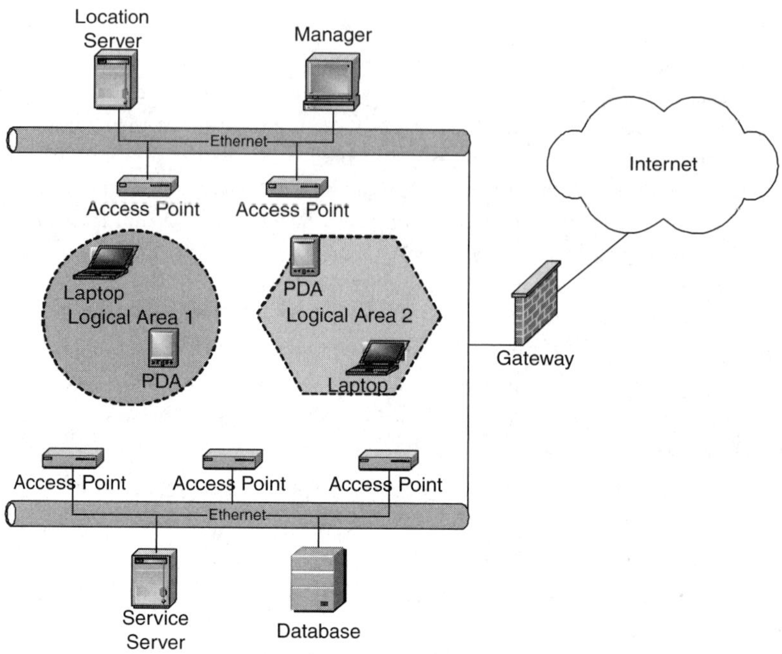

FIGURE 11.10 System architecture of the CSIE/NCTU tour guide system.

11.5 Conclusions

In this chapter, some fundamental techniques in positioning and location tracking have been discussed and several experimental systems reviewed. Location information may enable new types of services. Accuracy and deployment costs are two factors that may contradict each other, but both are important factors for the success of location-based services.

References

1. A. Savvides, C.C. Han, and M.B. Srivastava. Dynamic fine-grained localization in ad hoc networks of sensors. In *ACM/IEEE MOBICOM*, Rome, 2001, pp. 166–179.
2. M. Addlesee, R. Durwen, S. Hodges, J. Newman, P. Steggles, A. Ward, and A. Hopper. Implementing a sentient computing system. *Computer*, 34(8), 50–56, 2001.
3. P. Bahl, A. Balachandran, and V. Padmanabhan. Enhancements to the RADAR user location and tracking system. Technical report MSR-TR-00-12, Microsoft Research, 2000.
4. P. Castro, P. Chiu, T. Kremenek, and R.R. Muntz. A probabilistic room location service for wireless networked environments. In *Ubicomp*, 2001 pp. 18–34.
5. P. Enge and P. Misra. Special issue on GPS: the global positioning system. *Proc. IEEE*, 87, 3–15, 1999.
6. Y. Ko and N. Vaidya. Geocasting in mobile ad hoc networks: location-based multicast algorithms. In *IEEE Workshop Mobile Computing Syst. Applications (WMCSA)*, 101–110, 1999.
7. Y. Ko and N. Vaidya. GeoTORA: a protocol for geocasting in mobile ad hoc networks. In *8th Int. Conf. Network Protocols (ICNP)*, 2000, pp. 240–250.
8. W.-H. Liao, Y.-C. Tseng, K.-L. Lo, and J.-P. Sheu. GeoGRID: a geocasting protocol for mobile ad hoc networks based on GRID. *J. Internet Technol.*, 1(2), 23–32, 2000.
9. J.C. Navas and T. Imielinski. GeoCast — geographic addressing and routing. In *ACM/IEEE MOBICOM*, 66–76, 1997.
10. D. Niculescu and B. Nath. Ad hoc positioning system (APS) using AoA. In *IEEE INFOCOM*, San Francisco, 2003, pp. 1734–1743.

11. P. Bahl and V.N. Padmanabhan. RADAR: an in-building RF-based user location and tracking system. In *IEEE INFOCOM*, 2, 775–784, 2000.
12. N.B. Priyantha, A. Chakraborty, and H. Balakrishnan. The Cricket location-support system. In *ACM/IEEE MOBICOM*, 32–43, 2000.
13. N.B. Priyantha, A.K.L. Miu, H. Balakrishnan, and S.J. Teller. The Cricket compass for context-aware mobile applications. In *ACM/IEEE MOBICOM*, 1–14, 2001.
14. T.S. Rappaport. *Wireless Communications. Principles and Practice*, IEEE Press, 1996.
15. T. Roos, P. Myllymaki, H. Tirri, P. Misikangas, and J. Sievanen. A probabilistic approach to WLAN user location estimation. *Int. J. Wireless Inf. Networks*, 9(3), 2002.
16. S. Slijepcevic, S. Megerian, and M. Potkonjak. Characterization of location error in wireless sensor networks: analysis and applications. In *Int. Workshop Inf. Process. Sensor Networks* (IPSN), 2634, 593–608, 2003.
17. Y.-C. Tseng, S.-P. Kuo, H.-W. Lee, and C.-F. Huang. Location tracking in a wireless sensor network by mobile agents and its data fusion strategies. In *Int. Workshop Inf. Process. Sensor Networks (IPSN)*, 2634, 625–641, 2003.
18. Want, A. Hopper, V. Falcao, and J. Gibbons. The active badge location system. Technical report 92.1, Olivetti Research Ltd. (ORL), 1992.
19. M. Youssef, A. Agrawala, and U. Shankar. WLAN location determination via clustering and probability distributions. In *IEEE PerCom*, 2003.

12

Comparison of Data Processing Techniques in Sensor Networks

Vicente González-Millán
University of Valencia

Enrique Sanchis-Peris
University of Valencia

12.1 Sensor Networks: Organization and Processing 12-1
Evolution of Sensor Systems • Sensor Processing Systems
12.2 Architectures for Sensor Integration 12-3
Problems with High Data Rate • Introduction of Preprocessing Elements
12.3 Example of Architecture Evaluation in High-Energy Physics 12-18

12.1 Sensor Networks: Organization and Processing

In the scientific study of the natural phenomena that surround us, one of the first tasks to carry out consists of detailed analysis of the physical variables of the phenomenon to obtain the maximum information about it. To perform this analysis, capable sensors are used to measure the physical variables and to transform that measure into useful information for the study. A sensor is a device made to respond to a physical variable in a predictable form. Sensors can be mechanical, electric, electromechanical, electronic, magnetic, electromagnetic, or optic, to name some. The so-called sensor transfer function assures a well-known relationship between the physical variable and the sensor output.

Sensors can be of very varied form, even those that measure the same variable. However, any sensor can be studied under two aspects: physical and functional. The physical aspect refers to how the sensor is made or to what its form is. The term *physical sensor* refers to devices that sense the physical variable of interest, for example, a barometer, radar, a thermometer, etc. The functional aspect refers to what the sensor is supposed to do or which is its abstraction. The term *abstract* or *logical sensor* is used to refer to an abstraction of the reading taken by a particular sensor. Different possible abstractions exist. The reading of a sensor can be denoted as a simple number or as an interval in the real numbers set. In most cases, sensors are always associated with a transducer element that converts the sensor variations into useful electric signal.

Because the existence of a phenomenon implies a variation in some or all of the parameters associated with it, the electric signal obtained will present a certain variation with time, directly related with the variation of the measured magnitude. Therefore, in the study of the phenomenon a change takes place from an $(n + 1)$-dimensional space of physical magnitudes (n magnitudes and time) to n two-dimensional spaces of electric magnitudes (amplitudes and times), each corresponding to one of the measured physical magnitudes.

The advantage gained with this transformation is that, with electrical signals, an entire series of tools and technologies allow us their analysis and treatment, something that is not always possible directly on the physical magnitudes of the phenomenon. To treat the information obtained by the sensors, a processing system is needed. This system must operate appropriately with the data of each sensor, interpreting them and obtaining the desired result. This system will be more or less complex, depending on the number of sensors used and whether they are the same or different types, which in turn depends on the phenomenon studied.

12.1.1 Evolution of Sensor Systems

The evolution of the sensor systems can be described in five stages, each represented by one different type of sensor, although a new stage does not imply the disappearance of the sensors used in the previous one [1].

- *Single-sensor systems.* An example of a single sensor system is a radar system. This equipment sends a radio signal of a determined frequency and in a given direction, and receives the signal reflected back on the objects that are on the beam way. From the time difference between the emitted and received pulses, the system calculates the object distance. Another example of a single-sensor system is the sonar used to locate objects underwater. Because of the technology that was available, single-sensor systems were used before the microelectronic era. With only one sensor, system setup and data analysis were inexpensive and easy to perform. However, its simplicity was also a disadvantage because of the limited range of applications. For example, an autonomous mobile robot needs several sensors (tactile, cameras, CCD, etc.) and therefore cannot be built as a single-sensor system. Another drawback of this type of system is its robustness and the impossibility of using it in mission-critical applications. The third disadvantage comes from the fact that a single-sensor system cannot guarantee that the reading is always correct.
- *Replied sensors.* A solution to this third drawback is the use of several sensors, each one giving a reading on the phenomenon of interest. This strategy allows for validation of the reading using different techniques such as majority voting, average, or weighted average.
- *Different sensors.* When it comes to study of a complex phenomenon, it may be necessary to gather different types of information from it. For this purpose, we may use different kinds of sensors that will get different aspects of the phenomenon. For example, an autonomous mobile robot is equipped with different sensors needed to obtain a complete apprehension of the environment. The main advantages of the integration of different sensors are [2]:
  - Reliability increase
  - Improved fault tolerance
  - Improved detection and noise reduction

  This last advantage is explained if we realize that, observing the same signal of interest, the noise picked up by different sensors tends to be uncorrelated.
- *Spatially distributed sensors.* Some applications require that observations of an object are taken simultaneously from two or more points in space. Several degrees exist in the spreading of sensors: from a limited surface area to a region or even an entire country. The type of sensor used can be any of those previously seen, even a combination of them. The peculiarity of these systems is that, now, the information varies spatially and temporarily, so the processing system will be more complex.
- *Intelligent sensors.* If a high number of sensors, replied or different, are used, the volume of information to process may grow to a point at which the problem has a difficult solution. A possibility in this case is to use intelligent sensors in the measure of the physical variables. An intelligent sensor includes certain logical circuitry to abstract information with a bigger semantic content than the one obtained with the electrical signal of the physical variable. For example, a system to detect the passing of people in an enclosure can only offer an electric pulse when a person gets in, or have the necessary logic to offer a representative numeric value of the number of people that have passed through. In this case, the sensor becomes intelligent, offering more elaborate information than the purely electric one. The abstraction of the information may come with a reduction of the information that affects the design of the processing system, reducing the computational load and the necessary bandwidth.

12.1.2 Sensor Processing Systems

Sensor processing is a crucial issue for sensor systems. For its own nature, it requires knowledge of fields like physics, electronics, and computer science. No matter how they are implemented, each sensor processing system consists of four activities: acquisition, processing, integration, and analysis. For particular systems, some of the activities may be lightly or not implemented. Single-sensor systems do not need an integration phase, whereas for a replied sensor system, processing could be minimum, but integration is crucial. In different sensor systems, processing is important to make all readings compatible for the integration phase.

However, most of the sensor processing systems will include the four activities. The physical variable will be sensed in the acquisition activity and the data obtained will be appropriately processed (for example, scaled or formatted) before passing to the integration activity. The output of this activity goes to the analysis phase, where a decision is made. The mechanism of obtaining the decision can be deterministic, stochastic, or empiric. Several options exist to organize the sensor processing system, depending on the characteristics of the problem. The main types of sensor processing systems are [1]:

- Sensor collection — referred to a group of sensors set up in series, parallel, or mixed mode in which integration takes place progressively through the different sensors.
- Hierarchical systems — applied in cases in which the data volume is high so data sent to a central processor may require high bandwidth. A hierarchical distribution may help to reduce bandwidth and increase semantic contents of data as they go down the hierarchy. An important aspect of this organization is that its size does not grow linearly with the problem.
- Tree systems —organized like trees, with sensors in the different levels of the tree. The leaf nodes basically are sensors while the intermediate and root nodes carry out the local processing of data coming from the leaf nodes and on the data read by the sensors connected to them. In this way, at the top, the root node makes the decision with the data processed. The difference between this and hierarchical systems is that, in the latter, the entirety of the sensors is processed in the first level, while in the tree system sensors are progressively integrated.
- Multisensor integration — sensors are of different types and the integration is made at multiple levels, thus implying that the information from the different sensors must be processed to assure its compatibility.
- Distributed sensor processing — the four sensor processing activities take place in a distributed form. This means that not only is acquisition of data by the sensors geographically distributed, but also the processing, integration, and decision. This kind of system gives way to distributed sensor networks (DSN) in which multiple sensors of different types are geographically distributed. Examples of DSN are robotics, particle physics experiments, medical imaging, radar tracking, and flight navigation, to name a few. These systems and others of the same characteristics constitute the logical step in the evolution of sensor processing systems. The design and implementation of these types of networks would not have been possible without advances in technology, mainly in processors and communications.

12.2 Architectures for Sensor Integration

As mentioned earlier, in cases in which the volume of information to treat is large, the sensor processing system is organized in a hierarchical way and, generally, in three levels. The question now is how to compare the goodness of this solution to other types of architectures, for example a fully parallel one. In any case, the real implementation of the sensor processing system falls within one of the well-known Flynn classes for computer architecture [3]. This taxonomy divides the systems according to their number of instruction and data flow paths, dividing them between multiple and single paths. Figure 12.1 shows the Flynn taxonomy for computer architectures.

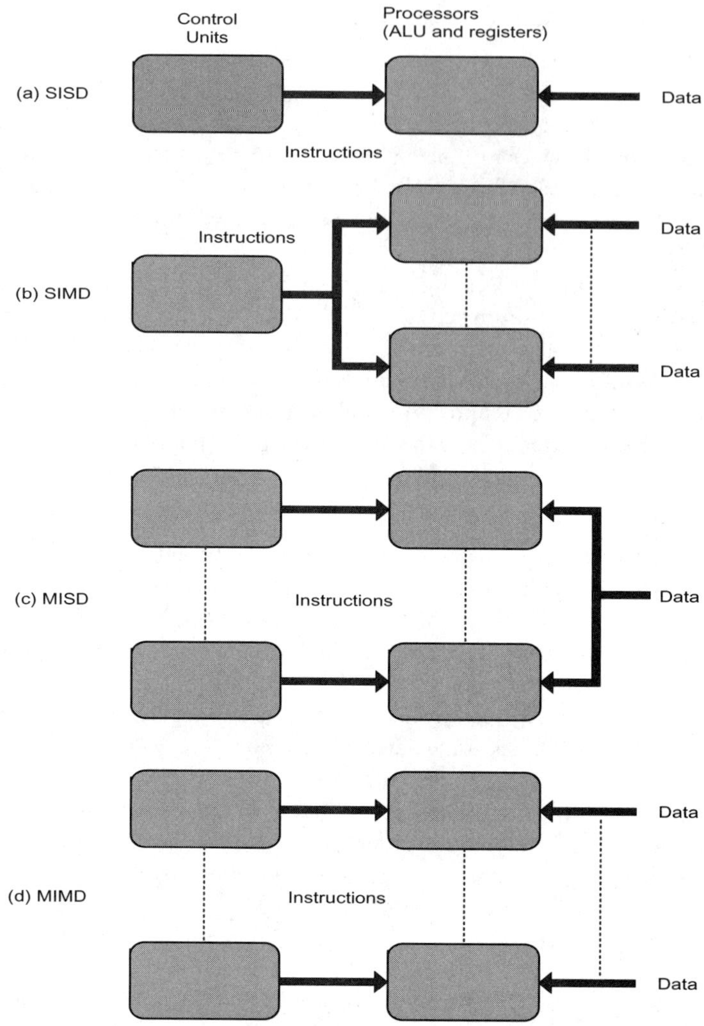

FIGURE 12.1 Flynn taxonomy for computer architectures.

- *SISD* (Figure 12.1a). This represents most of the available computers at present. The instructions are executed sequentially but they can be overlapped in the stages of execution (pipeline segmentation). An SISD computer can have more than a functional unit, but all are under the control of only a control unit.
- *SIMD* (Figure 12.1b). In this class one finds the matrix processors in which several units process different data, executing the same instructions, provided by one control unit. These systems are further classified in local or global memory SIMD according to their memory organization, particularly depending on whether the memory access is local to the processor or remote through an interconnection network.
- *MISD* (Figure 12.1c). This type of organization is characterized by the existence of several processors, each one executing a different instruction but on the same data flow. In this case, the output of a processor is the input of the following one.
- *MIMD* (Figure 12.1d). This category includes most of the multiprocessor and multicomputer systems. An MIMD computer implies interactions between several processors because all the data flows are derived from the same data space shared by all. If the data flows come from disjoint subspaces inside the shared memory, then one would have a multiple SISD system, or MSISD, that is really a group of independent SISD systems.

12.2.1 Problems with High Data Rate

Due to improvements in technology, it is possible currently to treat large quantities of data with a reasonable cost. High-resolution image processing is now possible, even in real time, thanks to increased bandwidth and processing power of the CPUs employed; video is in the same situation. In the field of high-energy physics, trace detectors are used to observe the trajectories of the particles; the lower the processing power is, the larger is the resolution in the determination of the traces, which implies an error in identifying the particle. Because of the increased computing power of processors, it is now possible to think about the construction of systems with better resolutions.

High-resolution image processing and trace detectors are representative examples of areas of investigation in which the volume of information provided by the sensors is high; however, they are not the only ones because one can also find these problems in areas like the robotics, aerospace control, or meteorological prediction. Traditionally, in cases in which the volume of information from the sensors is high, hierarchical architectures have been used for the processing. These architectures are implemented in three or more levels, each with an MIMD structure, so the global system can be viewed as an interconnected cluster of MIMD systems.

However, the option of a hierarchical architecture may not be always the best. Indeed, one can consider other solutions different from the hierarchical system, like a full parallel MIMD system. The best choice for the architecture may not be an easy one to make because it depends on factors such as the problem itself, available technology, reliability, complexity and performance of the solution, and, inevitably, budget. The analysis that follows will be centered in system performance, defining a merit factor that would allow comparison of the systems.

12.2.1.1 Merit Factor of a System

Generally, systems to compare can be implemented in completely different ways, so one needs a parameter independent of the particular implementation; on the other hand, it should somehow indicate which of the two systems will be more complex, difficult, and expensive to implement. Therefore, the merit factor (MF) is defined as the product of the bandwidth, BW, times the processing power, PC, needed to be able to solve a certain problem. That is to say,

$$MF(Mbytes/s \cdot MIPS) = BW_r \cdot PC_r \tag{12.1}$$

To be able to obtain the necessary expressions for the parallel and hierarchical systems, it is necessary to know the MF value of the association of a certain number of processors, each with its specific MF in serial or in parallel.

12.2.1.1.1 Merit Factor in a Parallel System

In this case, assume a number N of parallel connected processors, each with a certain MF_i value. To obtain the equivalent MF, evaluate separately the total bandwidth and computing power of the parallel system. Evidently, the bandwidth of the system is the sum of the individual bandwidths, BW_i, of each one of the processors. That is to say:

$$BW_p = \sum_{i=1}^{N} BW_i \tag{12.2}$$

The computing power is also the sum of the capacities, PC_i, of each one of the processors:

$$PC_p = \sum_{i=1}^{N} PC_i \tag{12.3}$$

Therefore, the merit factor of a parallel system in function of the MF of each processor is:

$$MF_p = BW_p \cdot PC_p = \sum_{i=1}^{N} BW_i \cdot \sum_{i=1}^{N} PC_i = BW_p \cdot \sum_{i=1}^{N} \frac{MF_i}{BW_i} = PC_p \cdot \sum_{i=1}^{N} \frac{MF_i}{PC_i} \tag{12.4}$$

12.2.1.1.2 *Merit Factor in a Series System*

Suppose now a case in which the processors are connected sequentially, i.e., the data output of one processor is the data input of the next one. In this case, for N processors, each one with its bandwidth, BW_i, and computing power, PC_i, the equivalent system with just a single processor will have the bandwidth of the first one of the series:

$$BW_s = BW_1 \tag{12.5}$$

On the other hand, the equivalent computing power will be determined by the total processing time, t_N, and the number of operations to carry out in this time. Therefore,

$$PC_s = \frac{D \cdot \sum_{i=1}^{N} op_i}{\sum_{j=1}^{N} t_j} = \sum_{i=1}^{N} PC_i \cdot \frac{t_i}{t_N} \tag{12.6}$$

where D is the number of data and op_i is the number of operations carried out by processor i. From the two expressions, the required quality to the equivalent system is:

$$MF_S = BW_S \cdot PC_S = BW_1 \cdot \sum_{i=1}^{N} PC_i \cdot \frac{t_i}{t_N} = \frac{BW_1}{t_N} \cdot \sum_{i=1}^{N} PC_i \cdot t_i \cdot \frac{BW_i}{BW_i} = \frac{BW_1}{t_N} \cdot \sum_{i=1}^{N} MF_i \cdot \frac{t_i}{BW_i} \tag{12.7}$$

12.2.1.2 Parameterization of Parallel and Hierarchical Architectures

A certain processing problem can be parameterized, indicating the required total bandwidth, BW, to read the data and the processing power necessary, PC, for their computation. Solving the problem by means of a parallel architecture as that of Figure 12.2 yields the following expressions:

$$\begin{aligned} BW_i &= \frac{BW}{N} \quad 1 \le i \le N \\ CP_i &= \frac{CP}{N} \quad 1 \le i \le N \end{aligned} \tag{12.8}$$

where N is the number of processors of the system.

However, it is common to have processing units (processors) with some particular processing power, $PC_i = PC_{pu}$, and bandwidth, $AB_i = AB_{pu}$. Therefore, the parameter to determine will be the number of units with those characteristics necessary to implement the system. This number, N, is:

$$N = \max\left(\frac{BW}{BW_i}, \frac{PC}{PC_i}\right) \tag{12.9}$$

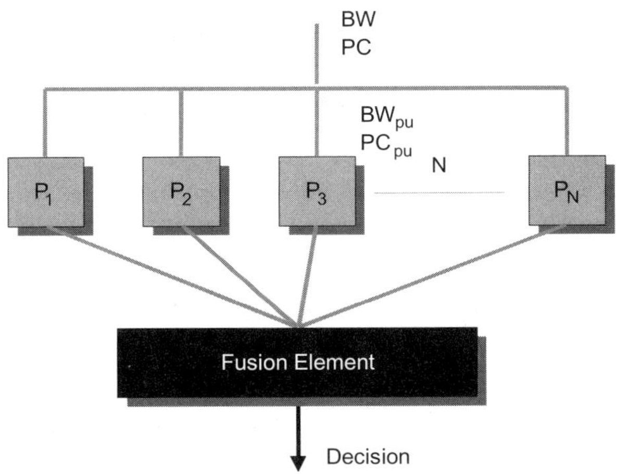

FIGURE 12.2 Parallel processing system. (From Gonzalez, V. et al., *IEEE Trans. Nucl. Sci.*, 49, 2002. With permission.)

The fusion element (fuser) picks up the decisions of the N processing elements and elaborates a final one. The ways of getting this decision are very varied. For this analysis, suppose that the system offers a yes/no binary decision elaborated performing the logical *and* of all the decisions.

The required bandwidth in the fuser depends on the size of the partial decisions of the N processing elements and on the time, T, to make the final decision. The processing power depends roughly on the number of *and* operations necessary to carry out the decision and of the time T necessary to carry them out. Using two input *and* operations, the number of necessary operations to obtain the result with N inputs is $N - 1$.

In this case, suppose that the system works in a pipeline way: first the system gets the N partial decisions in T seconds; after this phase, the fuser final decision is obtained in the same T time, on time to receive the next N partial ones. In this way, final and partial decisions are overlapped in time.

If S_{dec} is the size in bytes of the partial decision; S_{dat} the number of bytes for each datum coming from the sensors; ξ the relationship between the size of the partial decision and the size of the data from each sensor; D the number of data; and op the number of operations per datum; the bandwidth and processing power of the fuser can be expressed as:

$$BW_f = N \cdot \frac{S_{dec}}{T} = \left(\begin{array}{l} S_{dec} = \xi \cdot S_{dat} \\ BW = \dfrac{D \cdot S_{dat}}{T} \end{array} \right) = N \cdot \frac{\xi}{D} \cdot BW = N \cdot K_1 \cdot BW$$

$$PC_f = \frac{N-1}{T} = \left(PC = \frac{D \cdot op}{T} \right) = (N-1) \cdot \frac{1}{D \cdot op} \cdot PC = (N-1) \cdot K_2 \cdot PC \tag{12.10}$$

If the resolution of the problem is outlined by means of the employment of a hierarchical architecture such as the one shown in Figure 12.3, it will be necessary to introduce a new parameter to be able to obtain the expressions for BW_i and PC_i that will be a function of the level in the architecture. This parameter is the reduction factor in the data volume due to the extraction of the information from the received raw data from the sensors, in the case of a measurement system, or the proportion of data discarded by not completing certain requirements if it is a detection system.

It is necessary to notice that at each level, the bandwidth, BW_i, is reduced in a factor similar to the reduction due to the extraction of information or to the elimination of data not interesting (a factor p). In this case, suppose that, from a level to the next level, all data (because a "yes" decision was taken) or no data ("no" decision) pass. This decision has a probability p, so the time for the information arrival

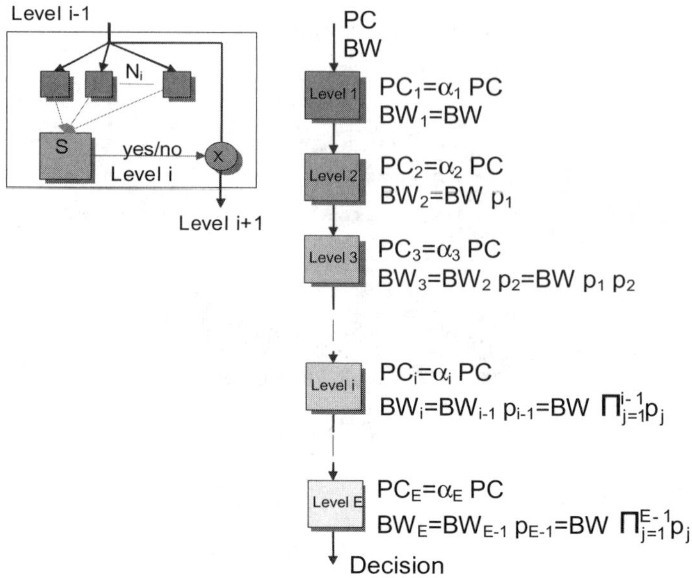

FIGURE 12.3 Hierarchical processing system. (From Gonzalez, V. et al., *IEEE Trans. Nucl. Sci.*, 49, 2002. With permission.)

between levels is increased by the inverse of that probability. Because all the data are sent, the effect is more time to send the same quantity of data, which implies a reduction of the bandwidth.

Assume that the processing power at each level i can be expressed as a proportion, α_i, of the total processing capacity PC. Each one of the levels in the hierarchical system is thought of as a parallel system with a fuser; the result of the decision makes the entirety of the data pass or not toward the next level.

12.2.1.3 Evaluation of Merit Factor for Parallel and Hierarchical Systems

With the considerations of the previous epigraph, one can now evaluate the MF for the parallel and hierarchical systems.

12.2.1.3.1 Parallel System

The MF for the parallel system has two terms. The first one, MF_{proc}, depends on the used processors while the second, MF_{fus}, is due to the fuser. According to Figure 12.2, the equivalent of the parallel system is an association of N parallel processors in series with a fuser.

Suppose that all the processors have the same characteristics; applying Equation 12.1 yields:

$$MF_{proc} = BW_{proc} \cdot PC_{proc} = \sum_{i=1}^{N} BW_i \cdot \sum_{i=1}^{N} PC_i = N \cdot BW_{pu} \cdot N \cdot PC_{pu} = N^2 \cdot MF_{pu} \tag{12.11}$$

where N is the number of units and MF_{pu} it is the MF of each processing unit.

The second term, related to the fuser, is:

$$MF_{fus} = BW_{fus} \cdot PC_{fus} = N \cdot (N-1) \cdot K_1 \cdot K_2 \cdot BW \cdot PC = N \cdot (N-1) \cdot K \cdot BW \cdot PC \tag{12.12}$$

The MF of the parallel system will be the series of both calculated, that is to say:

$$MF_{Paral} = \frac{MF_{proc}}{t+t} \cdot \sum_{i=1}^{2} MF_i \cdot \frac{t}{BW_i} = \frac{BW_{proc}}{2} \cdot \left[\frac{N^2 \cdot BW_{pu} \cdot PC_{pu}}{BW_{proc}} + \frac{N \cdot (N-1) \cdot K \cdot BW \cdot PC}{N \cdot K_1 \cdot BW} \right] =$$

$$= \frac{N \cdot b \cdot BW}{2} \cdot \left[\frac{N^2 \cdot b \cdot BW \cdot a \cdot PC}{N \cdot b \cdot BW} + \frac{N \cdot (N-1) \cdot K_1 \cdot K_2 \cdot BW \cdot PC}{N \cdot K_1 \cdot BW} \right] =$$

$$= BW \cdot PC \cdot \frac{N \cdot b}{2} \cdot \left[N \cdot a + (N-1) \cdot K_2 \right] \tag{12.13}$$

where a and b are the relationships between the processing capacity, PC_{pu}, and the bandwidth, BW_{pu}, respectively, of each processor and the total of the problem.

12.2.1.3.2 Hierarchical System

The hierarchical system is not more than a series of E sequential connected parallel systems. In each level, the system is implemented with N processors and a fuser. Therefore, the MF, MF_{Hier}, will be:

$$MF_{Hier} = \frac{BW_1}{\sum_{i=1}^{E} t_i} \sum_{i=1}^{E} MF_{Paral_i} \cdot \frac{t_i}{BW_i} \tag{12.14}$$

where MF_{Paral_i} is the MF of each level that is expressed as:

$$MF_{P_i} = \frac{BW_i}{t_i + t_i} \cdot \sum_{j=1}^{2} MF_{ij} \cdot \frac{t_j}{BW_j} = \frac{BW_{proc_i}}{2} \cdot \left(\frac{MF_{proc_i}}{BW_{proc_i}} + \frac{MF_{fus_i}}{BW_{fus_i}} \right) \tag{12.15}$$

assuming, as it was stated in the parallel association, that $t_j = t_i \ \forall i,j$ (that is, it takes the fuser the same time to get a decision as it takes the processors on the level to get theirs).

The MF for the processors on the level is the parallel association of N of them, that is:

$$MF_{proc_i} = N_i^2 \cdot a_i \cdot b_i \cdot BW \cdot PC \tag{12.16}$$

where a_i and b_i are defined the same way as in the parallel system, and N is defined as:

$$N_i = \max\left(\frac{BW_i}{BW_{pu_i}}, \frac{PC_i}{PC_{pu_i}} \right) = \max\left(\frac{BW \cdot \prod_{j=0}^{i-1} p_j}{BW_{pu_i}}, \frac{\alpha_i \cdot PC}{PC_{pu_i}} \right) \tag{12.17}$$

Looking at Figure 12.3

$$\begin{aligned} P_i &= \prod_{j=1}^{i-1} p_j \quad i \geq 2 \\ P_1 &= 1 \end{aligned} \tag{12.18}$$

one can calculate the MF of the fuser. For this, first calculate the bandwidth of the fuser for level *I*, which turns out to be:

$$\begin{aligned} BW_{fus_i} &= N \cdot \frac{S_{dec}}{t_i} = (t_i = T/P_i) = N \cdot \frac{S_{dec}}{T} \cdot P_i = \begin{bmatrix} S_{dec} = \xi \cdot S_{dat} \\ BW = \dfrac{D \cdot S_{dat}}{T} \end{bmatrix} = \\ &= N \cdot \frac{\xi}{D} \cdot P_i \cdot BW = N \cdot K_1 \cdot P_i \cdot BW \end{aligned} \tag{12.19}$$

On the other hand, the processing power necessary in the fuser of level *i* can be expressed as:

$$\begin{aligned} PC_{fus_i} &= \frac{N-1}{t_i} = (t_i = T/P_i) = \frac{N-1}{T} \cdot P_i = \left[PC = \frac{D \cdot op}{T} \right] = (N-1) \cdot \frac{1}{D \cdot op} \cdot P_i \cdot PC = \\ &= (N-1) \cdot K_2 \cdot P_i \cdot PC \end{aligned} \tag{12.20}$$

Therefore, the MF of the fuser of level *i* will be:

$$MF_{fus_i} = BW_{fus_i} \cdot PC_{fus_i} = N \cdot (N-1) \cdot K_1 \cdot K_2 \cdot P_i^2 \cdot BW \cdot PC \tag{12.21}$$

Substituting in Equation 12.15 yields:

$$\begin{aligned} MF_{P_i} &= \frac{BW_{proc_i}}{2} \cdot \left(\frac{N_i^2 \cdot a_i \cdot b_i \cdot BW \cdot PC}{N_i \cdot b_i \cdot BW} + \frac{N_i \cdot (N_i - 1) \cdot K \cdot P_i^2 \cdot BW \cdot PC}{N_i \cdot K_1 \cdot P_i \cdot BW} \right) = \\ &= BW \cdot PC \cdot \frac{N_i \cdot b_i}{2} \cdot \left(N_i \cdot a_i + (N_i - 1) \cdot K_2 \cdot P_i \right) \end{aligned} \tag{12.22}$$

Therefore, the MF of the hierarchical system, MF_{Hier}, is:

$$
\begin{aligned}
MF_{Hier} &= \frac{BW_1 \cdot PC}{\sum_{j=1}^{E} t_j} \cdot \sum_{i=1}^{E} \frac{N_i \cdot b_i}{2} \cdot \left(N_i \cdot a_i + \left(N_i - 1\right) \cdot K_2 \cdot \mathrm{P}_i\right) \cdot \frac{t_i}{BW_i} = \\
&= \frac{BW^2 \cdot PC}{\sum_{j=1}^{E} t_j} \cdot N_1 \cdot b_1 \cdot \sum_{i=1}^{E} \frac{N_i \cdot b_i}{2} \cdot \left(N_i \cdot a_i + \left(N_i - 1\right) \cdot K_2 \cdot \mathrm{P}_i\right) \cdot \frac{t_i}{N_i \cdot b_i \cdot BW} = \\
&= \frac{BW \cdot PC}{\sum_{j=1}^{E} t_j} \cdot N_1 \cdot b_1 \cdot \sum_{i=1}^{E} \frac{1}{2} \cdot \left(N_i \cdot a_i + \left(N_i - 1\right) \cdot K_2 \cdot \mathrm{P}_i\right) \cdot t_i = \qquad (12.23) \\
&= \frac{BW \cdot PC}{\sum_{j=1}^{E} t_1 / \mathrm{P}_j} \cdot N_1 \cdot b_1 \cdot \sum_{i=1}^{E} \frac{1}{2} \cdot \left(N_i \cdot a_i + \left(N_i - 1\right) \cdot K_2 \cdot \mathrm{P}_i\right) \cdot \frac{t_1}{\mathrm{P}_i} \Rightarrow \\
\Rightarrow MF_{Hier} &= BW \cdot PC \cdot \frac{N_1 \cdot b_1}{2} \cdot \sum_{i=1}^{E} \frac{\left(N_i \cdot a_i + \left(N_i - 1\right) \cdot K_2 \cdot \mathrm{P}_i\right)}{\mathrm{P}_i \cdot \sum_{j=1}^{E} 1/\mathrm{P}_j}
\end{aligned}
$$

Evidently, Equation 12.23 becomes identical to Equation 12.13, corresponding to the parallel system, when the number of levels E is one.

The following step is to carry out the comparison between the expressions for the parallel and the hierarchical systems and to try to obtain an analytic expression that allows the decision of the better solution for a given problem. However, if the MF of the parallel case is taken as a reference, one would have one equation and 4E variables, which would determine infinite solutions.

A different approach to compare both systems would be the following:

1. The parallel solution is determined and its MF is calculated.
2. A hierarchical system is designed and its MF is calculated.
3. Both results are compared. If one is interested in a hierarchical solution and its MF is bigger than the one for the parallel solution, the hierarchical parameters (a, b, and P) can be adjusted and the process repeated until the MF is smaller.

However, it is possible to obtain an analytic expression if what is known is the value of the MF of the parallel system and the values of the parameters of $E - 1$ levels of the hierarchical system of E levels. In this case, imposing the condition that, for example, the MF in the hierarchical system is smaller than the one in the parallel system, the following expression is obtained to calculate the values of the parameters of the last level:

$$
\frac{N_E \cdot a_E + \left(N_E - 1\right) \cdot K_2 \cdot \mathrm{P}_E}{\mathrm{P}_E \cdot \sum_{j=1}^{E} 1/\mathrm{P}_E} < \frac{N \cdot b}{N_E \cdot b_E} \cdot \left[N \cdot a + \left(N - 1\right) \cdot K_2\right] - \sum_{i=1}^{E-1} \frac{N_i \cdot a_i + \left(N_i - 1\right) \cdot K_2 \cdot \mathrm{P}_i}{\mathrm{P}_i \cdot \sum_{j=1}^{E} 1/\mathrm{P}_E} \qquad (12.24)
$$

12.2.2 Introduction of Preprocessing Elements

Up to now, the hierarchical and parallel systems for sensor processing have been studied. This subsection examines an improvement on the hierarchical system based on the introduction of preprocessing elements in the levels that will improve the system performance due to the reduction of their processing load.

12.2.2.1 Regions of Interest

When the number of sensor channels to process is very high, the required processing power and bandwidth in the levels of the system can be too high. In the field of image processing, hierarchical processing systems are used in which successive levels carry out the processing with higher resolutions [4]. Also in this field, particularly in analysis of video images, because of the great correlation that exists between frames, a technique is used based on the location of areas (*region of interest*, RoI) of the image that have changed from one frame to the next. Processing then takes place only on those regions with the consequent reduction in the processing time.

The idea is to apply the concept of RoI to a hierarchical processing sensor system with a great volume of information to release the computational load in the levels where this is possible. The RoI, Ω, can be defined as a group of sensor channels of the system. The RoI can represent a one-dimensional, two-dimensional, or three-dimensional space of measure of the physical environment. In the most general case of a three-dimensional space, the RoI is expressed as:

$$\Omega = \begin{bmatrix} c_{111} & \cdots & c_{11n} \\ \vdots & \vdots & \vdots \\ c_{1m1} & \cdots & c_{1mn} \\ c_{211} & \cdots & c_{21n} \\ \vdots & \vdots & \vdots \\ c_{2m1} & \cdots & c_{2mn} \\ \vdots & \vdots & \vdots \\ c_{p11} & \cdots & c_{p1n} \\ \vdots & \vdots & \vdots \\ c_{pm1} & \cdots & c_{pmn} \end{bmatrix} \tag{12.25}$$

where n, m, and p are the number of channels in each one of the dimensions x, y, and z.

The size of the region of interest will be:

$$\dim(\Omega) = n \cdot m \cdot p \tag{12.26}$$

In each data acquisition, a certain number of RoIs will be identified; call $\overline{n_{RoI}}$ the average number of RoIs of each data acquisition. In this case, the levels in the architecture will only process the channels of these regions and make the decision with only their information. In general, the average number of channels to process will be smaller than the total. The fraction of channels to process related to the total is:

$$N_{RB_i} = \max\left(\frac{\Upsilon_i \cdot PC}{PC_{pu_{RB}}}, \frac{BW \cdot P_i}{BW_{pu_{RB}}}\right) \tag{12.27}$$

where N_{ch} is the total number of channels of the sensor system (normally the number of sensors).

When introducing RoIs, it is necessary to add the necessary modules for their calculation to the architecture of processing. These modules are called RoI builders, or RB, and will be placed between levels of the hierarchy. To analyze how they would affect the MF, the RB must be modeled with a bandwidth and processing power (Figure 12.4). The bandwidth of each RB module located between two

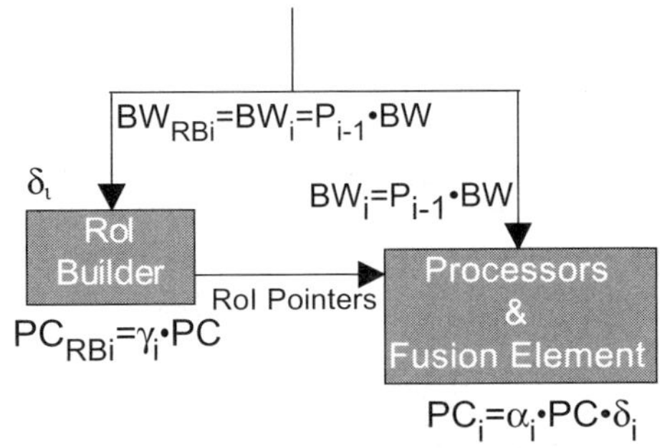

FIGURE 12.4 RoI builder placement in the hierarchical architecture. (From Gonzalez, V. et al., *IEEE Trans. Nucl. Sci.*, 49, 2002. With permission.)

levels is similar to that of the corresponding level, and its processing power can be expressed as a fraction γ_i of the total one of the problem.

Suppose that the RB in each level i is formed by a parallel system with units of bandwidth $BW_{pu_{RB}}$ and processing capacity $PC_{pu_{RB}}$, then the number of necessary units N_{RB_i} will be:

$$N_{RB_i} = \max\left(\frac{\gamma_i \cdot PC}{PC_{pu_{RB}}}, \frac{BW \cdot P_i}{BW_{pu_{RB}}}\right) \tag{12.28}$$

The RB MF will be expressed as:

$$\begin{aligned} MF_{RB_i} &= \sum_{j=1}^{N_{RB_i}} BW_{pu_{RB}} \cdot \sum_{j=1}^{N_{RB_i}} PC_{pu_{RB}} = \\ &= N_{RB_i}^2 \cdot BW_{pu_{RB}} \cdot PC_{pu_{RB}} = N_{RB_i}^2 \cdot a_{RB_i} \cdot b_{RB_i} \cdot BW \cdot PC \quad 1 \le i < E \end{aligned} \tag{12.29}$$

where, exactly as before, a_{RB_i} and b_{RB_i}, are the relationships between the processing power and the bandwidth of each RB unit and total of the problem. It is necessary to point out that there is no RB from level E to the $E + 1$, which explains the limits of index i in Equation 12.29.

The bandwidth of each level in the hierarchy is also modified because now it must also accept data relative to the RoI. However, this study rejects this contribution, supposing that the problem data rate is very high compared to the one due to this fact. Figure 12.5 shows the modified outline of the architecture with RoIs. Note that it is not always necessary to introduce RB between all levels of the hierarchy because it depends on the particular application.

Decreasing the number of channels to process will also decrease the processing power in the level of the hierarchy at which RoIs are used, although this does not apply to the bandwidth because each level should be capable of reading all the channels — not only those selected by the RB. The reduction factor is similar to the fraction δ, so the MF of that level is reduced by a factor $\delta < 1$, that is,

$$MF_i' = \delta \cdot MF \tag{12.30}$$

The change in the processing power may vary the number of units needed in each level, depending on which parameter (bandwidth or processing power) determined it. In general, when introducing RoI, the number of processing units, N'_i, is:

$$N_i' = máx\left(\frac{\delta_{i-1} \cdot PC_i}{PC_{pu_i}}, \frac{P_i \cdot BW}{BW_{pu_i}}\right) = máx\left(\frac{\delta_{i-1} \cdot \alpha_i \cdot PC}{PC_{pu_i}}, \frac{P_i \cdot BW}{BW_{pu_i}}\right) \tag{12.31}$$

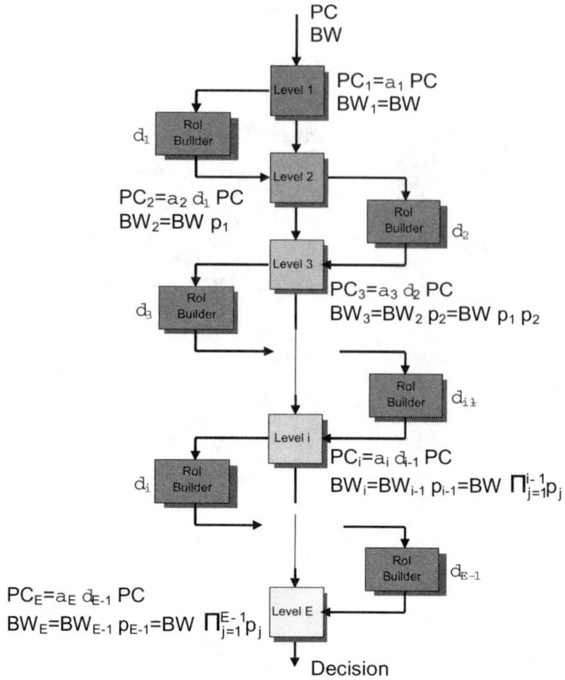

FIGURE 12.5 Hierarchical systems with RoIs. (From Gonzalez, V. et al., *IEEE Trans. Nucl. Sci.*, 49, 2002. With permission.)

for $i > 1$, with $\delta_0 = 1$ (because there is no RB from level 0 to level 1). In this calculation we supposed that the processing units used have the same characteristics as those used in the system without RoI.

The fuser MF is also modified by the use of RoI because it varies the number of processing units in the levels, although not the available time for the information processing and transmission.

Taking everything into account, the MF of each level of the hierarchy is modified and results in:

$$MF'_{P_i} = \frac{BW'_{proc_i}}{2 \cdot t_i + t_{RB_i}} \cdot \left(\begin{array}{l} \dfrac{N_{RB_i}^2 \cdot a_{RB_i} \cdot b_{RB_i} \cdot BW \cdot PC}{N_{RB_i} \cdot b_{RB_i} \cdot BW} \cdot t_{RB_i} + \\ + \dfrac{N_i'^2 \cdot a_i \cdot b_i \cdot BW \cdot PC}{N_i' \cdot b_i \cdot BW} \cdot t_i + \dfrac{N_i' \cdot (N_i' - 1) \cdot K \cdot \mathrm{P}_i^2 \cdot BW \cdot PC}{N_i' \cdot K_1 \cdot \mathrm{P}_i \cdot BW} \cdot t_i \end{array} \right) =$$

$$= BW \cdot PC \cdot \frac{N_i' \cdot b_i}{2 \cdot t_i + t_{RB_i}} \cdot \left(N_{RB_i} \cdot a_{RB_i} \cdot t_{RB_i} + N_i' \cdot a_i \cdot t_i + (N_i' - 1) \cdot K_2 \cdot \mathrm{P}_i \cdot t_i \right) = \left(K_{t_i} = t_{RB_i} / t_i \right)$$

$$= BW \cdot PC \cdot \frac{N_i' \cdot b_i}{2 + K_{t_i}} \cdot \left(N_{RB_i} \cdot a_{RB_i} \cdot K_{t_i} + N_i' \cdot a_i + (N_i' - 1) \cdot K_2 \cdot \mathrm{P}_i \right) \tag{12.32}$$

where K_{t_i} is the relationship between the processing time in the RB and the one of the level, and $K_{t_E} = 0$ because there is no RB between level E and $E + 1$. As in the case without RoIs, the processing and the fusion operate in a pipelined way.

The MF of the hierarchical system with RoIs is obtained from Equation 12.32:

$$MF_{H_{RoI}} = BW \cdot PC \cdot N_1 \cdot b_1 \cdot \sum_{i=1}^{E} \frac{\left(N'_i \cdot a_i + \left(N'_i - 1\right) \cdot K_2 \cdot \mathrm{P}_i + N_{RB_i} \cdot a_{RB_i} \cdot K_{t_i}\right)}{\left(2 + K_{t_i}\right) \cdot \mathrm{P}_i \cdot \sum_{i=1}^{E} 1/\mathrm{P}_i} \tag{12.33}$$

The MF of the hierarchical systems with and without RoIs can now be compared to decide when the system with RoIs has an MF smaller than the one of the system without RoIs. Comparing Equation 12.23 with Equation 12.33 for a certain level j, as a sufficient condition for the system with RoI to have an MF lower than the one without RoI, yields:

$$MF_{H_{RoI}} < MF_{Hier} \Rightarrow$$

$$\Rightarrow \frac{N'_j \cdot a_j + N_{RB_j} \cdot a_{RB_j} \cdot K_{t_j} + (N'_j - 1) \cdot K_2 \cdot \mathrm{P}_j}{\left(2 + K_{t_j}\right) \cdot \mathrm{P}_j \cdot \sum_{s=1}^{E} 1/\mathrm{P}_s} < \frac{N_j \cdot a_j + (N_j - 1) \cdot K_2 \cdot \mathrm{P}_j}{2 \cdot \mathrm{P}_j \cdot \sum_{s=1}^{E} 1/\mathrm{P}_s} \tag{12.34}$$

If $K_{t_j} \to 0$, i.e., the RB_j uses very little time in identifying the channels of the RoI compared to the processing time of the corresponding level, then the previous expression can be simplified, yielding:

$$N'_i \cdot a_j + \left(N'_i - 1\right) \cdot K_2 \cdot \mathrm{P}_j < N_j \cdot a_j + \left(N_j - 1\right) \cdot K_2 \cdot \mathrm{P}_j \tag{12.35}$$

Reordering,

$$\left(N_j - N'_j\right) \cdot \left(a_j + K_2 \cdot \mathrm{P}_j\right) > 0 \tag{12.36}$$

This condition is always true because a_j, K_2, and P_j are positive defined and $N_j > N'_j$ when using RoIs as the number of channels to process decreases. Therefore, in this case, the system with RoIs will have an MF always smaller than the one without regions.

In a case where $K_{t_j} \not\to 0$, one can obtain K_{t_j}:

$$K_{t_j} \lesseqgtr \frac{2 \cdot \left(N_j - N'_j\right) \cdot \left(a_j + K_2 \cdot \mathrm{P}_j\right)}{2 \cdot N_{RB_j} \cdot a_{RB_j} - \left(N_j \cdot a_j + \left(N_j - 1\right) \cdot K_2 \cdot \mathrm{P}_j\right)} \tag{12.37}$$

where the direction of the inequality (smaller than or bigger than) depends on the sign (positive or negative) of the denominator. If the denominator is negative, the condition of bigger than is always true because, as $N_j > N'_j$, the expression on the right of the inequality will be negative and, by definition, $K_{t_j} > 0$, assuring that the system with RoIs will have a smaller MF than the system without RoIs.

The denominator can now be evaluated to see when it is positive or negative:

$$N_{RB_j} \cdot a_{RB_j} \lesseqgtr \frac{N_j \cdot a_j + \left(N_j - 1\right) \cdot K_2 \cdot \mathrm{P}_j}{2} \tag{12.38}$$

The expression on the left of the inequality is the relationship between the total processing capacity of the RB and the total one of the problem. The right of the expression is the half-sum of the relative processing power of the processors of the level and of the fuser referred to the total one.

Therefore, if the relative processing power of the RB is smaller than half of that of the level (considering the processors and the fuser), the denominator will be negative and it will be true, for the reason mentioned before, that the system with regions of interest has an MF smaller than the one for the system with RoIs. If, on the contrary, the relative processing power of the RB is bigger than half of that of the level, it will depend on the value of K_{t_j} whether the MF of the hierarchical system with RoIs is smaller or bigger than the one without RoIs.

If the number of processing units of the RB is limited by their processing power, Equation 12.38 transforms to offer the value of γ, the ratio of the RB processing capacity and the total one of the problem:

$$\gamma_j \begin{smallmatrix} < \\ > \end{smallmatrix} \frac{N_j \cdot a_j + (N_j - 1) \cdot K_2 \cdot \mathrm{P}_j}{2} \tag{12.39}$$

If the number of units of the RB is limited by the bandwidth, the relationship between the relative processing power and the relative bandwidth of each unit of the RB is obtained:

$$\frac{a_{RB_j}}{b_{RB_j}} \begin{smallmatrix} < \\ > \end{smallmatrix} \frac{1}{\mathrm{P}_j} \cdot \frac{N_j \cdot a_j + (N_j - 1) \cdot K_2 \cdot \mathrm{P}_j}{2} \tag{12.40}$$

Equation 12.38 through Equation 12.40 are equivalent and the values obtained allow the calculation (Equation 12.37) of the new number of processing units at each level with the value of K_{t_j} that, in turn, can be evaluated once the processing power of the units of the RB is known. The new value for the number of units of the level is related to the reduction factor due to the employment of RoIs through Equation 12.31.

Thus, it is demonstrated that, under certain conditions, it is possible to find a hierarchical system with RoIs with lower technical requirements (lower MF) than the hierarchical system without them.

12.2.2.2 Data Clustering

Another of the improvements that can be introduced in the system is to try to avoid the dispersion of the data to process in each processing element at each level in the hierarchy. The use of an RoI suggests that all its channels should be processed in a combined way because the elaboration of the decision will be made on the basis of existing relationships among the values of the channels. If the levels of the hierarchy are implemented like parallel systems, it can happen that the channels of the RoI may be distributed among several processors, making intercommunication necessary, and thus increasing the time necessary for processing and reducing the performance of the system.

A solution to this problem is to try to gather the data so that one can maximize the probability that all the channels of an RoI are sent to only one processing unit of the parallel system inside the level. The way to do this is to study the problem to discover channels that will be part of an RoI with bigger probability. This implies that the physical process has a certain bias and is present more probably in certain subspaces of the measure space.

If this is not the case, then two options exist:

- To use the information of the RoIs to carry out a dynamic routing of the data. This solution is good but it needs the implementation of a channel multiplexing system and the use of delays to prepare the information of the RoI before routing the channels.
- To gather the channels in a static form, but following some relationship with the physical phenomenon that is being observed. This solution implies, on occasion, the necessity to exchange data among the processors, but the introduced delay will be smaller than in the case of dynamic routing.

The case in which all RoI data are supposed to be in the correct processing unit is the one studied previously. In the second case, the sharing of data will introduce delays in the processing. If one wants to keep the total processing time at the level of the hierarchy constant, the effect of the delay would be to increase the processing capacity units or its number if the processing capacity stays constant.

Taking the hierarchical system with regions of interest as a reference, the total processing power, PC_i, of the parallel system in the level i of the hierarchy can be expressed as:

$$PC_i = \delta_{i-1} \cdot \alpha_i \cdot \frac{D \cdot op}{t_i} \quad 1 \le i \le E \tag{12.41}$$

where D is the number of data to process; op is the number of operations to carry out per datum; t_i is the time to carry out the processing at level i; and δ is the reduction factor due to the employment of RoIs.

For each of the N_i units of this parallel system, the processing power is:

$$PC_{pu_i} = \delta_{i-1} \cdot \alpha_i \frac{D/N_i \cdot op}{t_i} \quad 1 \le i \le E \tag{12.42}$$

If we now have the possibility of exchanging data, this will modify the number of operations to carry out in function of the probability that this exchange will occur. However, if one wants to maintain the number of units, the processing power of each one of them will be increased. The relationship between the new processing power, PC'_{pu_i}, and the previous one is:

$$PC'_{pu_i} = \delta_{i-1} \cdot \alpha_i \cdot \frac{D/N_i \cdot op + p_{ai} \cdot D/N_i \cdot op_a}{t_i} = PC_{pu_i} \cdot \left[1 + p_{ai} \frac{op_a}{op} \right] \quad 1 \le i \le E \tag{12.43}$$

where op_a is the number of operations to carry out to get the data from other units of level i and p_{ai} is the probability of having to make this access; to simplify the calculation, this probability is supposed equal for all the units.

As observed, the increment in the computing power is proportional to the ratio between the number of total operations carried out when accessing other units and the number of operations when this access is not required. If one wants to maintain the processing power, then it is necessary to increase its number. To calculate this increment, we make:

$$PC'_{pu_i} = \delta_{i-1} \cdot \alpha_i \cdot \frac{D/N_i'^{(PC)} \cdot op + p_{ai} \cdot D/N_i'^{(PC)} \cdot op_a}{t_i} = \delta_{i-1} \cdot \alpha_i \cdot \frac{D \cdot op}{N_i^{(PC)} \cdot t_i} \quad 1 \le i \le E \tag{12.44}$$

Solving to get $N'^{(PC)}_i$, the new number of units according to the processing power,

$$N_i'^{(PC)} = N_i^{(PC)} \cdot \left[1 + p_{ai} \frac{op_a}{op} \right] \quad 1 \le i \le E \tag{12.45}$$

where $N_i^{(cp)} = PC_i / PC_{pu_i}$.

In either of the two cases, the required bandwidth is increased because now it must cope not only with the data of the channels but also with the data transfers between processors. To simplify the problem, suppose that the data request between units is uniformly distributed and that the request probability for data exchange is equal for all the units. If this is not the case, then the analysis would get more complicated with the introduction of the probability distribution functions of the requests, as well as that of their destination.

Assuming the simplest case, each unit carries out $D{\cdot}p_a$ accesses distributed among the $N-1$ remaining units, and receives the same amount from all those $N-1$ other units, where p_a is the probability of requesting external data. In this way, the required bandwidth will be:

$$BW'_{pu_i} = \mathrm{P}_i \cdot \frac{D\cdot S_{dat}/N_i'^{(BW)} + p_{ai}\cdot D\cdot S_{dat}/N_i'^{(BW)} + \cdot p_{ai}\cdot D\cdot S_{dat}/N_i'^{(BW)}}{t_1} =$$

$$= \mathrm{P}_i \cdot \frac{D\cdot S_{dat}/N_i'^{(BW)} + 2\cdot p_{ai}\cdot D\cdot S_{dat}/N_i'^{(BW)}}{t_1} = \frac{D\cdot S_{dat}\cdot \mathrm{P}_i}{t_1\cdot N_i'^{(BW)}}\cdot\left(1+2\cdot p_{ai}\right) = \frac{D\cdot S_{dat}\cdot \mathrm{P}_i}{t_1\cdot N_i^{(BW)}} \quad 1\le i\le E \tag{12.46}$$

From here, one can get the new number of units depending on the bandwidth, $N_i^{(BW)}$:

$$N_i'^{(BW)} = N_i^{(BW)}\cdot\left(1+2\cdot p_{ai}\right) \tag{12.47}$$

with $N_i^{(BW)} = BW_i/BW_{pu_i}$.

Therefore, the number of necessary units, N_i'', will be the biggest of the results of Equation 12.45 and Equation 12.47. That is,

$$N_i'' = \max\left(N_i'^{(PC)}, N_i'^{(BW)}\right) \tag{12.48}$$

As can be seen, when introducing a cluster of static data, we modify the number of units of the system and therefore its MF.

The new expression of the MF of the hierarchical system with data clustering is obtained by simply substituting in Equation 12.33 the number of units for the resulting value of Equation 12.48:

$$MF'_{H_{RoI}} = BW\cdot PC\cdot N_1\cdot b_1\cdot \sum_{i=1}^{E}\frac{\left(N_i''\cdot a_i + \left(N_i''-1\right)\cdot K_2\cdot \mathrm{P}_i + N_{RB_i}\cdot a_{RB_i}\cdot K_{t_i}\right)}{\left(2+K_{t_i}\right)\cdot \mathrm{P}_i\cdot \sum_{j=1}^{E} 1/\mathrm{P}_j} \tag{12.49}$$

If, in addition to data clustering, another type of data processing is used (formatting, detection and correction of errors, etc.), this would be reflected as a sequential element with the processors and the MF would increase. Depending on the particular case, the final result of the MF for the system with RoI could be greater than for the system without RoI.

12.3 Example of Architecture Evaluation in High-Energy Physics

High energy physics experiments try to confirm theories by detecting and measuring particle properties. For this kind of experiment, accelerators and particle detectors are used. These last are organized as a distributed sensor network with thousands, or even millions, of sensors of different types whose information must be processed in a short time, which leads to high data rates. Traditionally, hierarchical data acquisition systems have been employed for data taking because, of the total amount of data acquired, only a few are of interest. The reason is that not all the particles produced by the accelerator are of interest.

CERN, the European Laboratory for Particle Physics, is the most important laboratory in the world for the study of particle physics. It holds the biggest accelerator in construction nowadays — the LHC (large hadron collider [5]) — in which two beams of protons will collide with an energy near 14 TeV (tera electron-volts) to study the origin of mass by searching a new particle, the Higgs boson.

For analysis of the collision results, two big detectors, ATLAS [6] and CMS [7], are being constructed. ATLAS will be a huge toroid, 22 m long and 32 m high, with more than 170 million electronic channels to read, coming from sensors inside the detector. All these channels sum a total of 1.3 Mbytes to be read every 25 ns, which gives a rate of 50 TBytes/s. The total processing capacity needed to perform all the operations is estimated at $5{\cdot}10^{10}$ MIPS.

TABLE 12.1 Main Parameters of Hierarchical System Levels

	BW_{input} (MB/s)	PC (MIPS)	p_i	N_i	MF (MB/s × MIPS)
Level 1	$14.84{\cdot}10^6$	$45{\cdot}10^6$	1/400	$1.12{\cdot}10^6$	$1.24{\cdot}10^{11}$
Level 2	10^4–10^5	$32{\cdot}10^6$	1/100	$32{\cdot}10^4$	$3.5{\cdot}10^{13}$
Level 3	10^3–10^4	10^6	1/10	10^4	$1.1{\cdot}10^{14}$

TABLE 12.2 Comparison of MF for the Second Level with and without RoIs

	BW_{input} (MB/s)	PC (MIPS)	N_i	MF (MB/s × MIPS)
Without RoI	10^4–10^5	$32{\cdot}10^6$	$32{\cdot}10^4$	$3.5{\cdot}10^{13}$
With RoI	10^4–10^5	$144{\cdot}10^3$	1440	$1.548{\cdot}10^{11}$

A parallel solution for this problem would require, assuming 100 MIPS and 200 MB/s processors, 500 million processing units according to Equation 12.9. For the calculation of the MF, one needs the value of K_1 and K_2, which can be estimated [6] as $K_1 = 3'33{\cdot}10^{-9}$ and $K_2 = 6'67{\cdot}10^{-18}$. The a and b values are $a = 100/5{\cdot}10^{10} = 2{\cdot}10^{-9}$ and $b = 200/5{\cdot}10^7 = 4{\cdot}10^{-6}$. All these data compute a total MF of $2.5{\cdot}10^{21}$ MIPS × Mbytes/s.

A three-level hierarchical solution can be implemented using 40 MIPS and 200 Mbytes/s hardware processors for the first level and 100 MIPS, 200 Mbytes/s processors for the two other levels. The other parameters for this solution are, from ATLAS Collaboration [6], $a_1 = 8{\cdot}10^{-10}$; $a_2 = a_3 = 2{\cdot}10^{-9}$; and K_1 and K_2 equal to the values used in the parallel solution because they do not depend on the architecture but on the characteristics of the problem. Table 12.1 summarizes the MF for each level. The total result of the series of the three levels is $1.45{\cdot}10^{14}$.

The introduction of RoIs improved the hierarchical system. For this case, only level 2 will include RoI and RB. Simulations made [8] showed that the average number of RoIs per acquisition in level 2 would be 5, each one with 135,000 channels [9]. This leads to a reduction factor of $\delta_1 = 4.5{\cdot}10^{-3}$. The RB is estimated as a processing system of 500 units with a total processing capacity of $12{\cdot}10^3$ MIPS [10], which make the γ_1 parameter equal to $2.4{\cdot}10^{-7}$. From Brawn et al. [10], the K_{T2} parameter can be estimated to a value of $K_{T2} = 0.0875$. With these data, the new number of processing units at level two is 1440. The MF for this second level, where RoIs have been applied, is then recalculated. Table 12.2 shows the differences with and without RoIs.

References

1. Iyengar, S.S., Prasad, L., and Min H., *Advances in Distributed Sensor Integration. Application and Theory*, 1st ed., Prentice Hall, Englewood Cliffs, NJ, 1995, chap. 2.
2. Durrant–Whyte, H.F., Sensor models and multisensor integration, *Int. J. Robotics Res.*, 7(6), 97, 1988.
3. Flynn, M.J., Very high speed computing systems, *Proc. IEEE*, 54, 1901, 1966.
4. Nagin, P.A. et al., Region relaxation in a parallel hierarchical architecture, in *Real-Time Parallel Computing Image Analysis*, 1st ed., Onoe, M., Ed., Plenum Press, New York, 1981, 37.
5. LHC Study Group, *The Large Hadron Collider Accelerator Project*, CERN, Switzerland, 1993.
6. ATLAS Collaboration, *ATLAS Technical Proposal*, CERN, Switzerland, 1994.
7. CMS Collaboration, *CMS Technical Proposal*, CERN, Switzerland, 1994.

8. Kozlov, V., Functional simulation of detector, front-end and read-out parts of a LHC-like DAQ architecture, Report CERN/RD13-120, CERN, Switzerland, 1994.
9. Bock, R. and LeDu, P., Detector and readout specifications for the level-2 trigger demonstrator program, Report ATLAS/DAQ-NO-53, CERN, Switzerland, 1996.
10. Brawn, I.P. et al., The level-1 calorimeter trigger system for ATLAS, Report ATLAS-DAQ-NO-30, CERN, Switzerland, 1995.

13

Cooperative Computing in Sensor Networks

13.1 Introduction ... **13**-1
13.2 The Cooperative Computing Model ... **13**-3
13.3 Node Architecture ... **13**-4
Admission Manager • Code Cache • Virtual Machine • Tag Space
13.4 Smart Messages ... **13**-5
Smart Message Life Cycle • Smart Message Self-Routing
13.5 Programming Interface ... **13**-7
13.6 Prototype Implementation and Evaluation ... **13**-8
Cost of SM Migration • Cost of Tag Space Operations
13.7 Applications ... **13**-12
SPIN Using Smart Messages • Directed Diffusion Using Smart Messages
13.8 Simulation Results ... **13**-14
13.9 Related Work ... **13**-15
13.10 Conclusions ... **13**-18

Liviu Iftode
Rutgers University

Cristian Borcea
Rutgers University

Porlin Kang
Rutgers University

13.1 Introduction

As the cost of embedding computing becomes negligible compared to the actual cost of goods, a trend toward incorporating computing and wireless communication capabilities in most of the consumer products occurs. Therefore, the next generation of computing systems will be embedded, in a virtually unbounded number, and dynamically connected. Although these systems will penetrate every possible domain of daily life, the expectation is that they will operate outside normal cognizance, requiring far less attention from human users than today's desktop computers.

The first illustration of these systems that has received considerable interest in the last couple of years is sensor networks [11–13]. These networks have severe resource limitations in terms of processing power, amount of available memory, network bandwidth, and energy. However, during the next decade sensor networks will become part of a larger class of networks of embedded systems (NES) that have sufficient computing, communication, and energy resources to support distributed applications. For instance, already some companies propose computer systems embedded into cars or video cameras that are able to communicate with each other [1, 4].

For some of these networks, such as networks of intelligent cameras performing object tracking over a large geographical area, it might be beneficial to perform local computations and to cooperate in order to execute a global task. They may perform sophisticated filtering of data at a node that acquired an image or even distributed object tracking, rather than running a centralized algorithm at a server. The challenge is how to program NES, namely, to determine the appropriate computing model and the system support necessary to execute distributed applications in these networks.

NES pose a unique set of challenges that make traditional distributed computing models difficult to employ in programming them. The number of devices working together to achieve a common goal is orders of magnitude greater than those seen so far. These systems are heterogeneous in their hardware architectures because each embedded system is tailored to perform a specific task. Unlike the Internet, NES are typically deployed in environments void of human attention, where it is unacceptable to expect a human to hit a "reset" button to recover from a failure. NES are inherently fragile, with node and connection failures the norm rather than the exception. The availability of nodes may vary greatly over time; they can become unreachable due to mobility, depletion of energy resources, or catastrophic failures.

The nodes in NES communicate through wireless network interfaces. Thus, they can communicate directly only with nodes within their transmission range. Similarly to most ad hoc networks, the separation between hosts and routers disappears (i.e., each node must perform routing). However, the scale and heterogeneity encountered in NES as well as different application requirements preclude the existence of a common routing support. Therefore, the flexibility to use multiple routing algorithms in the same network is desirable.

The applications running in NES target specific data or properties within the network, not individual nodes. From an application point of view, nodes with the same properties are interchangeable. Fixed naming schemes, such as IP addressing, are inappropriate in most situations. The need to target specific data or properties within the network raises the issue of a different naming scheme with dynamic bindings between names and node addresses. A naming scheme based on content or properties is more appropriate for NES than a fixed naming scheme [10].

This chapter presents distributed computing model, cooperative computing, and a software architecture for NES based on execution migration. Cooperative computing applications consist of migratory execution units, called smart messages (SMs), working together to accomplish a distributed task. SMs are user-defined distributed programs (composed of code, data, and execution control state) that migrate through the network searching for nodes of interest (i.e., nodes on which the program needs to run) and execute their own routing at each node in the path. Distributed computing based on execution migration is more suitable for NES than data migration (message passing) due to the volatility and dynamic binding of names to nodes specific to these networks. Cooperative computing provides flexible support for a wide variety of applications, ranging from data collection and dissemination to content-based routing and object tracking.

Nodes in the network support SMs by providing: a name-based shared memory (tag space) for inter-SM communication and access to the host system; and an architecturally independent environment (virtual machine) for SM execution. SMs are self-routing, namely, they are responsible for determining their own paths through the network. SMs name the nodes of interest by properties and self-route to them using other nodes as "stepping stones." Applications in cooperative computing are able to adapt to adverse network conditions by changing their routing dynamically.

To validate the cooperative computing model, the authors have designed and implemented a prototype by modifying Sun Microsystem's Java KVM (kilobyte virtual machine) [3]. Microbenchmark results are reported for this prototype running over a test bed consisting of Linux-based HP iPAQs equipped with 802.11 cards for wireless communication. These results indicate that cooperative computing is a feasible solution for programming real-world applications.

For larger scale evaluation, a simulator has been developed that executes SMs and allows one to account for execution as well as communication time. In this simulator, two previously proposed applications for data collection and data dissemination in sensor networks have been implemented: directed diffusion [13] and SPIN [11]. The simulation results show that this model is able to provide high flexibility for user-defined distributed applications while limiting the increase in response time to, at most, 15% over traditional nonactive communication implementations.

The next section describes cooperative computing; Section 13.3 presents the node architecture for the model. In Section 13.4, details of smart messages are discussed, and Section 13.5 presents the API for cooperative computing. Section 13.6 shows microbenchmark results for the prototype implementation.

Section 13.7 describes the applications implemented using SMs and their simulation results are presented in Section 13.8. Section 13.9 discusses related work and the chapter concludes with Section 13.10.

13.2 The Cooperative Computing Model

Cooperative computing is a distributed computing model for large-scale, ad hoc NES. In this model, distributed applications are defined as dynamic collections of migratory execution units, called SMs, that cooperate in achieving a common goal. The execution of an SM is described in terms of computation and migration phases. The execution performed at each step is determined by the particular properties of that node. On nodes that present interest to the current computation, the SM may read and process data; on intermediate nodes, the SM executes only its routing algorithm. During migrations, SMs carry mobile data, the code missing at destination, and a lightweight execution state.

Nodes in the network cooperate by providing an architecturally independent programming environment (virtual machine) for SM execution and a name-based shared memory (tag space) for inter-SM communication and interaction with the host system. SMs, along with the system support provided by nodes, form the cooperative computing infrastructure, which allows programming user-defined distributed applications in NES.

In this model, a new distributed application can be developed without *a priori* knowledge about the scale and topology of the network or the specific functionality of each node. Placing intelligence in SMs provides this flexibility and also obviates the issue of implementing a new application or protocol in NES, which is difficult or even impossible using conventional approaches [10].

SMs are resilient to network volatility. Over time, certain nodes may become unavailable due to mobility or energy depletion, but SMs are able to adapt by controlling the routing. These messages can carry multiple routing procedures and choose the most appropriate one based on the conditions encountered in the network. Using this feature, SMs can discover routes to nodes of interest even in adverse network conditions.

Moving the execution to the source of data improves the performance for applications that need to process large amounts of data. For example, instead of transferring large size images through the network for an object tracking application, an SM can perform the analysis of the images at the nodes that acquired them. Thus, it reduces the network bandwidth and energy consumption, and in the same time, it improves the user-perceived response time. The impact of transferring code on performance can be limited by caching code at the nodes.

Figure 13.1 shows a simple application that illustrates the novel aspects of computation and communication in cooperative computing. The application performs object tracking over a large area (e.g., a campus, airport, or urban highway system) using a network of mobile robots with attached cameras [17].

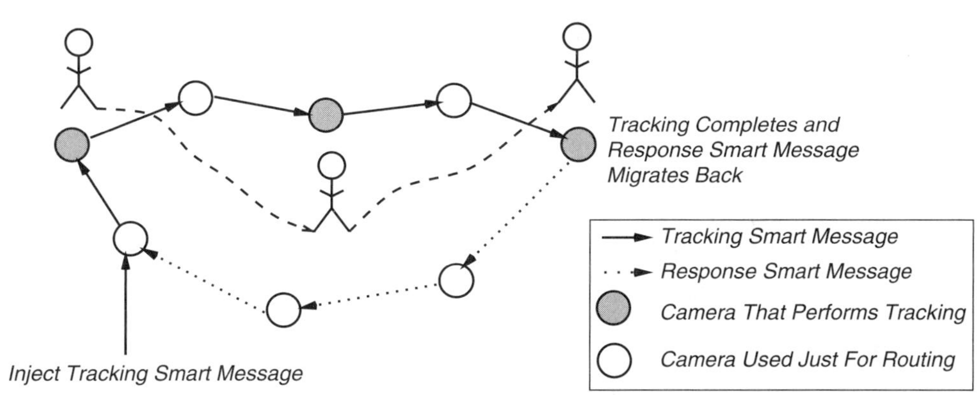

FIGURE 13.1 Distributed object tracking using cooperative computing.

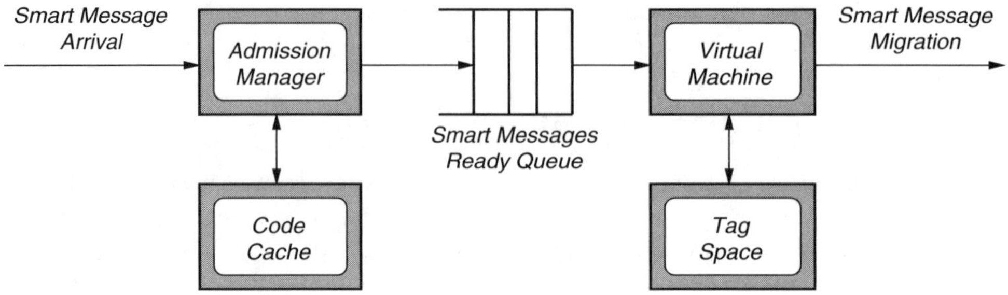

FIGURE 13.2 Node architecture.

In the figure, the target is represented by a person moving across a given geographical region. A user can inject the tracking SM into any node of the network.

The SM migrates to a node that acquired an image of a possible target object, analyzes this image, and then may decide to follow the object. The network maintains no routing infrastructure, and the SM is responsible for determining its path to cameras that detected the object. The smart message can use the direction of motion and geographical information to "chase" the object. Once the SM arrives at a new node that has a picture of the object, it generates a task to analyze the object and its motion further. The SM may migrate to neighbor nodes to obtain pictures of the object from a different angle or lighting conditions. When the tracking completes, the SM generates a response SM that will transport the gathered information back to the user node.

13.3 Node Architecture

The goal of the SM software architecture is to keep the support required from nodes in the network to the minimum, placing intelligence in SMs rather than in individual nodes. Figure 13.2 shows the common system support provided by nodes for cooperative computing. The admission manager receives incoming SMs, decides whether to accept them, and stores these messages into the SM-ready queue. The code cache stores frequently used codes to reduce the amount of traffic in the network. The virtual machine (VM) acts as a hardware abstraction layer for scheduling and executing tasks generated by incoming SMs. The tag space is a name-based shared memory that stores data objects persistent across SM executions and offers a unique interface to the host's OS and I/O system.

13.3.1 Admission Manager

To prevent excessive use of its resources (energy, memory, bandwidth), a node needs to perform admission control. Each SM presents its resource requirements within a resource table. The admission manager is responsible for receiving incoming messages and storing them in the SM ready queue, subject to admission restrictions.

13.3.2 Code Cache

Commonly, the applications executing in NES have a localized behavior, exhibiting spatial and temporal locality. Therefore, frequently used SM codes are cached in order to amortize over time the initial cost of transferring the code through the network.

13.3.3 Virtual Machine

The VM schedules, executes, and migrates SMs. To migrate an SM, the VM captures the execution state and sends it along with the code and data to the next hop. The VM at the destination will resume the

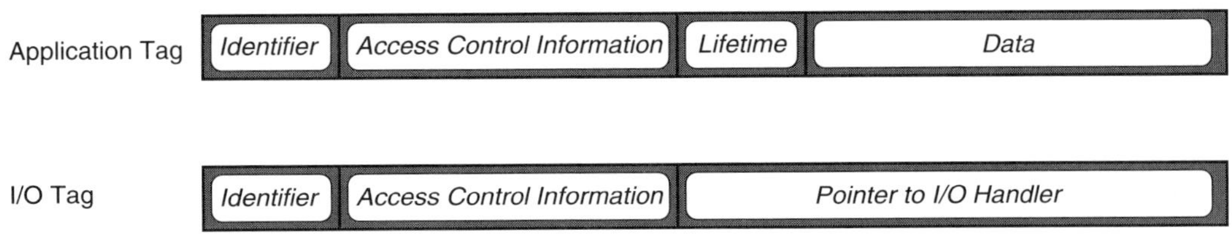

FIGURE 13.3 Structure of application and I/O tags.

SM from the instruction following the migration invocation. The VM also ensures that an SM conforms to its declared resource estimates; otherwise, the SM can be removed from the system.

13.3.4 Tag Space

Each node that supports SMs manages a name-based shared memory, called tag space, consisting of tags that are persistent across SM executions. The tag space contains two types of tags: application tags, which are created by SMs, and I/O tags that are provided by the system. The I/O tags define the basic hardware of the node and provide SMs with a unique interface to the local OS and I/O system. SMs are allowed to read and write both types of tags, but they can create or delete only application tags.

Figure 13.3 illustrates the structure of application and I/O tags. The identifier is the name of the tag and is similar to a file name in a file system; it is used by SMs for content-based node naming. The access of SMs to tags is restricted, based on the access control information associated with each tag. For application tags, the VM associates the access control information carried by the SM that created the tag (i.e., the owner of the tag). For I/O tags, the owner of the device sets the access control information.*

Application tags and I/O tags differ in terms of functionality and lifetime. Application tags offer persistent memory for a limited lifetime (i.e., application tags are still "alive" for a certain amount of time after the SMs that created them have finished the execution at the local node); after this time interval, the tags expire, and the node reclaims their memory. I/O tags, on the other hand, are permanent and provide a pointer to an I/O handler (i.e., a system call or an external process) capable of serving I/O requests. The list of all the possible utilizations of tags consists of:

- *Naming*. SMs name the nodes of interest using tag identifiers.
- *Data storage*. An SM can store data in the network by creating its own tags.
- *Data exchange and data sharing*. Exchanging data through the tag space is the only communication channel among different SMs.
- *Routing*. SMs may create routing tags at visited nodes to store routing information in the data portion of these tags.
- *Synchronization*. An SM can block on a specific tag pending a write of this tag by another SM. Once the tag is written, all SMs blocked on it are waked up and made ready for execution.
- *Interaction with the host system*. An SM can issue commands to or request data from the host OS and I/O devices using I/O tags.

13.4 Smart Messages

SMs are execution units that migrate through the network to execute on nodes of interest and route themselves at each node in the path toward a node of interest. SMs comprise code and data sections (referred to as "bricks"), a lightweight execution state, and a resource table. The code and data bricks can be dynamically used to assemble new, possibly smaller SMs. The ability to incorporate only the necessary code and data bricks in the new SMs can reduce their size and, consequently, the amount of

*More information about access control, protection domains, and SM security in general can be found in Xu et al. [29].

traffic in the network (i.e., the code and data carried by SMs are divided into bricks solely for this purpose). The execution state contains the execution context necessary to resume the SM after a successful migration. The resource table consists of resource estimates: execution time, tags to be accessed or created, memory requirements, and network bandwidth. These resource estimates set a bound on the expected needs of an SM at a node; they are used by the admission manager to make the admission decision.

The SM computation is embodied in tasks. During its execution, a task may modify the data bricks of the SM as well as the local tags to which it has access. It can also migrate, create new SMs, or block on tags of interest. A collection of SMs cooperating toward a common goal forms a distributed application.

13.4.1 Smart Message Life Cycle

Each SM has a well-defined life cycle at a node: (1) it is subject to admission control; (2) upon admission, a task is generated out of the SM's code and data bricks and scheduled for execution; and (3) after completion at a node, the SM may terminate or may decide to migrate to other nodes of interest.

13.4.1.1 Admission

To avoid unnecessary resource consumption, the admission manager executes a three-way handshake protocol for transferring SMs between neighbor nodes. First, only the resource table is sent to the destination for admission control. If the SM admission fails, the task will be informed, and it can decide on subsequent actions. If the SM is accepted, the admission manager checks, using the code bricks' IDs (computed off-line by applying a hash function on the code), whether the code bricks belonging to this SM are cached locally. Then, it informs the source to transfer only the missing code bricks. (These code bricks will also be cached upon arrival.)

13.4.1.2 Scheduling and Execution

Upon admission, an SM becomes a task scheduled (in FIFO order) for execution. The execution is nonpreemptive; new SMs can be accepted, but they will not be dispatched for execution until the current SM terminates. An executing SM can yield the VM, however, by blocking on a tag. The execution time is bounded by the estimated running time presented during admission (i.e., the VM may terminate an SM that does not respect the admission contract).

Nonpreemptive scheduling is used for three reasons. First, the execution time of SMs is usually short (many times a node is used merely as a "stepping stone" en route to a node of interest). Thus, context switching would incur too much overhead with respect to the total execution time of the SM. Second, it is not necessary to support multiprogramming for interactive programs (unlike traditional computer systems, embedded systems commonly operate unattended). Third, the communication always terminates the current SM (i.e., the only form of communication in cooperative computing is a migration invocation) and, consequently, the idea of using multiple threads in one application to overlap communication and computation does not make sense for SM programs. On the other hand, nonpreemptive scheduling makes inter-SM synchronization and sharing particularly simple to implement.

13.4.1.3 Migration

If the current computation does not complete at the local node, the task may continue its execution at another node. The current execution state is captured and migrated along with the code and data bricks. Because a task accesses only mobile data and tags, an efficient migration has been implemented in which only a small part of the entire execution context is saved and transferred through the network. Essentially, the instruction and stack pointers are transferred for all the stack frames corresponding to the current task. It is important to notice that migration is explicit (i.e., the programmers call a "migration" primitive when needed), and that data transferred during a migration are specified by the programmer as data bricks.

13.4.2 Smart Message Self-Routing

SMs are self-routing, i.e., they are responsible for determining their own paths through the network. SMs require no system support for routing; the entire process takes place at application level. An SM names its destinations in terms of tag identifiers and executes its routing algorithm at each node in the path. SMs may create routing tags at intermediate nodes in the network to store routing information. If routing information is not locally available, an SM may create other SMs for route discovery and block on a routing tag. A write on this tag unblocks the SM, which will resume its migration. Because tags are persistent for their lifetime, the routing information, once acquired, can be used by subsequent SMs, thus amortizing the route discovery effort.

Each SM must include at least one *routing brick* among its code bricks. A single routing algorithm, however, might not always reach a node of interest in the presence of highly dynamic network configurations. Therefore, an SM can carry multiple routing algorithms and change them during execution according to the current network conditions. For instance, an SM can use a proactive routing algorithm in a stable and relatively dense network and an on-demand algorithm in a volatile and sparse network. In this way, the SM may complete even if network conditions change significantly during its execution. Borcea and colleagues [6] offer a complete description of the self-routing mechanism.

13.5 Programming Interface

The API for the cooperative computing model is given in Table 13.1. It provides simple, yet powerful, primitives. SMs can access the tag space, dynamically create new SMs, synchronize on tags, and migrate to nodes of interest.

createTag, deleteTag, readTag, and writeTag. These operations allow SMs to create, delete, or access existing tags. As mentioned in Section 13.3, these operations are subject to access control. The same interface is used to access the I/O tags. SMs can issue commands to I/O devices by writing into I/O tags or can get I/O data by reading I/O tags.

createSMFromFiles, createSM, and spawnSM. An SM is created by injecting a program file at a node; this program calls createSMFromFiles with a list of program file names to build the new SM structure. An SM may use createSM during execution to assemble a new SM from a subset of its code and data bricks. A createSM call is commonly used to create a route discovery SM when routing information is not locally available. An SM that needs to clone itself calls spawnSM; this primitive returns true in the "parent" and false in the "child" SMs. Typically, spawnSM is invoked when the current computation needs to migrate a copy of itself to nodes of interest while continuing the execution at the local node. A newly created SM is inserted into the SM ready queue.

blockSM. This primitive implements the update-based synchronization mechanism. An SM blocks on a tag waiting for a write. To prevent deadlocks, blockSM takes a timeout as parameter. If nobody writes the tag in the timeout interval, the VM returns the control to the SM. A typical example is an SM that blocks on a routing tag while waiting for a route discovery SM to bring a new route.

TABLE 13.1 Cooperative Computing API

Category	Primitives
Tag space operations	createTag(tag_name, lifetime, data); deleteTag(tag_name); readTag(tag_name); writeTag(tag_name, value);
SM creation	createSMFromFiles(program_files); createSM(code_bricks, data_bricks); spawnSM();
SM synchronization	blockSM(tag_name, timeout);
SM migration	migrateSM(tag_names, timeout); sys_migrate(next_hop);

```
Typical_SM(tag){
  do
    migrateSM(tag, timeout);
    <do computation>
  until (<quality of result>);
  migrateSM(back, timeout);
}
```

FIGURE 13.4 Code skeleton for typical smart message.

migrateSM and sys_migrate. The migrateSM primitive implements a high-level content-based migration, provided usually as a library function. It allows applications to name the nodes of interest by tag names and to bound the migration time. When migrateSM returns normally (no timeout), the SM is guaranteed to resume its execution at a node of interest. In case of timeout, the SM regains control at one of the intermediate nodes in the path. Figure 13.4 presents an example of a typical SM that uses migrateSM. For instance, this SM can be used in the object tracking application described in Section 13.2. The SM migrates to nodes hosting the tag of interest and executes on these nodes until a certain quality of result is achieved. When this is done, the SM migrates back to the node that injected it into the network.

The migrateSM function implements routing using routing tags, the low level primitive called sys_migrate, and possibly other SMs for route discovery. An SM can choose among multiple migrateSM functions that correspond to different routing algorithms. The sys_migrate primitive is used to migrate SMs between neighbor nodes. The entire migration protocol of capturing the execution state and sending the SM to the next hop is implemented in sys_migrate.

13.6 Prototype Implementation and Evaluation

The authors have implemented their SM prototype in Java over Linux, thus harnessing well-developed and supported Java application development tools and knowledge base.* Specifically, Sun Microsystem's KVM (Kilobyte Virtual Machine) [3] has been modified because it has a small memory footprint (i.e., as little as 160 KB, which makes it suitable for resource-constrained devices) and its source code is publicly available.

The SM API is encapsulated in two Java classes: *SmartMessage* and *TagSpace*; for efficiency, the API was implemented as Java native methods. The authors have also implemented their own serialization mechanism because KVM does not support serialization. In addition to the KVM interpreter thread, two additional threads have been introduced for admission control and local code injection. The design of the SM computing platform is not specific to any hardware or software environment. It can be implemented on any VM (e.g., Mate [20], Scylla [27]), language, or underlying operating system.

Next, microbenchmark results for this SM prototype are reported. Specifically, the cost of one-hop migration and the cost of tag space operations have been measured. The test bed consists of HP iPAQs 3870 running Linux 2.4.18-rmk3-hh24. Each iPAQ contains an Intel StrongARM 1110 206-Mhz RISC processor, 32-MB flash memory, and 64-MB RAM memory. For communication, Lucent Orinoco 802.11b Silver PC Cards are used in ad hoc mode. To factor out the cost of Java method call overhead (approximately 6 μs), the code for measuring costs has been inserted inside the native methods associated with the SM API.

13.6.1 Cost of SM Migration

The one-hop migration has three phases: execution capture at source, SM transfer, and execution resumption at destination. The SM is converted into a machine-independent representation to allow state capture

*The SM software distribution is freely available at http://discolab.rutgers.edu.

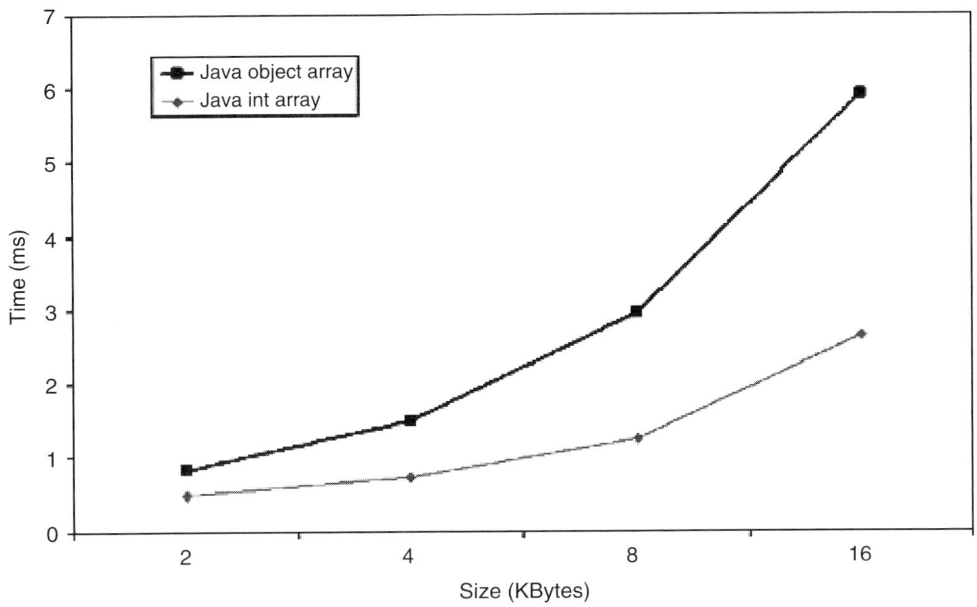

FIGURE 13.5 Cost of data brick serialization.

and resumption. Because the code bricks are already in machine-independent Java class format, only the data bricks and execution state need to be converted. This conversion is done using the authors' simple object serialization mechanism. The serialization of the execution state does not have a significant impact because only the execution control state is captured and transferred, not the local variables. Therefore, the important factors that determine the cost of one-hop migration are the data brick serialization, the SM transfer, and data brick deserialization.

13.6.1.1 Data Brick Serialization and Deserialization

To study the effect of data brick serialization, a fixed-size code brick (1197 bytes) has been used and the data brick size has been varied from 2 to 16 KB. The stack frames have also been kept constant (131 bytes for two activation records). The cost of serializing these two stack frames is 0.235 ms. Commonly, the data bricks in an SM consist of a mixture of objects and primitive types. Two types of data bricks have been used in this evaluation; they represent a practical lower and upper bound for typical data bricks: an array of integers and an array of objects. The object array represents an upper bound because each of its elements causes a call to the top-level VM serialization method, while the integer array represents a lower bound because there is only one call to the top level VM serialization method.

Figure 13.5 shows that the serialization cost is below 6 ms for data bricks as large as 16 KB. Commonly, the SMs process the data at source and therefore they carry small size data. The applications developed by the authors carry less than 2 KB, which costs less than 1 ms to serialize. Figure 13.6 presents the cost of deserialization for the same data bricks. Observe that this cost is as much as 30% larger than the cost of serialization — an increase caused by the memory allocation costs during object deserializations.

13.6.1.2 SM Transfer

To evaluate the total cost of migrating an SM (serialization, transfer, deserialization), two sets of experiments were performed. In the first, the code brick size was varied while data brick size and stack frame size were kept fixed at 53 and 131 bytes, respectively. In the second, data brick size was varied while keeping the code brick size and stack frame size fixed at 1197 and 131 bytes, respectively.

Figure 13.7 and Figure 13.8 show the results of these two experiments for two cases: when the code is not cached and when the code is cached. In Figure 13.7, the time to transfer the SM when the code is

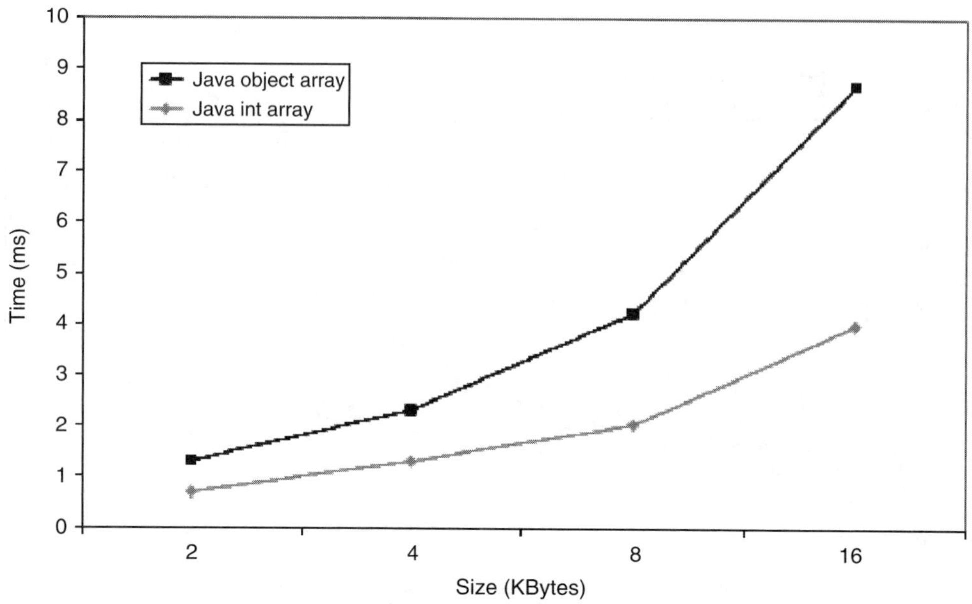

FIGURE 13.6 Cost of data brick deserialization.

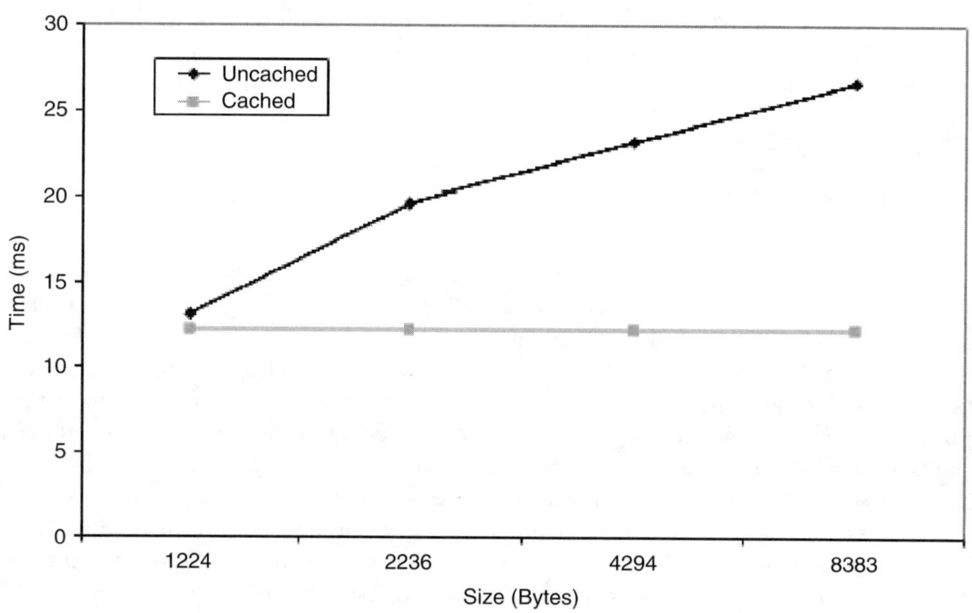

FIGURE 13.7 Effect of code brick size on single-hop migration.

cached represents, essentially, the overhead of the three-way handshake protocol because the sizes of the data bricks and stack frames are small. Figure 13.8 demonstrates that the data brick size contributes significantly to the total cost of migration. Thus, it is important to have a serialization mechanism with minimal space overhead.

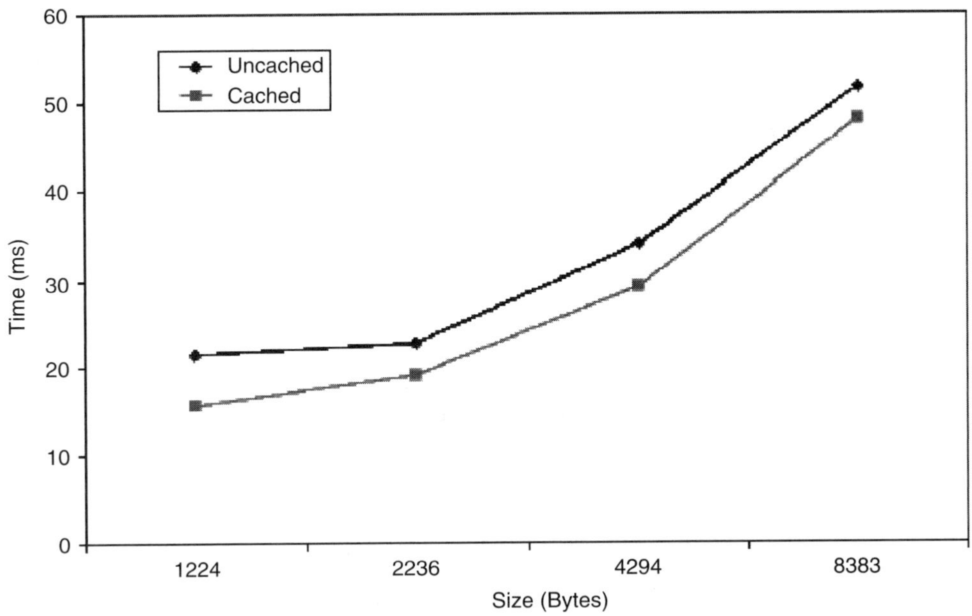

FIGURE 13.8 Effect of data brick size on single-hop migration.

13.6.2 Cost of Tag Space Operations

Table 13.2 shows the cost of the tag space operations for application tags. The *readTag* primitive has the lowest cost because it performs the least number of operations. When an SM reads a tag, the VM interpreter acquires a lock, performs a lookup in the tag space, and returns the data to the SM. The *writeTag* operation costs are slightly higher because the interpreter must check and unblock any SMs blocked on the tag. The *createTag* primitive involves an additional step to register a timer for the tag lifetime, while *blockSM* needs to append the SM to the queue and suspend the current task. The *deleteTag* primitive has the highest cost because the interpreter needs to wake up all SMs blocked on the tag, remove the timer for the tag lifetime, and remove the tag structure from the tag space.

Table 13.3 presents the access time to several I/O tags that are currently implemented in our prototype: GPS location query; neighbor discovery; camera image capture; light sensor; and system status inquiry (battery lifetime, system time, and amount of free memory). A typical node with a video camera and a GPS receiver attached is shown in Figure 13.9. The *gps_location* is updated by a user-level process that

TABLE 13.2 Time for Tag Space Operations

Tag Space Operation	Time (μs)
createTag	43.4
deleteTag	55.9
readTag	20.8
writeTag	31.7
blockSM	45.8

TABLE 13.3 Cost of Reading I/O Tags

Tag Name	Time (ms)
gps_location	0.20
neighbor_list	0.34
image_capture (32-KB)	341.23
light_sensor	0.11
battery_lifetime	25.63
system_time	0.09
free_memory	0.12

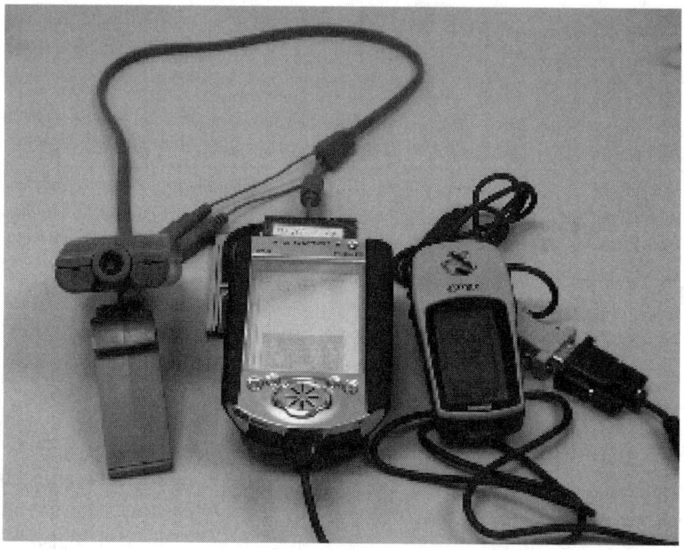

FIGURE 13.9 Prototype node with video camera and GPS receiver attached.

reads from the GPS serial interface. The location of the neighbors along with their identifiers can be returned by reading the *neighbor_list* tag, which is typically used by geographical routing algorithms carried and executed by SMs. To get the information about neighbor nodes, a neighbor discovery protocol has been implemented that maintains a cache of known neighbors. For the *image_capture* tag, the system also performs YUYV to RGB format conversion on the captured image before returning it to the tag reader. All the other tag values are obtained directly from Linux using system calls.

13.7 Applications

To prove that virtually any protocol or application can be written using SMs, two previously proposed applications — SPIN [11] and directed diffusion [13] — have been implemented. They present different paradigms for content-based communication and computation in sensor networks; SPIN is a protocol for data dissemination and directed diffusion implements data collection.

13.7.1 SPIN Using Smart Messages

SPIN is a family of adaptive protocols that disseminates information among nodes in a sensor network. The implementation of SPIN-1 is a three-stage handshake protocol for data dissemination. Each time a node obtains new data, it disseminates them in the network by sending an advertisement to its neighbors.

```
1  DisseminateSM(String tag, int timeout){
2     int timestamp;
3     Data data;
4     String tagData=tag+"data";
5     String tagTimestamp=tag+"timestamp";
6     Address src, dest;
7     while(true){ // SM at source
8        TagSpace.blockSM(TagData, timeout);
9        timestamp = TagSpace.readTag(tagTimestamp);
10       if (!SmartMessage.spawnSM()){ // child SM
11          while(true){ // SM at every node
12             src = SmartMessage.getLocalAddress();
13             SmartMessage.sys_migrate(all); // migrate to all neighbors
14             if (timestamp.CompareTo((Integer)TagSpace.readTag(tagTimestamp))<=0){
15                System.exit(0); // the same or more recent data exists at this node
16             }
17             TagSpace.writeTag(tagTimestamp, timestamp);
18             dest = SmartMessage.getLocalAddress();
19             SmartMessage.sys_migrate(src); // migrate back to source
20             data = TagSpace.readTag(tagData);
21             SmartMessage.sys_migrate(dest); // bring data to destination
22             TagSpace.writeTag(tagData, data);
23          }
24       }
25    }
26 }
```

FIGURE 13.10 SPIN with smart messages.

The node receiving the advertisement checks whether it has already received or requested those data. If not, it sends a request message to the sender asking for the advertised data. The initiator sends the requested data and then the process is executed recursively for the entire network.

As an example of a cooperative computing program, Figure 13.10 presents the code for the authors' implementation of SPIN using SMs. The tag space at each node hosts two tags: the value of the most recent data received (*tagData*) and the timestamp associated with these data (*tagTimestamp*).

The protocol is initiated by injecting a *Disseminate SM* into a node that produces data. This SM blocks on tagData (line 8) waiting for new data. Each time new data are produced, the SM reads the tagTimestamp and spawns itself (lines 9 and 10). The child SM migrates to the neighbors to advertise the new data (line 13). If a destination node does not have these or more recent data, the child SM updates the tagTimestamp and migrates back to the source to bring the data (lines 14 to 22). Upon data arrival, the child SM executes recursively the same algorithm until the data are disseminated in the entire network.

13.7.2 Directed Diffusion Using Smart Messages

In directed diffusion, a sink node requests data by sending "interests" for named data. Data matching an interest are then drawn from source nodes toward the sink node. Intermediate nodes can cache and aggregate data; they may also direct interests based on previously cached data. At the beginning, the sink may receive data from multiple paths, but after a while it will reinforce the path providing the best data rate. All future data will arrive on the reinforced path only.

For the implementation of directed diffusion using SMs, the tag space at each node hosts three tags: the most recent data value (*tagData*); the best data rate available at that node (*tagDataRate*); and the best next hop toward the source (*tagBestRoute*). Directed diffusion is initiated by injecting an SM at the sink. The execution of this SM has two main phases: (1) *exploration* starts at the sink and floods the network to find data of interest; and (2) *reinforcement* chooses the best path and brings data from source to sink.

If the information of interest is not locally available (no tagDataRate value), the *explore SM* spawns itself; the child SM migrates to all neighbors, while the parent SM blocks on tagDataRate. This operation is performed recursively at every node until an SM reaches a node containing the tagDataRate. At this point, the child SM migrates back to its parent carrying the discovered data rate. If the new data rate is better than the value stored in tagDataRate, the SM updates tagDataRate with the new value and tagBestRoute with its source as the best node in the path toward the source of data. This update unblocks the parent SM, which will carry the data rate one hop back. Eventually, the sink node is reached and the reinforcement phase begins.

During the reinforcement phase, a *collect SM* migrates to the best next hop starting from the sink. At each intermediate node, this SM spawns; the child SM migrates to the best next hop, while the parent SM blocks, waiting for data. When the SM reaches the source, it spawns new SMs to carry the data one hop back at the promised data rate. Recursively, a blocked SM is awakened by the data arrival, and it will carry the data back until it reaches the sink.

13.8 Simulation Results

For large-scale evaluation, the authors have developed an event-driven simulator, similar to ns-2 [21], extended with support for SM execution. The simulator is written in Java to allow rapid prototyping of applications. To get accurate results, the communication and the execution times must be accounted for. The simulator provides accurate measurements of the execution time by counting, at the VM level, the number of cycles per VM instruction. To account for the execution time, each node has been simulated with a Java thread, and a new mechanism has been implemented for scheduling these threads inside JVM. The communication model used in this simulator is "generic wireless," with contention solved at the message level. Before any transmission, a node "senses" the medium and backs off in case of contention.

The main goal in conducting the simulation experiments was to quantify the data convergence time for the authors' implementations of SPIN and directed diffusion using SMs and to compare these results with those for traditional message-passing implementations. Data convergence time is defined as the time when a certain percentage of the total number of nodes has received the data (SPIN), or the data rate (directed diffusion). In both cases, due to flooding, all nodes end up receiving the data and the data rate. SPIN completes after all nodes have received the data; directed diffusion will start the reinforcement phase after all nodes have received the data rate. The same network configuration is used for all experiments. The network has 256 nodes distributed uniformly over a square area, and each node has the same transmission range. The average number of neighbors per node is four.

The first set of experiments evaluated the data convergence time when only one SM is injected in the network. Figure 13.11 presents the data convergence time for a single directed diffusion SM, with the sink and source located at the diagonal corners of the square region. The data convergence time for three different cases of the same SM and a base case that uses passive communication (no SM) are plotted. The top curve shows the time when code caching is not used. The second curve shows a more than fourfold improvement in performance when code caching is activated during the first execution of the SM in the network. The code is cached when an SM visits a node for the first time and will be used by subsequent SMs during the same execution. The effects of caching are very important in this case because the SMs visit a node multiple times in directed diffusion; they travel the network forward (looking for the source) and backward (diffusion of data rate).

In the third curve, a 30% decrease can be observed in completion time when the code is already cached at all nodes. The fourth curve shows the data convergence time for a traditional implementation: the protocol is implemented at each node; only data are transferred through the network; and the execution time is not accounted for. Observe that the degradation in performance for this implementation, when the code is cached at all nodes, compared to the traditional implementation is only 5%. This is a reasonable price for the flexibility to program any user-defined distributed application in NES.

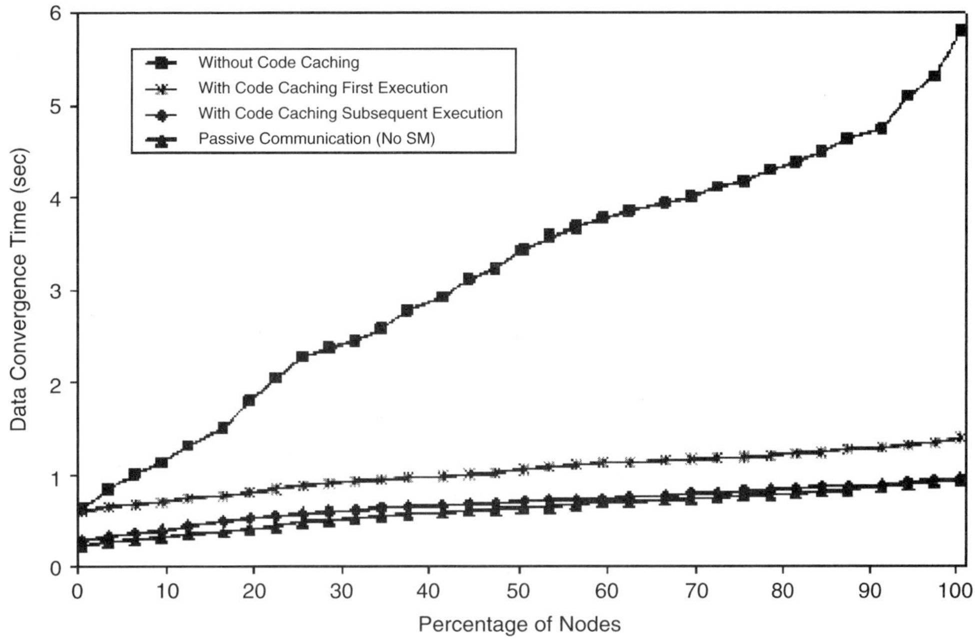

FIGURE 13.11 Directed diffusion using smart messages.

Figure 13.12 plots the same curves for a single SPIN SM launched in the network at a node located in a corner of the square area. During the first execution, code caching leads to a threefold improvement in performance (i.e., reducing the size of SMs is essential for a protocol based on flooding and three-stage communication). The third curve shows a 30% decrease in the completion time (similar to directed diffusion) when the code is already cached at all nodes. The completion time increases from 10 to 15% compared to the traditional implementation.

The second set of experiments quantified the performance of these applications when multiple SMs run simultaneously in the network. Figure 13.13 and Figure 13.14 show the data convergence time for directed diffusion and SPIN with the code already cached at nodes. For these experiments, data convergence time is the time when a certain percentage of nodes have received the data (or data rate) for all the SMs running in parallel. The nodes at which the SMs start are distributed uniformly in the network.

The results show that data convergence time increases with the number of SMs, but only during the initial flooding phase because of increased contention in the network. After that, the shapes of the curves are the same, independent of the number of SMs. The results also indicate that SPIN completes faster than directed diffusion in all cases (i.e., 2.3 s compared to 3.4 s for the top curves in the figures). The cause is that SPIN floods only the neighbors and then brings the data to them, while directed diffusion needs to flood the entire network until it finds the source and then brings the data rate back to all nodes. In the initial phase, directed diffusion generates more messages in the network leading to higher contention, but its performance will increase as soon as the reinforcement phase begins.

13.9 Related Work

SMs have been influenced by the design of mobile agents for IP-based networks [9, 18, 22, 28]. A mobile agent may be viewed as a task that explicitly migrates from node to node assuming that the underlying network assures its transport between them. SMs apply the general idea of code migration, but focus more on flexibility, scalability, reprogrammability, and the ability to perform distributed computing over unattended NES.

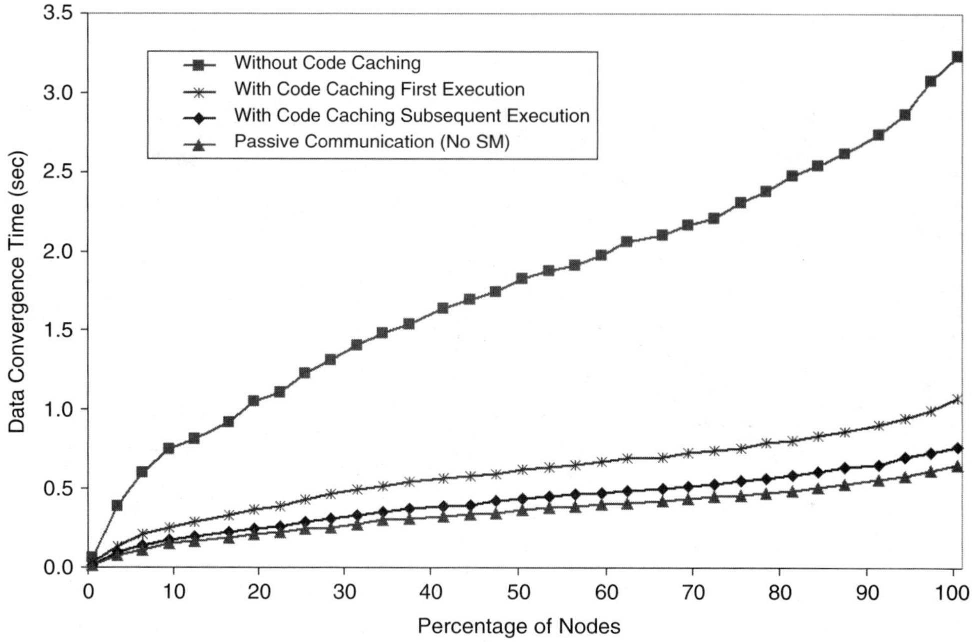

FIGURE 13.12 SPIN using smart messages.

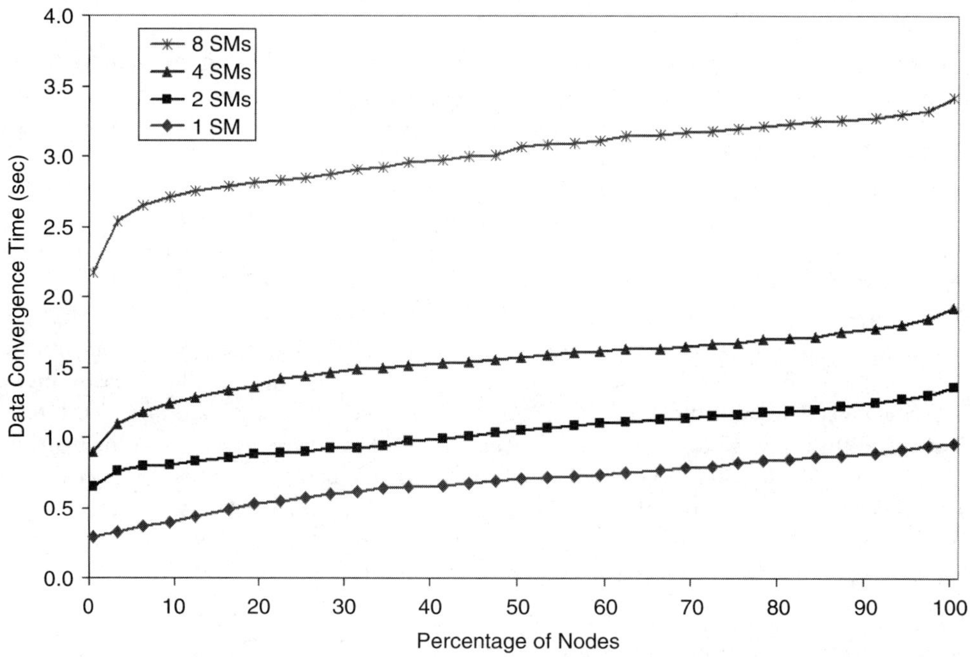

FIGURE 13.13 Directed diffusion: multiple smart messages.

Unlike mobile agents, SMs are defined to be responsible for their own routing in a network. A mobile agent names nodes by fixed addresses and commonly knows the network configuration *a priori*, while an SM names nodes by content and discovers the network configuration dynamically. Furthermore, the SM software architecture defines the common system support that each node must provide. The goal of this architecture is to reduce the support required from nodes in NES because they possess limited resources.

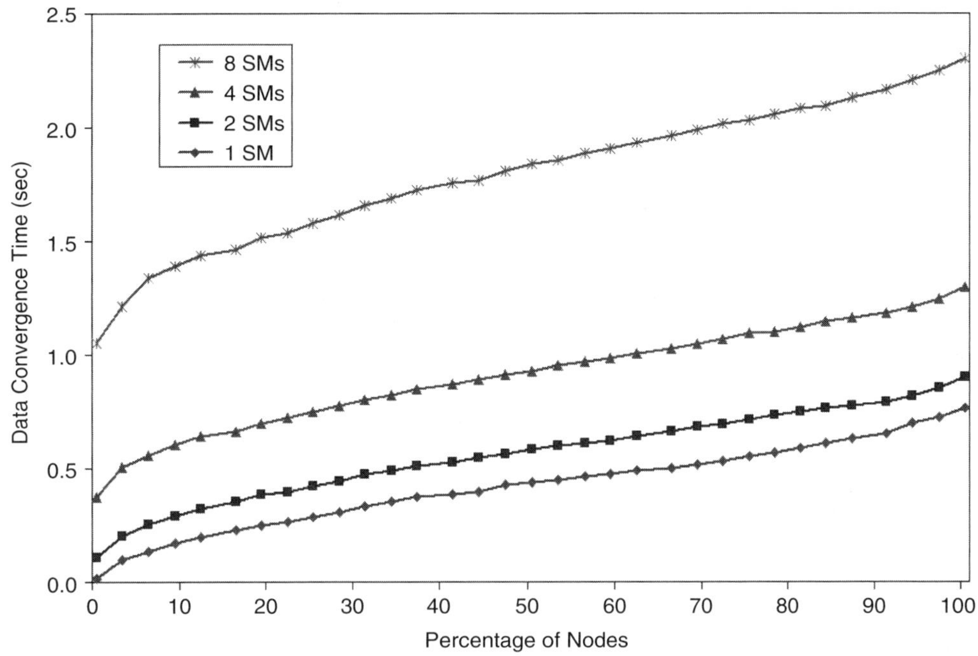

FIGURE 13.14 SPIN: multiple smart messages.

The SM self-routing mechanism shares some of the design goals and leverages work done in active networks (ANs) [8, 23, 26]; however, SMs differ from ANs in several key features. The main difference between them is in terms of programmability. Unlike ANs, which target faster communication in IP-based networks, cooperative computing defines a distributed computing model for NES whereby several SMs can cooperate, exchange data, and synchronize with each other through the tag space. Additionally, the AN model does not contain the migration of execution state as the authors' model does. The migration of execution state for SMs trades overhead for flexibility in programming sophisticated tasks that require cooperation and synchronization among several entities. For example, this execution state allows SMs to make routing decisions based on the results of computation done at previously visited nodes.

Research in mobile ad hoc networking [14, 15, 19, 24] has resulted in numerous routing protocols. These protocols have generally been designed for IP-based networks and have primarily targeted traditional mobile computing applications over networks of mobile personal computers. Some of these protocols have been leveraged into routing algorithms used by the SM self-routing mechanism.

Sensor networks represent the first step toward large networks of embedded systems. Most of the research in this area has focused on hardware [16, 25]; operating systems [12]; or network protocols [5, 11, 13]. Cooperative computing provides a solution for developing user-defined distributed applications in sensor networks, a crucial issue that has been tackled only marginally so far. As demonstrated, cooperative computing provides enough flexibility to enable implementation of previously proposed protocols over this computing platform.

SensorWare [7] is similar to cooperative computing in that both are frameworks for programmable NES based on code migration. Therefore, both are suitable to reprogram the network. However, SensorWare supports mobile control scripts and accesses the resources through virtual devices, whereas cooperative computing supports mobile Java code (i.e., Java is supported on many embedded systems today [2]), execution state migration, and uniform access to resources through tags.

Mate [20] is an efficient VM for sensor networks that can significantly simplify code development and dissemination efforts. The main difference between cooperative computing and this research is that Mate targets only the reprogrammability of the network, but the programming model is still the traditional message passing. SMs, on the other hand, are based on execution migration. An SM transfers not only the code, but also the execution state through the network.

13.10 Conclusions

This chapter has described a programming model for large-scale networks of embedded systems, in which distributed applications are implemented as collections of smart messages. The model overcomes the scale, heterogeneity, and connectivity issues encountered in these networks by using execution migration, content-based naming, and self-routing. The experimental results for this prototype implementation demonstrate the feasibility of cooperative computing. The implementation and simulation results for two sensor network applications show that this model represents a flexible, yet simple, solution for programming large networks of embedded systems.

Acknowledgments

This work was supported in part by the NSF under grant ANI-0121416. The authors thank Ulrich Kremer and Chalermek Intanagonwiwat for our frequent discussions regarding the design of cooperative computing. We also thank Philip Stanley–Marbell, Deepa Iyer, and Akhilesh Saxena for their contributions at various stages of this project.

References

1. Axis Communications. http://www.axis.com.
2. Java 2 Platform, Micro Edition (J2ME). http://java.sun.com/j2me/.
3. K Virtual Machine. http://java.sun.com/products/cldc/.
4. Sensoria Corporation. http://www.sensoria.com.
5. Blum, B., Nagaraddi, P., Wood, A., Abdelzaher, T., Son, S., and Stankovic, J. An entity maintenance and connection service for sensor networks. In *1st In. Conf. Mobile Syst., Applications, Services (MobiSys)* (May 2003), 201–214.
6. Borcea, C., Intanagonwiwat, C., Saxena, A., and Iftode, L. Self-routing in pervasive computing environments using smart messages. In *Proc. 1st IEEE Int. Conf. Pervasive Computing Commun. (PerCom)* (March 2003), 87–96.
7. Boulis, A., Han, C., and Srivasta, M. Design and Implementation of a framework for programmable and efficient sensor networks. In *Proc. 1st Int. Conf. Mobile Syst., Applications, Services (MobiSys).* (May 2003), 187–200.
8. Wetheral, D. Active network vision reality: lessons from a capsule-based system. In *Proc. 17th ACM Symp. Operating Syst. Principles (SOSP'99)* (1999), 64–79.
9. Gray, R.S., Cybenko, G., Kotz, D., and Rus, D. Mobile agents: motivations and state of the art. In *Handbook of Agent Technology*, J. Bradshaw, Ed. AAAI/MIT Press, 2001.
10. Heideman, J., Silva, F., Intanagonwiwat, C., Govindan, R., Estrin, D., and Ganesan, D. Building efficient wireless sensor networks with low-level naming. In *Proc. 18th ACM Symp. Operating Syst. Principles (SOSP)* (October 2001), 146–159.
11. Heinzelman, W.R., Kulik, J., and Balakrishnan, H. Adaptive protocols for information dissemination in wireless sensor networks. In *Proc. 5th Annu. ACM/IEEE Int. Conf. Mobile Computing Networking (MobiCom)* (August 1999), 174–185.
12. Hill, J., Szewczyk, R., Woo, A., Hollar, S., Culler, D., and Pister, K. System architecture directions for networked sensors. In *Proc. 9th Int. Conf. Architectural Support Programming Languages Operating Syst. (ASPLOS)* (November 2000), 93–104.
13. Intanagonwiwat, C., Govindan, R., and Estrin, D. Directed diffusion: a scalable and robust communication paradigm for sensor networks. In *Proc. 6th Annu. ACM/IEEE Int. Conf. Mobile Computing Networking (MobiCom)* (August 2000), 56–67.

14. Li, J., Janotti, J., De Couto, D., Karger, D.R., and Morris, R. A scalable location service for geographic ad hoc routing. In *Proc. 6th Annu. ACM/IEEE Int. Conf. Mobile Computing Networking (MobiCom)* (August 2000), 120–130.
15. Johnson, D.B. and Maltz, D.A. *Dynamic Source Routing in Ad Hoc Wireless Networks.* In *Mobile Computing*, T. Imielinski and H. Korth (Eds.). Kluwer Academic Publishers, 1996, 153–181.
16. Juang, P., Oki, H., Wang, Y., Martonosi, M., Peh, L.-S., and Rubenstein, D. Energy-efficient computing for wildlife tracking: design trade-offs and early experiences with ZebraNet. In *Proc. 10th Int. Conf. Architectural Support Programming Languages Operating Syst. (ASPLOS)* (October 2002), 96–107.
17. Jung, B. and Sukhatme, G.S. Cooperative tracking using mobile robots and environment-embedded, networked sensors. In *2001 IEEE Int. Symp. Computational Intelligence Robotics Automation.* (July 2001), 206–211.
18. Karnik, N. and Tripathi, A. Agent server architecture for the Ajanta mobile-agent system. In *Proc. 1998 Int. Conf. Parallel Distributed Process. Tech. Applications (PDPTA'98)* (July 1998), 66–73.
19. Karp, B. and Kung, H. Greedy perimeter stateless routing for wireless networks. In *Proc. 6th Annu. ACM/IEEE Int. Conf. Mobile Computing Networking (MobiCom)* (August 2000), 243–254.
20. Levis, P. and Culler, D. Mate: a virtual machine for tiny networked sensors. In *Proc. 10th Int. Conf. Architectural Support Programming Languages Operating Syst. (ASPLOS)* (October 2002), 85–95.
21. McCanne, S. and Floyd, S. ns Network Simulator. http://www.isi.edu/nsnam/ns/.
22. Milojicic, D., LaForge, W., and Chauhan, D. Mobile objects and agents. In *Proceedings of the 4th USENIX Conf. Object-Oriented Technol. Syst.* (1998), 1–14.
23. Moore, J.T., Hicks, M., and Nettles, S. Practical programmable packets. In *Proc. 20th Annu. Joint Conf. IEEE Computer Commun. Soc. (INFOCOM'01)* (April 2001), 41–50.
24. Perkins, C.E., Royer, E., and Das, S.R. Ad hoc on demand distance vector (AODV) routing. In *2nd IEEE Workshop Mobile Computing Syst. Applications (WMCSA'99)* (February 1999), 90–100.
25. Priyantha, N.B., Miu, A.K.L., Balakrishnan, H., and Teller, S. The Cricket compass for context-aware mobile applications. In *Proc. 7th Annu. ACM/IEEE Int. Conf. Mobile Computing Networking (MobiCom)* (2001), 1–14.
26. Schwartz, B., Jackson, A.W., Strayer, W.T., Zhou, W., Rockwell, R.D., and Partridge, C. Smart packets for active networks. *ACM Trans. Computer Syst.* (February 2000), 397–413.
27. Stanley–Marbell, P. and Iftode, L. Scylla: A smart virtual machine for mobile embedded systems. In *3rd IEEE Workshop Mobile Computing Syst. Applications, WMCSA2000* (December 2000).
28. White, J. *Mobile Agents.* J.M. Bradshaw (Ed.), MIT Press, 1997.
29. Xu, G., Borcea, C., and Iftode, L. Toward a security architecture for smart messages: challenges, solutions, and open issues. In *Proc. 1st Int. Workshop Mobile Distributed Computing (MDC'03)* (May 2003).

14

Dynamic Power Management in Sensor Networks

14.1 Introduction 14-1

14.2 Idle Power Management 14-2
Multiple Shutdown States • Sensor Node Architecture • Sleep State Transition Policy

14.3 Active Power Management 14-5
Variable Voltage Processing

14.4 System Implementation 14-6
DVS Circuit • Idle Power Management Hooks • Processor Power Modes • OS Architecture • Sensor-Specific Application Programming Interface Extensions

14.5 Results 14-12

Amit Sinha
Engim, Inc.

Anantha Chandrakasan
Engim, Inc.

14.1 Introduction

Wireless distributed sensor networks have gained importance in a wide spectrum of civil and military applications [1]. Advances in microelectricalmechanical systems (MEMS) technology, combined with low-power, low-cost DSPs and RF circuits have resulted in cheap wireless sensor networks becoming feasible. A distributed, self-configuring network of adaptive sensors has significant benefits. They can be used to monitor inhospitable and toxic environments remotely. Large classes of benign environments also require the deployment of a large number of sensors such as intelligent patient monitoring, object tracking, assembly line sensing, etc. Their distributed nature provides wider resolution as well as increased fault tolerance compared to a single sensor node. Several projects, such as the MIT μAMPS project [2], that demonstrate the feasibility of sensor networks are underway.

A wireless sensor node is typically battery operated and is thus energy constrained. To maximize the lifetime of the sensor node after its deployment, all aspects, including circuits, architecture, algorithms and protocols, must be made energy efficient. Once the system has been designed, additional energy savings can be obtained by using dynamic power management concepts [3] whereby the sensor node is shut down if no interesting events occur or slowed down during periods of reduced activity. Such event-driven power consumption is critical for obtaining maximum battery life from the sensor node. In addition, it is desirable that the node has graceful energy quality scalability so that, if the application demands, the user is able to extend the mission lifetime at the cost of sensing accuracy. Energy-scalable algorithms and protocols have been proposed for such energy-constrained situations [4].

Sensing applications present a wide range of requirements in terms of data rates, computation, average transmission distance, etc. As such, protocols and algorithms will need to be tuned to each application.

Therefore, embedded operating systems and software will be critical ingredients in such sensor networks because programmability will be a necessary requirement. This chapter proposes operating system (OS)-directed dynamic power management (DPM) techniques to improve the energy efficiency of sensor nodes. DPM is an effective tool to reduce system power consumption without significantly degrading performance. The embedded OS is used to facilitate active and idle power management.

The basic idea behind idle power management is to shut down devices when they are not needed and wake them when necessary. Formulating an optimum shutdown policy, in general, is a nontrivial problem. If the energy and performance overheads in transitioning to sleep states were negligible, a simple greedy algorithm that makes the system go into the deepest sleep state as soon as it is idle would be perfect. However, in reality, transitioning to a sleep state has the overhead of storing the processor state and shutting off the power supply. Waking also takes a finite amount of time. Therefore, implementing the right policy for transitioning to various sleep states is critical for effective idle power management.

Although shutdown techniques can yield substantial energy savings in the idle states of the system, additional energy savings are possible by optimizing the performance of the sensor node in its active state. Dynamic voltage scaling (DVS) is a very effective active power management technique for reducing processor energy consumption [5]. Most microprocessor-based systems are characterized by a time-varying computational load. Simply reducing the operating frequency during periods of reduced activity results in linear decrease in power consumption but does not affect the total energy consumed per task. Reducing the operating voltage implies greater circuit delays that, in turn, mean that peak performance is compromised. Significant energy benefits can be achieved by recognizing that peak performance is not always required and therefore the operating voltage and frequency of the processor can be dynamically adapted based on instantaneous processing requirements. The goal of DVS is to adapt the power supply and operating frequency to match the workload so that the visible performance loss is negligible. Pering and colleagues have conducted an evaluation of some DVS algorithms on portable benchmarks [6].

14.2 Idle Power Management

Efficient DPM in idle mode requires power-differentiated states and optimal OS policies to transition to and from various states.

14.2.1 Multiple Shutdown States

It is not uncommon for a device to have multiple power modes. For example, the StrongARM SA-1100 processor has three power modes: “run,” “idle,” and “sleep” [7]. Each of these modes is associated with a progressively lower level of power consumption. The run mode is the normal operating mode of the processor; all power supplies are enabled, all clocks are running, and every on-chip resource is functional. The idle mode allows the software to halt the CPU when not in use while continuing to monitor interrupt service requests. The CPU clock is stopped and the entire processor context is preserved. When an interrupt occurs, the processor switches back to run mode and continues operating exactly at the point at which it stopped. The sleep mode offers greatest power savings and minimum functionality. The power supply is cut off to a majority of circuits and the sleep state machine watches for a preprogrammed wake-up event. Similarly, a Bluetooth radio has four different power consumption modes: “active,” “hold,” “sniff,” and “park.”

Most power-conscious devices support multiple power-down modes offering different levels of power consumption and functionality. An embedded system with multiple such devices can have a set of power states based on various combinations of device power states. In fact, an open interface specification called the advanced configuration and power management interface (ACPI), jointly promoted by Intel, Microsoft, and Toshiba [8], standardizes how the OS can interface with devices characterized by multiple power states to provide dynamic power management. ACPI supports a finite state model for system resources and specifies the hardware/software interface that should be used to control them. ACPI controls

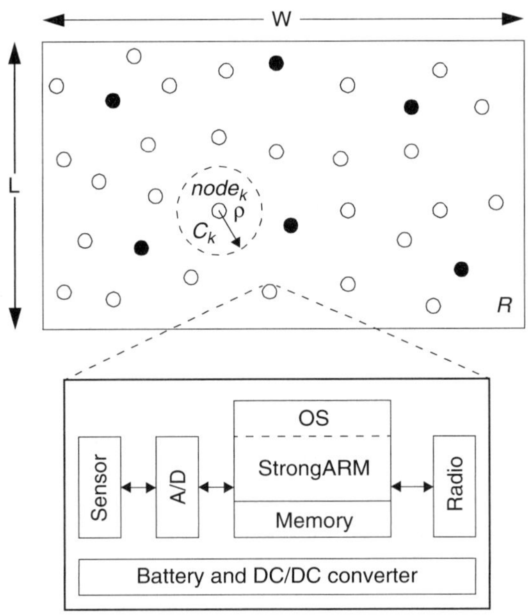

FIGURE 14.1 Sensor network and node architecture. (From Sinha, A. and Chandrakasan, A., *IEEE Des. Test Comp.* 62–74, 2001. With permission.)

the power consumption of the whole system as well as the power state of each device. An ACPI-compliant system has five global states — SystemStateS0 (working state) and SystemStateS1 to SystemStateS4 — corresponding to four different levels of sleep states. Similarly, an ACPI-compliant device has four states: PowerDeviceD0 (the working state) and PowerDeviceD1 to PowerDeviceD3. The sleep states are differentiated by the power consumed, the overhead required in going to sleep, and the wake-up time.

14.2.2 Sensor Node Architecture

Figure 14.1 illustrates the basic sensor node architecture. Each node consists of the embedded sensor, A/D converter, a processor with memory (in this case, the StrongARM SA-11x0 processor), and the RF circuits. Each of these components is controlled by the OS through primitive device drivers. An important function of the OS is to enable power management (PM). Based on event statistics, the OS decides which devices to turn off or on. The sensor network essentially consists of η homogeneous sensor nodes distributed over a rectangular region R with dimensions $W \times L$ with each node having a visibility radius ρ. There is no particular reason for the rectangular topology.

Table 14.1 enumerates the component power modes corresponding to five different useful sleep states for the sensor node. Each of these node-sleep modes corresponds to an increasingly deeper sleep state and is therefore characterized by an increasing latency and decreasing power consumption. These sleep

TABLE 14.1 Useful Sleep States for the Sensor Node

State	StrongARM	Memory	Sensor, A/D	Radio
s_0	active	active	on	tx, rx
s_1	idle	sleep	on	rx
s_2	sleep	sleep	on	rx
s_3	sleep	sleep	on	off
s_4	sleep	sleep	off	off

Source: From Sinha, A. and Chandrakasan, A., *IEEE Des. Test Comp.* 62–74, 2001. With permission.

states are chosen based on working conditions of the sensor node, e.g., it does not make sense to have the memory in the active state and everything else completely off:

- State s_0 is the completely active state of the node where it can sense, process, transmit, and receive data.
- In state s_1, the node is in sense and receive mode while the processor is in standby.
- State s_2 is similar to state s_1 except that the processor is powered down and is waked up when the sensor or the radio receives data.
- State s_3 is the sense-only mode in which everything except the sensing front-end is off.
- State s_4 represents the completely off state of the device.

The design problem is to formulate a policy of transitioning between states based on observed events so as to maximize energy efficiency. It can be seen that the power-aware sensor model is similar to the system power model in the ACPI standard. The sleep states are differentiated by the power consumed, the overhead required in going to sleep, and the wake-up time. The deeper the sleep state is, the lesser the power consumption and the longer the wake-up time.

14.2.3 Sleep State Transition Policy

Assume an event is detected by a sensor node at some time t_0; it finishes processing it at time t_1; and the next event occurs at time $t_2 = t_1 + t_i$. At time t_1, the node decides to transition to a sleep state s_k from the active state s_0 as shown in Figure 14.2. Each state s_k has a power consumption P_k, and the transition time to it from the active state and back is given by $\tau_{d,k}$ and $\tau_{u,k}$, respectively. By the definition of node-sleep states, $P_j > P_i$, $\tau_{d,i} > \tau_{d,j}$ and $\tau_{u,i} > \tau_{u,j}$ for any $i > j$. The power consumption between the sleep modes is modeled as a linear ramp between the states. When the node transitions from state s_0 to, say, state s_k, individual components such as the radio, memory, and processor are progressively powered down. This results in a stepped variation in power consumption between the states. The linear ramp is analytically simpler to handle and approximates the process reasonably well.

Now a set of sleep time thresholds $\{T_{th,k}\}$ corresponding to the states $\{s_k\}$ will be derived such that transitioning to a sleep state s_k from state s_0 will result in a net energy loss if the idle time $t_i < T_{th,k}$ because of the transition energy overhead. This assumes that no productive work can be done in the transition period, which is usually true (e.g., when a processor wakes up, the transition time is the time required

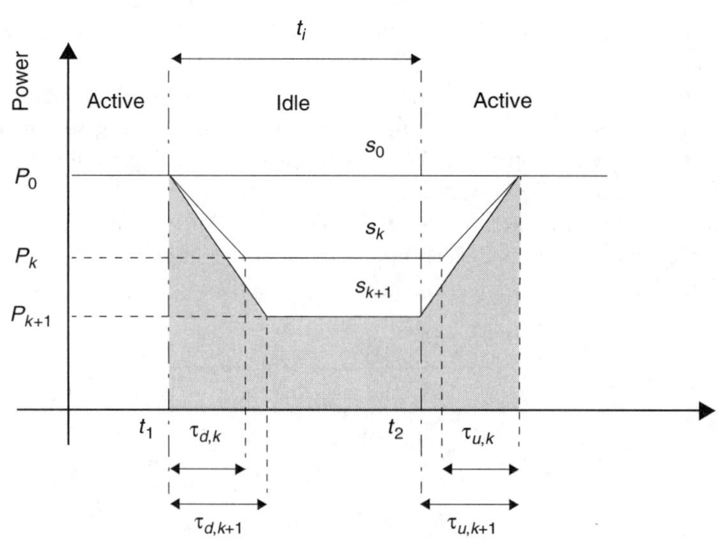

FIGURE 14.2 Dynamic voltage and frequency scaling. (From Sinha, A. and Chandrakasan, A., *IEEE Des. Test Comp.* 62–74, 2001. With permission.)

TABLE 14.2 Sleep State Power, Latency, and Threshold

State	P_k (mW)	τ_k (ms)	$T_{th,k}$
s_0	1040	–	–
s_1	400	5	8
s_2	270	15	20
s_3	200	20	25
s_4	10	50	50

Source: From Sinha, A. and Chandrakasan, A., *IEEE Des. Test Comp.* 62–74, 2001. With permission.

for the PLLs to lock, the clock to stabilize, and the processor context to be restored). The energy savings because of state transition are given by the difference in the area under the graphs shown in Figure 14.2 and are computed as follows:

$$E_{save,k} = \left(P_0 - P_k\right)t_i - \left(\frac{P_0 - P_k}{2}\right)\tau_{d,k} - \left(\frac{P_0 + P_k}{2}\right)\tau_{u,k}$$

Such a transition is only justified when $E_{save,k} > 0$. This leads to the following energy gain threshold:

$$T_{th,k} = \frac{1}{2}\left[\tau_{d,k} + \left(\frac{P_0 + P_k}{P_0 - P_k}\right)\tau_{u,k}\right]$$

This implies that the longer the delay overhead of the transition is, the higher the energy gain threshold, and the more the difference between P_0 and P_k is, the smaller the threshold.

Table 14.2 lists the power consumption of a sensor node described in Figure 14.1 designed using off-the-shelf components in different power modes and the corresponding energy gain thresholds. One can see that the thresholds are in the order of milliseconds. The OS shutdown policy is based on event interarrival statistics and energy gain thresholds and can be formulated as an optimization problem. If the events are modeled as a Poisson process, the probability of at least one event in time t_i is given by

$$p_E\left(t\right) = 1 - e^{\lambda t_i}$$

In this case, a simple algorithm that updates the average events per unit time, λ, as they happen computes the probability of an event happening within the thresholds, $T_{th,k}$, and chooses the deepest sleep state based on a minimum probability threshold that would be effective. Sinha and Chandrakasan have described the energy savings obtained from such an algorithm [9].

14.3 Active Power Management

The OS can be used to manage active power consumption in an energy-constrained sensor node. It reduces the operating frequency and voltage to a level just enough for the sensing application so that no visible loss is observed in performance while the energy consumption is reduced.

14.3.1 Variable Voltage Processing

Dynamic voltage scheduling is a very effective technique for reducing CPU energy. Several sensor systems are characterized by a time-varying computational load. Simply reducing the operating frequency during periods of reduced activity results in a linear decrease in power consumption but does not affect the total energy consumed per task, as shown in Figure 14.3(a) (the shaded area represents energy). Reduced operating frequency implies that the operating voltage can also be reduced. Because the switching power

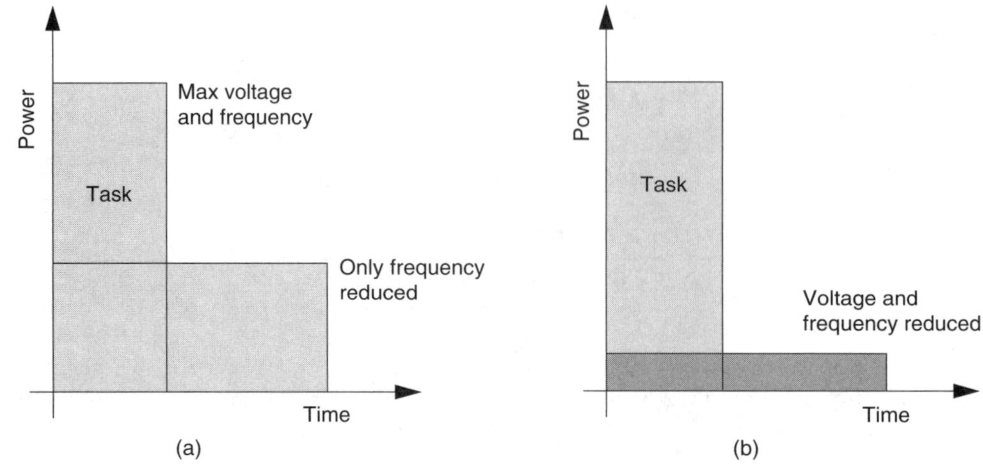

FIGURE 14.3 μAMPS processor board.

consumption scales linearly with frequency and quadratically with supply voltage, quadratic energy reduction can be obtained as shown in Figure 14.3(b). Significant system energy savings can be realized by recognizing that peak performance is not always required and therefore the operating voltage and frequency of the processor can be dynamically adapted based on instantaneous processing requirement.

14.4 System Implementation

Figure 14.4 shows the first generation μAMPS sensor node, which is based on the StrongARM SA-1110 processor and has 1 MB of on-board SRAM and flash memory. The board runs at a nominal battery (single lithium primary cell) power supply of about 4.0 V. The on-board power supply circuits generate a 3.3-V supply for all digital circuits. A separate analog power supply is also generated to isolate the digital power supply noise from the analog circuits. The 3.3-V digital power supply also powers the I/O pads of the StrongARM SA-1110 processor. The core power supply is generated through a DVS circuit that can regulate the power supply from 0.925 V to a maximum of 2.0 V with a conversion efficiency of about 85%.

The radio module is on a similarly sized board and consists of a dual power, 2.4 GHz-radio for 10- and 100-m ranges. The 16-b bus interface connector will allow the radio module to be stacked onto the processor board. In addition, the connector allows a different sensor board (e.g., a seismic sensor) to be stacked. The processor board has an RS-232 and a USB connector for remote debugging and connecting to a debug PC. The board features a built-in acoustic sensor (a microphone, some opamps, and A/D circuit) that talks to the StrongARM processor using the synchronous serial port (SSP). The opamp gains are programmable and processor controlled. An envelop detect mechanism has also been incorporated into the sensor circuit, which bypasses the A/D circuit and wakes the processor when the signal energy crosses a certain programmable threshold. Using this feature can significantly reduce power consumption in the sense mode and facilitates event-driven computation.

14.4.1 DVS Circuit

The basic variable core power supply schematic is shown in Figure 14.5. The MAX1717 step-down controller is used to regulate the core supply voltage dynamically through the 5-b digital-to-analog converter (DAC) inputs over a 0.925 to 2 V range. The converter works on the following principle. A variable duty cycle pulse width modulated (PWM) signal alternately turns on the power transistors M1 and M2. This produces a rectangular wave at the output of the transistors with duty cycle D. The LC low-pass filter passes a desired DC output equal to $DV_{battery}$ while attenuating the AC component to an acceptable ripple. The duty cycle D is controlled using the DAC pins (D0:D4), which results in 30 voltage

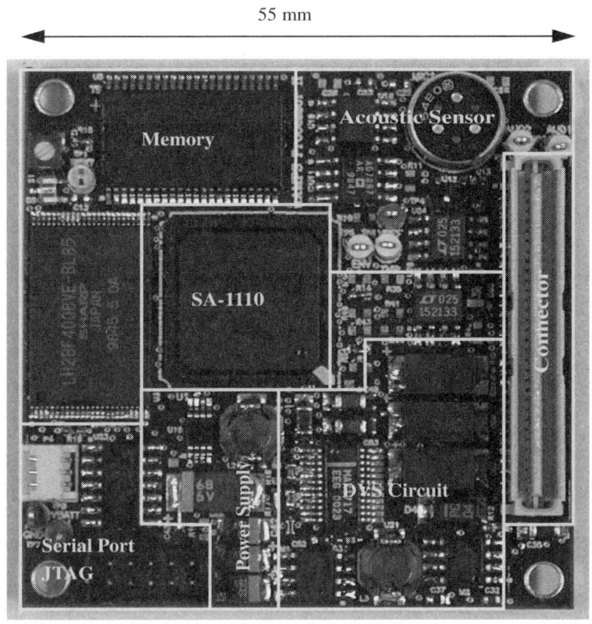

FIGURE 14.4 DVS circuit schematic.

levels (two combinations are not allowed). A two-wire remote sensing scheme compensates for voltage drops in the ground bus and output voltage rail. The StrongARM sets the DVS enable pin on the voltage regulator depending on whether DVS capability is desired or not. A feedback signal from the regulator lets the processor know if the output core voltage is stabilized. This is required for error-free operation during voltage scaling.

The processor clock frequency change involves updating the contents of the core clock configuration register (CCF) of the SA-1110 [10]. The core clock is derived by multiplying the reference crystal oscillator clocks using a phase-locked loop (PLL) based on CCF register settings, as shown in Table 14.3. The core clock (CCLK) can be driven using the fast CCLK or the memory clock (MCLK), which runs at half the frequency of the CCLK. The core clock uses CCLK normally except when waiting for fills to complete after a cache miss. Core clock switching between CCLK and MCLK can be disabled by setting a control register appropriately.

The sequence of operations during a voltage and frequency update depends on whether one is increasing or decreasing the processor clock frequency, as shown in Figure 14.6. When the clock frequency is increased, it is first necessary to increase the core supply voltage to the minimum required for that particular frequency. The optimum voltage frequency pairs are stored in a lookup table. Once the core voltage is stabilized, the frequency update can proceed. The first step involves recalibrating the memory timings. This is done by setting an appropriate value in the MSC control register. Before CCLK frequency is increased, clock switching between CCLK and MCLK is disabled to avoid an inadvertent switch of the core clock. CCLK frequency is changed by setting the CCF register. Once this is done, core clock switching between CCLK and MCLK is enabled.

The sequence of operations is somewhat reversed when reducing frequency. First, the core clock frequency is updated (following the three basic steps mentioned previously). Before one can reduce the core voltage, it is necessary to recalibrate the memory timing. This is required because, once the core clock frequency is reduced, memory read–write will result in errors unless the memory timing is adjusted (e.g., when reading the voltage–frequency lookup table). Subsequently, the core voltage is reduced and normal operation is started once it stabilizes. To ensure correct operation, the entire voltage frequency update must be done is an atomic fashion. For example, if an interrupt occurs while frequency is updated and memory has not been recalibrated, execution errors might occur.

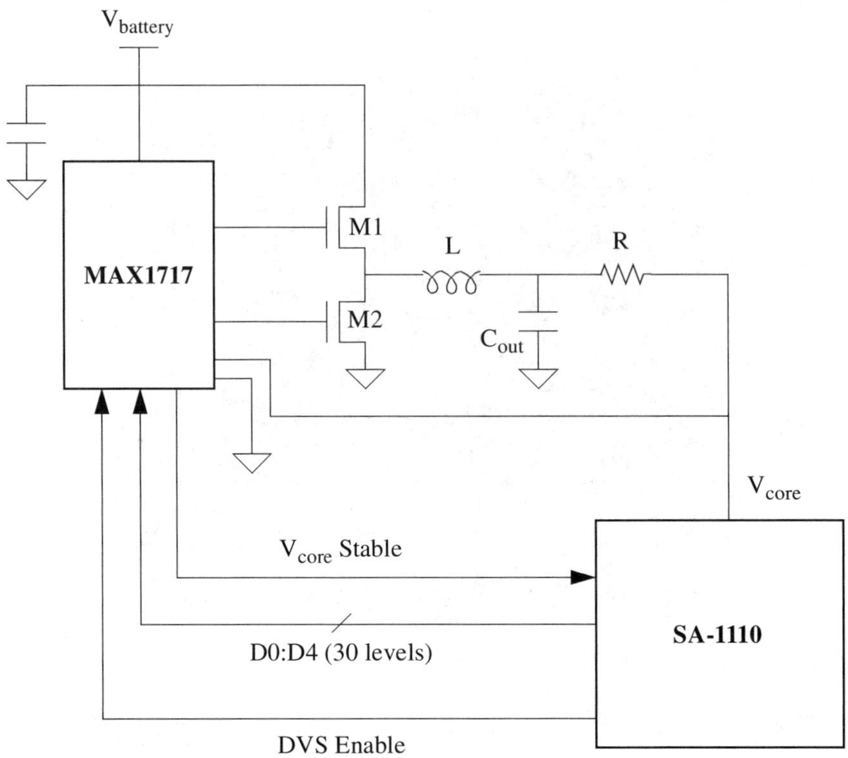

FIGURE 14.5 Idle power management hooks on the sensor processor board.

TABLE 14.3 SA-1110 Core Clock Configurations and Minimum Core Supply Voltage

CCF(4:0)	Core Clock Frequency (CCLK) in MHz 3.6864 MHz Oscillator	 3.6864 MHz Oscillator	Core Voltage (V) [3.6864 MHz Osc]
00000	59.0	57.3	1.000
00001	73.7	71.6	1.050
00010	88.5	85.9	1.125
00011	103.2	100.2	1.150
00100	118.0	114.5	1.200
00101	132.7	128.9	1.225
00110	147.5	143.2	1.250
00111	162.2	157.5	1.350
01000	176.9	171.8	1.450
01001	191.7	186.1	1.550
01010	206.4	200.5	1.650
01011	221.2	214.8	1.750
01100-11111	–	–	

14.4.2 Idle Power Management Hooks

The sensor node has been designed specifically to allow a set of sleep states similar to the one described earlier; in addition, it has hardware support for event-driven computation. The overall schematic is shown in Figure 14.7. The general purpose I/O (GPIO) pins on the StrongARM are used to generate and receive various signals from the peripherals. The SA-1110 features 28 GPIO pins, each of which can be configured as an input or an output. In addition, the GPIO pins can be configured specifically to detect a rising or

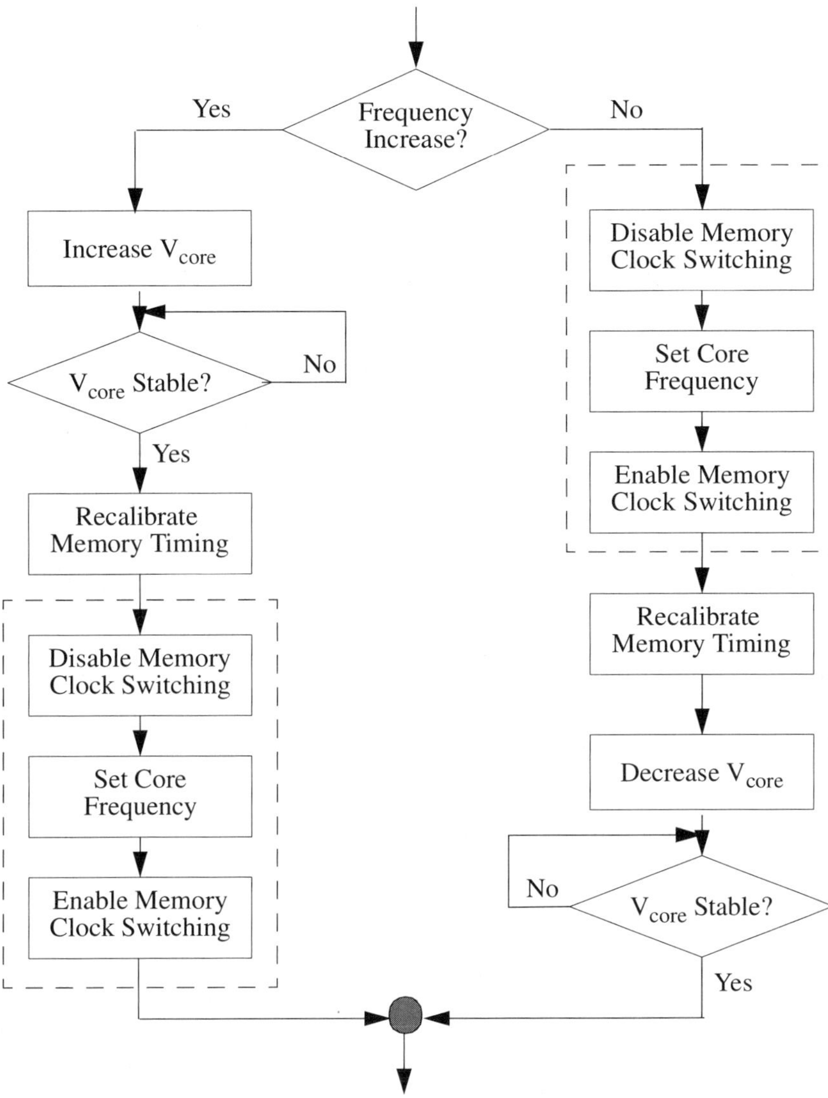

FIGURE 14.6 System-level power savings from active power management using DVS.

falling edge. In implementation, four GPIO pins are dedicated to power supply control in the system. The entire analog power supply can be switched off when no sensing is required. Alternately, only the power supply to the low pass filter (LPF) can be switched off and the envelop energy sensing circuit could be used to trigger a signal to the processor. When this happens, the processor could enable the LPF and start reading data off the A/D converter using the SSP (synchronous serial port). The signal detection threshold is also programmable using other GPIO pins and similar power supply control is available for the radio module; the processor can turn off the radio when it is not required.

14.4.3 Processor Power Modes

The SA-1110 contains power management logic that controls the transition among three different modes: run, idle, and sleep. Each of these modes corresponds to a reduced level of power consumption.

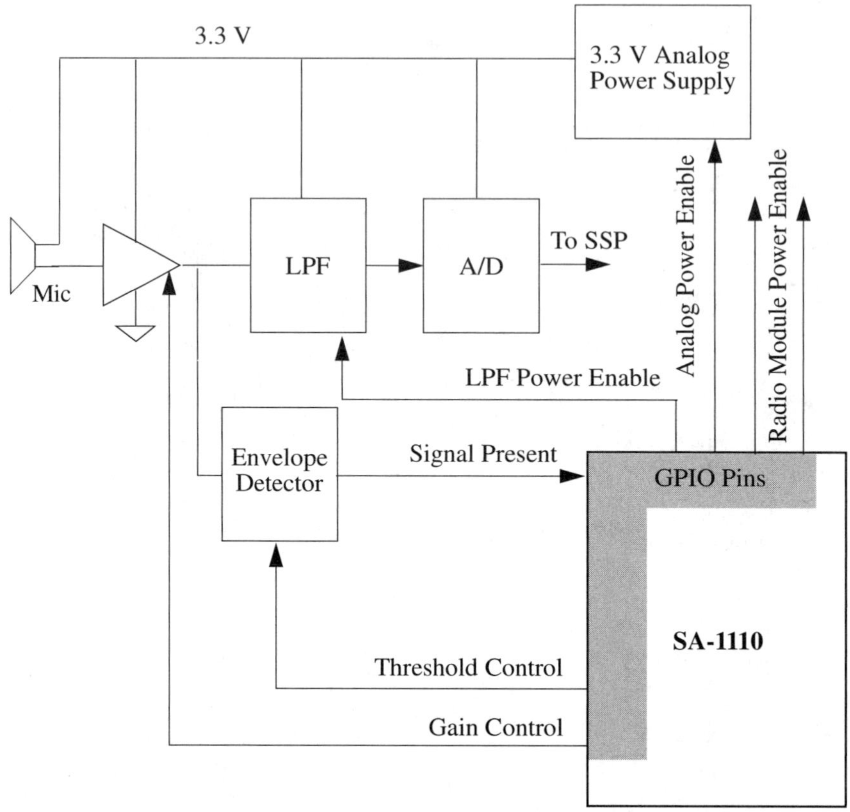

FIGURE 14.7 Degradation in DVS savings with increase in workload variance.

- *Run mode.* This is the normal mode of operation for the SA-1110. All on-chip power supplies are on, all clocks are on, and every on-chip resource is available. Under usual conditions, the processor starts up in the run mode after a power-up or reset.
- *Idle mode.* This mode allows an application to stop the CPU when not in use while continuing to monitor interrupt requests. The CPU clock is stopped and, because the SA-1110 is a fully static design, all state information is saved. When normal operation is resumed, execution is started exactly where it stopped. During idle mode, all on-chip resources (real-time clock; OS timer; interrupt controller; GPIO; power manager; DMA and LCD controllers; etc.) are on. The PLL also remains in lock so that the processor can be brought in and out of the idle mode quickly.
- *Sleep mode.* Sleep mode offers greatest power savings for the processor and, consequently, lowest functionality. When transitioning from run/idle to sleep mode, the SA-1110 performs an orderly shutdown of on-chip activity, applies an internal reset to the processor, and negates the power enable (PWR_EN) pin, thus indicating to the external system that the power supply can be turned off. Running off the 32.768 KHz crystal oscillator, the sleep state machine watches for a preprogrammed wake-up event to occur. Sleep mode is entered in one of two ways: through software control or through a power supply fault. Entry into sleep mode is accomplished by setting the force sleep bit in the power manager control register (PMCR). This bit is set by software and cleared by hardware during sleep so that, when the processor wakes, it finds the bit cleared. The entire sleep shutdown sequence takes about 90 ms.

Table 14.4 shows the power consumption in various modes of the SA-1110 processor at two different frequencies and the corresponding voltage specification [10]. Note that the minimum operating voltage

TABLE 14.4 Power Consumption of the SA-1110 Processor

		Power Consumption Modes		
Frequency	Supply Voltage (V)	Normal (mW)	Idle (mW)	Sleep (μA)
133	1.55	<240	<75	<50
206	1.75	<400	<100	<50

required (as shown in Table 14.3) at the two frequencies is slightly lower than what is shown in Table 14.4. Although the idle mode results in about 75% power reduction, the sleep mode saves almost all the power.

14.4.4 OS Architecture

The sensor OS is based on Redhat eCos, an open-source, real-time operating system for embedded applications [11]. It meets the requirements of the embedded space that Linux cannot yet reach. Linux currently scales from a minimal size of around 500 KB of kernel and 1.5 MB of RAM, all before taking into consideration application and service requirements. eCos can provide the basic runtime infrastructure necessary to support devices with memory footprints in the tens to hundreds of KB, with real-time options.

The original eCos OS is designed to be completely scalable across platforms as well as within a given platform. Essentially, source level configuration allows the user to add or remove packages from a source repository based on system requirements. For example, the user might choose to remove math libraries and the resulting kernel will be leaner. The core eCos system consists of a number of different components, such as the kernel, the C library, an infrastructure package, etc. Each of these provides a large number of configuration options, allowing application developers to build a system that matches the requirements of their particular applications.

To manage the potential complexity of multiple components and lots of configuration options, eCos has a component framework: a collection of tools specifically designed to support configuring multiple components. Furthermore, this framework is extensible, allowing additional components to be added to the system at any time. The eCos component description language (CDL) lets the configuration tools check for consistency in a given configuration and point out any dependencies that have not been satisfied.

At the core of the eCos kernel is the scheduler. This defines the way in which threads are run, and provides the mechanisms by which they may synchronize. It also controls the means by which interrupts affect thread execution. To allow threads to cooperate and compete for resources, it is necessary to provide mechanisms for synchronization and communication. The classic synchronization mechanisms are mutexes/condition variables and semaphores. These are provided in the eCos kernel, together with other synchronization/communication mechanisms that are common in real-time systems, such as event flags and message queues.

The kernel also provides exception handling. An exception is a synchronous event caused by the execution of a thread. These include the machine exceptions raised by hardware (such as divide-by-zero, memory fault, and illegal instruction) and machine exceptions raised by software (such as deadline overrun). The simplest, and most flexible, mechanism for exception handling is to call a function. This function needs context in which to work, so access to some working data is required. The function may also need to be handed some data about the exception raised — at least the exception number and some optional parameters. As opposed to exceptions, which are synchronous in nature, interrupts are asynchronous events caused by external devices. They may occur at any time and are not associated in any way with the currently running thread and are harder to deal with. The ways in which interrupt vectors are named, how interrupts are delivered to the software, and how interrupts are masked are all highly hardware specific. On the SA-1110, two kinds of interrupts are supported: fast interrupts (FIQ) and regular interrupts (IRQ); both can be masked.

TABLE 14.5 Typical Power Management Functions

Function Type	Functions Available
DVS	UAMPS_ENABLE_DVS(), UAMPS_SET_VOLTAGE(), UAMPS_SET_PROC_CLOCK(), UAMPS_CHECK_VCORE_STABLE(), UAMPS_SET_PROC_CLOCK(), uamps_set_proc_rate(), uamps_dvs_scheduler()
Shutdown	UAMPS_PERIPHERAL_POWER_ON(), UAMPS_PERIPHERAL_POWER_OFF(), UAMPS_V3_STANDBY_ON(), UAMPS_V3_STANDBY_OFF(), SA11X0_PWR_MGR_WAKEUP_ENABLE, SA11X0_PWR_MGR_GENERAL_CONFIG, SA11X0_PWR_MGR_CONTROL, uamps_set_proc_idle(), uamps_set_proc_sleep(), uamps_goto_sleep_state()

The kernel also provides a rich set of timing utilities such as counter, clocks, alarms, and timers. The counter objects provided by the kernel provide an abstraction of the clock facility that is generally provided. Application code can associate alarms with counters, where an alarm is identified by the number of ticks until it triggers, the action to be taken on triggering, and whether the alarm should be repeated. Clocks are counters associated with a stream of ticks that represent time periods. Clocks have a resolution associated with them, whereas counters do not.

14.4.5 Sensor-Specific Application Programming Interface Extensions

Table 14.5 illustrates some sample functions in the power management API. The functions are available to the μAMPS application developer to enhance the power efficiency of the sensing application. Programs written for the sensor node do not need to satisfy any unusual requirements, but some differences always exist between programs written for real-time OS platforms and those written for a time-sharing, virtual memory system like UNIX or Windows.

14.5 Results

Figure 14.8 shows the power consumption of the sensor node in the fully active state (all modules on) as a function of the operating frequency of the SA-1110. The figure shows the power consumption using DVS and only frequency scaling (done at a fixed core voltage of 1.65 V). The system power supply was 4.0 V. In the active mode, DVS is the primary source of power management. When running at the maximum operating voltage and operating frequency, the power consumption of the system is almost 1 W. Active power management using DVS results in about 53% maximum system-wide power savings. The actual savings depend on the workload requirement.

With DVS, minimum energy consumption results when the processing rate variation is minimized because of the convexity of the energy workload model. Figure 14.9 plots the relative battery life improvement as a function of the variance in workload. Each workload profile is Gaussian with a fixed average workload. Although the average workload might be constant, the battery life improvement from DVS will degrade as the fluctuations in workload increase.

Table 14.6 shows the measured power consumption of the sensor node in various modes of operation. The sensor node can be classified as a processor power-dominated architecture. The radio module follows the processor in power requirement (estimated at about 70 mA at 3.3 V). DVS can reduce system power consumption by 53%. Shutting down each of the components (analog power supply, radio module, and the processor) results in another 44% power savings, i.e., idle power management accounts for about 97% of system-wide power savings. Figure 14.10 shows overall power savings attributed to various power management hooks.

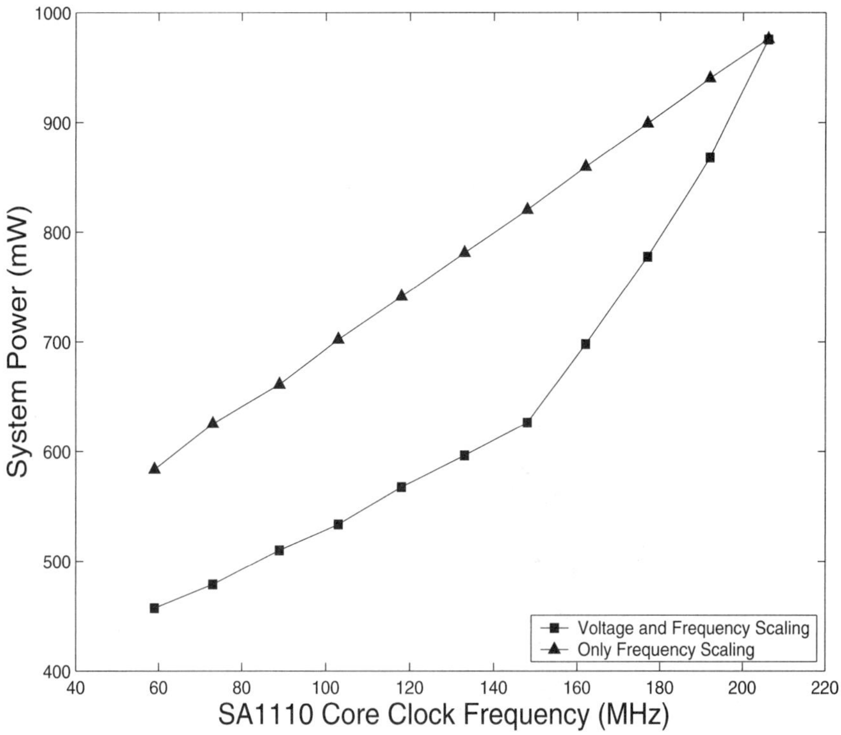

FIGURE 14.8 System-level power savings distribution.

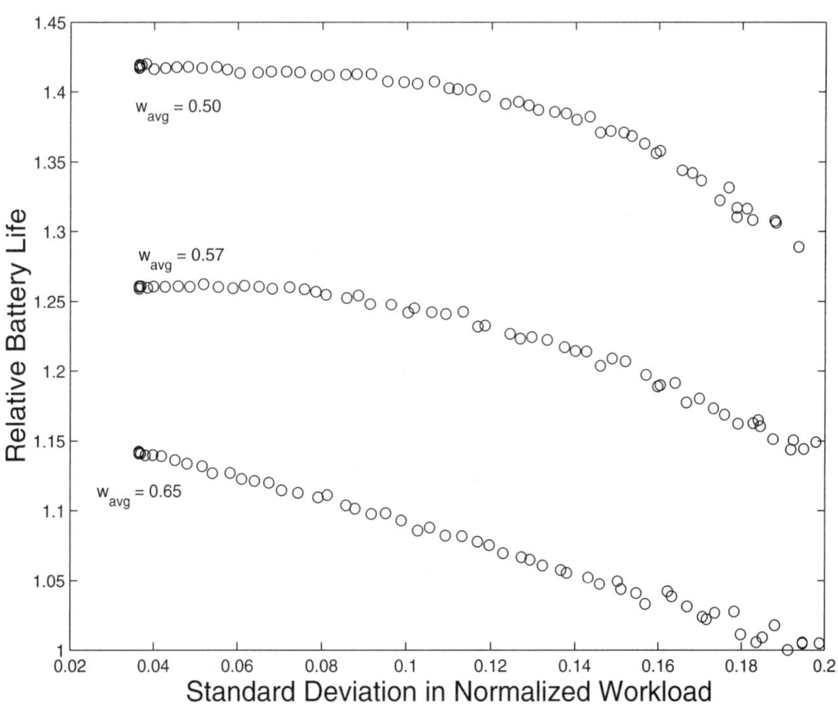

FIGURE 14.9 Battery life improvement as a function of duty cycle and active workload in a sensor node compared to a node with no power management.

TABLE 14.6 Measured Power Consumption in Various Sensor Modes

	System Mode	Component Modes: Processor	Radio	Analog	Power (mW)
Active	Active	Max Freq	on	on	975.6
States	Low Active	Min Freq	on	on	457.2
	Idle	idle	on	on	443.0
Sleep	Receive	idle	on	off	403.0
States	Sense	idle	off	off	103.0
	Sleep	sleep	off	off	28.0

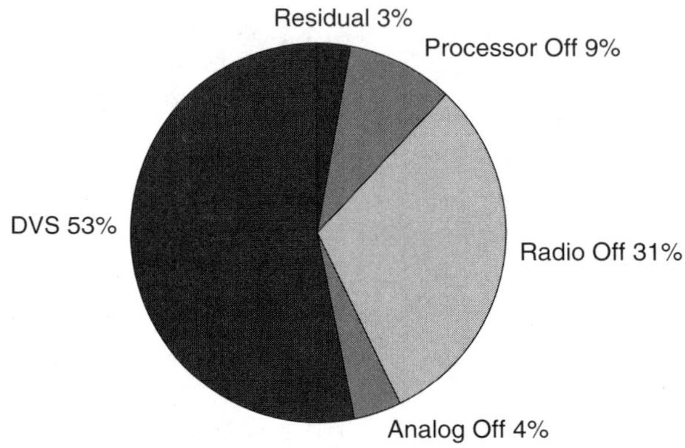

FIGURE 14.10 System level power savings distribution.

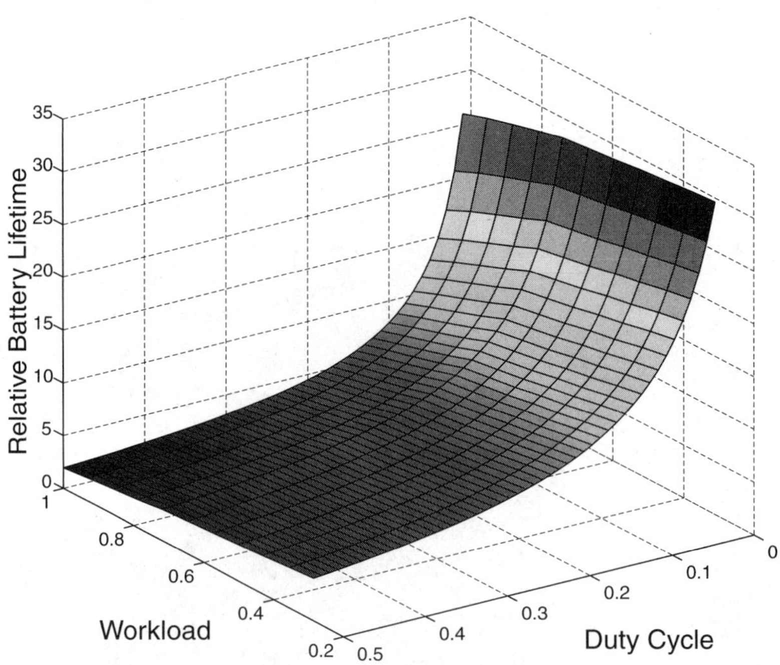

FIGURE 14.11 Battery life improvement in the sensor node compared to a node with no power management as a function of duty cycle and active workload.

Actual energy savings in the field depend significantly on processing rate requirements and event statistics. To estimate the energy savings from active mode power consumption, one would need an estimate of the workload variation on the system. If it is assumed that the average workload requirement is 50%, with slow variation, the estimated energy savings are about 30%.

Idle mode energy savings, on the other hand, can be significant. If it is assumed that the operational duty cycle is 1%, the estimated energy savings are about 96%. This implies that sensor node battery life can be improved by a factor of over 27 (i.e., a node that lasts for a day with no power management will now last for almost a month). With a 10% duty cycle, the battery life improvement is by a factor of about 10. The important point is that the system is energy scalable, i.e., it has the right hooks to tune energy consumption based on computational load and sensing requirements. Figure 14.11 shows the factor by which battery life of the sensor node can be enhanced by using power management techniques as a function of the workload and duty cycle requirement.

References

1. A.P. Chandrakasan et al., Design considerations for distributed microsensor systems, *Proc. Custom Integrated Circuits Conf.*, San Diego, CA, May 1999, 279–286.
2. The MIT μAMPS Project, http://www-mtl.mit.edu/research/icsystems/uamps/.
3. L. Benini and G.D. Micheli, *Dynamic Power Management: Design Techniques and CAD Tools*, Norwell, MA, Kluwer, 1997.
4. A. Sinha, A. Wang, and A.P. Chandrakasan, Algorithmic transforms for efficient energy scalable computation, *Int. Symp. Low Power Electron. Design*, Italy, July 2000.
5. V. Gutnik and A.P. Chandrakasan, An embedded power supply for low-power DSP, *IEEE Trans. VLSI Syst.*, 5(4), Dec. 1997, 425–435.
6. T. Pering, T. Burd, and R. Broderson, The simulation and evaluation of dynamic voltage scaling algorithms, *Int. Symp. Low Power Electron. Design*, 1998, 76–81.
7. Intel StrongARM SA-1100 Microprocessor Developer's Manual, http://developer. intel.com/design/ strong/manuals/278088.htm.
8. Advanced configuration and power interface specification, http://www.teleport. com/~acpi.
9. A. Sinha and A. Chandrakasan, Operating system and algorithmic techniques for energy-scalable wireless sensor networks, *Proc. 2nd Int. Conf. Mobile Data Manage.*, Hong-Kong, Jan 2001.
10. Intel Corp., Intel StrongARM SA-1110 Microprocessor — Advanced Developer's Manual, June 2000.
11. The eCos Operating System, http://www.redhat.com/ecos.

15

Design Challenges in Energy-Efficient Medium Access Control for Wireless Sensor Networks

Duminda Dewasurendra
Virginia Polytechnic Institute and State University

Amitabh Mishra
Virginia Polytechnic Institute and State University

15.1 Introduction 15-1
15.2 Unique Characteristics of Wireless Sensor Networks... 15-2
Why Are MAC Layer Design Issues Important?
15.3 MAC Protocols for Wireless Ad Hoc Networks..... 15-4
IEEE 802.11 • Bluetooth • Energy-Conserving Medium Access Control (EC-MAC) Protocol for Wireless ATM Networks • Power-Aware Multiple Access (PAMAS) Protocol
15.4 Design Challenges for Wireless Sensor Networks....... 15-10
Why Existing Methods for Wireless Ad Hoc Networks Cannot Be Used • Communication and Application Types in Sensor Networks
15.5 Medium Access Protocols for Wireless Sensor Networks..... 15-13
Sensor MAC (SMAC) • Self-Organizing MAC for Sensor Networks (SMACS) and Eavesdrop and Register (EAR) Algorithms • Traffic Adaptive Medium Access Protocol (TRAMA) • Power-Efficient and Delay-Aware Medium-Access Protocol for Sensor Networks (PEDAMACS) • Comparison
15.6 Open Issues..... 15-22
15.7 Conclusions 15-24

15.1 Introduction

Wireless sensor networks (WSNs) are an emerging paradigm posing new challenges for researchers in wireless communications [1]. This new class of networks closely resembles the behavior of wireless ad hoc networks. Nevertheless, they have a few unique differences; the principal one is the small size of nodes constituting a WSN. Although smaller nodes make WSNs suitable for several existing and emerging applications related to information sensing, this also implies that the nodes have limited resources, i.e., CPU speed, memory, battery, and radio interface. Because the nodes are resource constrained, they require network designs that can be customized for different types of application environments, thus placing significant demands on algorithm design, protocol specification, and technologies.

This chapter focuses on medium access control (MAC) schemes for WSNs. Unique features of these networks will be briefly discussed in order to highlight the issues demanding special attention during the design of MAC schemes. Significant research efforts currently underway in this context will be studied along with MAC schemes for generic wireless ad hoc networks (WAHNs) and wireless local area networks (WLANs). Finally, the challenges and open issues related to MAC algorithm design for the effective deployment of future WSNs will be discussed.

15.2 Unique Characteristics of Wireless Sensor Networks

WSNs consist of large numbers of distributed nodes that organize themselves to form a multihop wireless network. Each node consists of one or more types of sensors, an embedded processor, small memory, and a low-power radio transceiver. Generally, these nodes are battery powered and coordinate among themselves to achieve a common task. Compared to nodes in a generic WAHN operating under IEEE 802.11 [2] or Bluetooth [3, 4] protocols, these nodes are extremely small in size and possess limited energy resources. The transmitting power and thus the communication range are much lower, which is largely compensated by a higher density of nodes in most cases. WSNs can have distributed, hierarchical, or clustered architectures, as illustrated in Figure 15.1.

The lack of centralized control is common to WSNs as well as to other WAHNs. Nevertheless, the behavior of a WSN is largely governed by the application for which it is used. Even considering a single application, the desired role of nodes would be different from time to time. For example, in a battlefield application, it may be employed to monitor the ambient data patterns silently and generate alarms if the specified deviations are observed. The same network may be used to track the movement of a detected vehicle at another time. Such dynamic changes of network objectives and the corresponding change in node behavior are uncommon in most of the other generic WAHNs.

Furthermore, nodes of a sensor network are mostly unattended after deployment, permitting neither upgrade of energy sources nor troubleshooting. The node hardware is designed so that the overall cost is extremely low and nodes can be abandoned once the power sources are exhausted. Voids of discarded nodes may be filled with redundant nodes due to high node density, and perhaps by the deployment of additional nodes if the need arises. It is necessary that the network should accommodate such losses and new additions with least effort. These are unique issues pertaining to WSNs compared to generic WAHNs in which the nodes are mostly attended; energy sources are high capacity and can be recharged or replaced; and nodes have direct and individual interaction with users. Comparison between a WSN and a generic ad hoc network is summarized in Table 15.1.

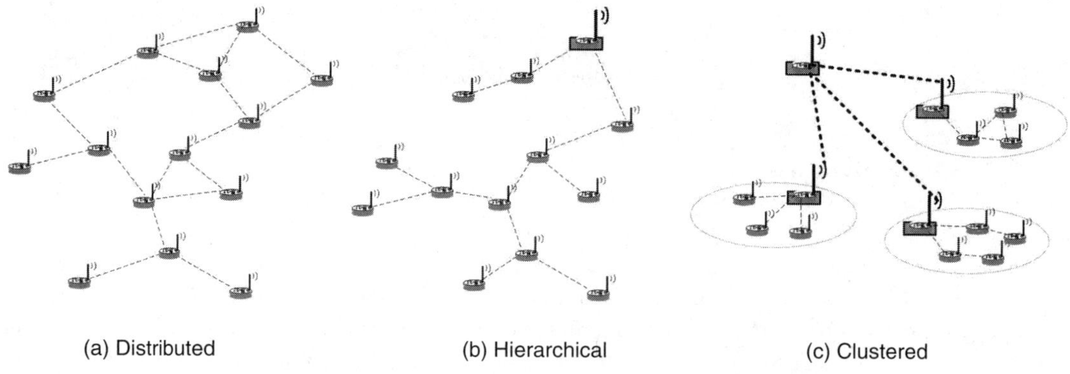

FIGURE 15.1 Sensor network architectures.

TABLE 15.1 Comparison of Features for WSNs and WAHNs

Wireless Sensor Network	Wireless Ad Hoc Network
Nodes involved in sensing the environment; events occurring in the environment can initiate certain communication in the network	No sensing behavior; network communication governed by user applications
Nodes are smaller in size	Larger nodes (e.g., PDAs, laptops)
Small and limited capacity power sources	High-capacity power sources
Inexpensive nodes	Relatively expensive nodes
Nodes unattended after deployment and designed for a prolonged lifetime with no maintenance or troubleshooting	Node troubleshooting and battery replacement possible
Node lifetime depends on the usage of attached power source	Node lifetime does not depend on energy resources because power sources are replaceable or rechargeable
Higher node density/highly redundant networks	Low node density/less redundant networks
Shorter transmission range (3 to 30 m)	Longer transmission range (10 to 500 m)
Limited processing and memory capacity	Higher processing power and memory
Nodes may stay in sleep mode for a significant amount of time	Nodes will be listening to the wireless medium most of the time
Data-centric communication; packet destination will depend on attributes of gathered data	Communication mostly occurs between specific nodes according to user requirements
Traffic profiles likely have statistically correlated properties comprising bursty traffic in case of event detection and low, continuous traffic during other times	Mostly continuous traffic, e.g., multimedia data streams
Low bandwidth (1–100 kb/s)	High bandwidth (e.g., 1 to 54 Mbps in IEEE 802.11-based WAHNs)
Network operation can be task oriented	Operation similar for all applications

15.2.1 Why Are MAC Layer Design Issues Important?

In all wireless networks, nodes must share a single medium for communication. Network performance largely depends upon how efficiently and fairly the nodes can share this common medium. Note that the packet transmission is directly handled by the MAC layer. Compared to a wired medium, a significant portion of the node's energy is spent on radio transmissions and on listening to the medium for anticipated packet reception. On the other hand, wireless networks always have restricted power sources; thus, careful design of the MAC scheme is necessary for the optimal performance and extended lifetime of the network.

In the context of WSNs, this requirement is extremely critical. According to the characteristics highlighted previously, nodes of a WSN carry extremely low energy resources and remain unattended after deployment; therefore, the node lifetime depends entirely on how energy is conserved during communication. Although some exhausted nodes could be compensated using redundant neighboring nodes, certain situations may arise rendering a part of the network completely inactive due to low connectivity and insufficient coverage, or making that part of the network inaccessible and isolated from the other parts. Such scenarios could be averted by avoiding unnecessary transmissions and longer listening periods — activities that consume the highest amount of power in nodes.

Another related issue is the high node density in WSNs. Although the transmission ranges are lower, a fairly high number of nodes can contend for the medium, at least in certain portions of the network. By the same token, transmissions from each node would increase the background noise for a large number of nodes, which may disrupt their own receptions. Thus, the MAC schemes for WSNs should be carefully designed to achieve the optimum performance toward the intended application. Previous surveys [1, 5] discuss some issues related to medium access in WSNs and WAHNs.

15.3 MAC Protocols for Wireless Ad Hoc Networks

The closest types of networks rendering a similar behavior to WSNs are WAHNs, although they have marked differences as highlighted in our discussion. Properly standardized MAC protocols designed to cater to the ad hoc and distributed nature of WAHNs have been developed and are in commercial use. Also, some of them focus on energy savings, mainly for mobile applications. These features are highly sought after in WSNs as well.

Currently available MAC protocols for wireless ad hoc networks are of two major types: *contention based* (CSMA) and *scheduling based* (TDMA, FDMA, or CDMA). In contention-based MAC schemes, the nodes compete among each other for channel access, whereas in scheduling-based methods, a specific schedule of channel access is used in time, frequency, or code domains. This section will briefly discuss several important medium access schemes belonging to both categories, including IEEE 802.11 [2]; Bluetooth [3, 4]; energy-conserving MAC (EC-MAC) [6]; and the power aware multiple access (PAMAS) [7]. Their merits, drawbacks, and suitability for WSNs will be highlighted.

15.3.1 IEEE 802.11

IEEE 802.11 is a standard developed for wireless LAN (WLAN) applications intended to replace conventional wired LANs so that the same applications can run seamlessly with media in 802.3 and 802.5 standards. Nodes in such networks would be mostly laptops and other typical equipment connected to a LAN. The distributed coordination function (DCF) in IEEE 802.11 is the access method used to support asynchronous data transfer on a best-effort basis when the network functions in an ad hoc mode. DCF can also coexist with an infrastructure network.

This is a contention-based protocol based on MACA [8] and MACAW [9] schemes proposed as improvements to the original CSMA scheme developed in Kleinrock and Tobagi [10]. It uses carrier sense multiple access with collision avoidance (CSMA/CA). Collision detection (CD) is not used because a node is unable to listen to the channel for collisions while transmitting. The scheme attempts to avoid the hidden terminal and exposed terminal problems in the original CSMA scheme.

15.3.1.1 Operation

Each node maintains a backoff counter controlling the channel access. Before a node starts data transmission, it senses the wireless medium. If the medium appears to be idle for a specified period of time (distributed interframe space — DIFS), it starts decrementing the backoff counter. If the carrier is detected during this time, the backoff counter is frozen; otherwise, it starts transmission once the backoff counter reaches zero. The sender and receiver exchange short request-to-send (RTS) and clear-to-send (CTS) control frames to establish a session. Data transmission is followed by an acknowledgment (ACK) frame to confirm successful reception. The gaps among RTS, CTS, DATA, and ACK frames are specified by short interframe space (SIFS). Duration of SIFS is relatively shorter than DIFS, thereby giving priority to the ongoing transmission. Contention for the channel access between two nodes, N1 and N2, is illustrated in Figure 15.2. Initially N1 is transmitting frame 1 followed by ACK reception. After waiting for the DIFS period, it starts decrementing the backoff counter in an attempt to transfer another packet. Because the backoff counter of N2 reaches zero first, it captures the medium and transmits a frame while N1 senses the medium is busy. Following the transmission of N2, N1 recaptures the medium for transmission of its second frame.

The DCF adopts a slotted binary exponential backoff mechanism to select the random backoff interval in case of unsuccessful transmission or after the completion of a successful transmission. This random number is drawn from a uniform distribution over the interval [0, CW-1], where CW is the contention window. After an unsuccessful transmission, CW is doubled; once CW reaches a maximum value (CW_{max}), it will remain there. In the case of a successful transmission, the CW value is reset to a minimum value (CW_{min}).

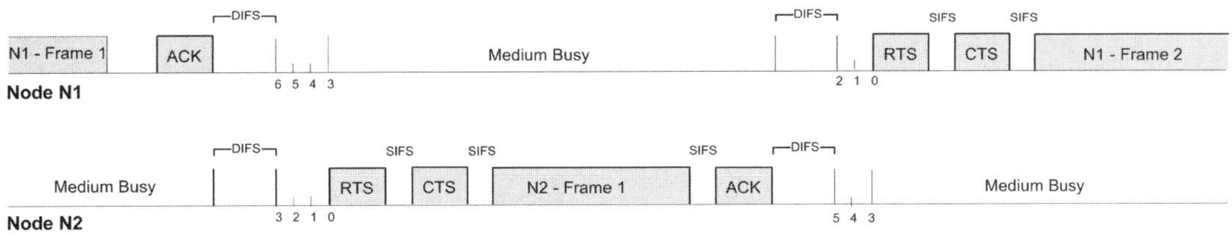

FIGURE 15.2 Contention between nodes N1 and N2 in IEEE 802.11 DCF.

The control frames RTS and CTS, as well as the data frames, include a parameter indicating the expected data transfer duration for the current session, which is used by the other nodes to update their network allocation vector (NAV). NAV is used to maintain a timer at each node, thus avoiding unnecessary transmission attempts before the current transmission is completed. This is termed a *virtual carrier sensing*. During the backoff period and the NAV timer active period, the node will be in idle mode listening to the channel with no transmission attempts.

15.3.1.2 Power-Saving Mode in IEEE 802.11

IEEE 802.11 standard also defines a power saving (PS) mode in which certain nodes can "go to sleep." Under DCF operation, PS nodes "wake up" periodically for a short interval called the Ad Hoc Traffic Indication Map (ATIM) window. It is assumed that hosts are fully connected and synchronized so that the ATIM windows of all PS hosts will start at about the same time. During this window, each node will contend to send a beacon frame. Any successful beacon serves the purpose of synchronizing node clocks and also inhibits other hosts from sending their beacons. After receiving the beacon, an active node can send a direct ATIM frame to a node in PS mode. These transmissions are also contention based and use the same DCF access procedure described earlier. On reception of the ATIM frame, the PS node will reply with an ACK and remain active for the rest of the period. Data transfer will take place after that.

15.3.1.3 Merits, Drawbacks, and Implications for WSNs

Recent work has shown that the energy consumption using IEEE 802.11 MAC protocol is significantly high because the nodes are listening to the channel most of the time. Although the 802.11 standard defines the PS mode, it provides very limited policy about when nodes should go to sleep. PS mode is designed for single-hop networks in which all nodes can hear each other. When used in multihop networks, IEEE 802.11 may have problems in clock synchronization, neighbor discovery, and network partitioning — thereby degrading the performance. Clock synchronization in a multihop WAHN is difficult because there is no centralized control; also, the synchronization packets exchanged among neighbors have variable delays due to unpredictable node mobility and radio interference.

PS mode is typically supported by letting low-power nodes wake up only at specific times. Without precise clocks, a host may not be able to know when other PS hosts will wake up to receive packets. Furthermore, a host may not be aware of a PS host at its neighborhood because a PS host will reduce its transmitting and receiving activities so that it cannot be detected. Such incorrect neighbor information may be detrimental to most routing protocols because the route discovery procedure may incorrectly report that there is no route even when routes exist with some PS hosts in the middle. Tseng et al. [11] proposed three sleep schemes to improve the PS mode in the IEEE 802.11 for its operation in multihop networks.

Requirements for clock synchronization and the suboptimal power saving makes this scheme an improper candidate for medium access in WSNs. Nevertheless, the idea of having a portion of nodes sleeping in the network while others are active may be an applicable concept to WSNs. The presence of redundant nodes in a WSN implies that all the nodes need not be active all the time because other nodes in the neighborhood can perform sensing and communication tasks covering the target area. Therefore, properly chosen redundant nodes can be put into sleep mode to achieve network-wide power savings. Open issues to be explored include the selection of redundant nodes and wake-up and connection reestablishment procedure for these nodes.

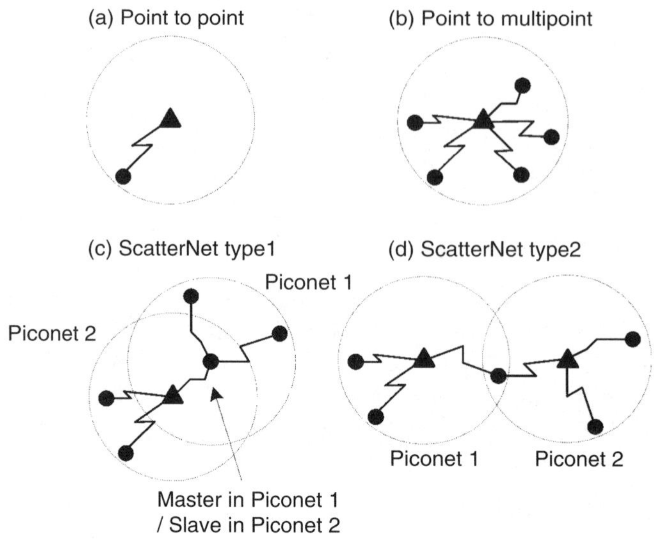

FIGURE 15.3 Piconet configurations in Bluetooth.

15.3.2 Bluetooth

Bluetooth [3, 4] is a short-range wireless networking for electronic consumer devices (mobile phones, pagers, PDAs, etc.). It uses a TDMA and CDMA hybrid scheduling-based MAC scheme. The topology is a star network in which several slave nodes are attached to and synchronized with a master node to form a piconet. The number of nodes in a piconet is limited to eight in order to keep a high-capacity link among all the units and to limit the overhead required for addressing. Basic piconet configurations are shown in Figure 15.3(a) and (b). Along with the basic TDMA scheme, Bluetooth uses frequency hopping code division multiple access (FH-CDMA), which uses a large number of pseudorandom hopping sequences. Interpiconet communication is achieved by forming ScatterNets as shown in Figure 15.3(c) and (d). A single node can be a master in one piconet while it is a slave in another. Also, a node can be a slave in two piconets.

15.3.2.1 Operation

The master node determines hopping sequence, provides clock synchronization information for each slave node, and also controls the traffic in the piconet. The master/slave role is only attributed to a unit for the duration of the piconet. When a piconet is cancelled after a certain period of time, the master and slave roles are also cancelled and new piconets will be formed. Any node can become a master or slave. By definition, the unit that establishes the piconet becomes the master. Mechanisms are in place for multiple piconets to interconnect and form a multihop topology.

The time slots are alternately used for master and slave transmissions. The master transmission includes slave address of the unit for which the information is intended. In order to prevent collisions on the channel due to multiple slave transmissions, a polling technique is used: for each slave-to-master slot, the master decides which slave is allowed to transmit. This decision is performed on a per-slot basis: only the slave addressed in the master-to-slave slot directly preceding the slave-to-master slot is allowed to transmit in this slave-to-master slot. If the master has information to send to a specific slave, it is polled implicitly and the slave can return information. If the master has no information to send, it must poll the slave explicitly with a short poll packet. Because the master node schedules the traffic in the uplink and the downlink, intelligent scheduling algorithms that take into account the slave characteristics must be used. The master node control effectively prevents collisions among the participants of the piconet. Independent collocated piconets may interfere when they occasionally use the same hop carrier.

15.3.2.2 Merits, Drawbacks, and Implications for WSNs

Compared to contention-based MAC schemes, TDMA schemes have a natural advantage of energy conservation because the duty cycle of the radio is reduced and there are no contention-introduced overheads or collisions. Nodes can be put to sleep to save energy during the off intervals of the duty cycle, thereby making this an obvious candidate for WSNs.

Use of a TDMA protocol usually requires the nodes to form real communication clusters such as the piconets described here. Nodes in such clusters are restricted to communicate within the cluster, except for the master node and possible gateway nodes. Managing intercluster communication and interference is not an easy task. Moreover, when the number of nodes within a cluster changes, it is not easy for a TDMA protocol to change its frame length and time slot assignment dynamically. Thus, its scalability is normally not as good as that of a contention-based protocol.

In Bluetooth, nodes within a piconet must be synchronized to use the TDMA scheme. Achieving local synchronization within the cluster is not a difficult task. However, network-wide synchronization will be almost impractical, especially in WSNs. Thus, proper mechanisms need to be developed for intercluster communication, perhaps based on contention-based schemes.

15.3.3 Energy-Conserving Medium Access Control (EC-MAC) Protocol for Wireless ATM Networks

This particular MAC protocol is briefly described here because of its significant contribution toward minimizing the power consumption of nodes in wireless and mobile ATM networks. Goals of this access protocol are to conserve battery power; to support multiple traffic classes; and to provide different levels of service quality through bandwidth allocation. Although the IEEE 802.11 and Bluetooth standards address energy efficiency, this was not one of the central design issues in developing these protocols. The EC-MAC protocol [6], on the other hand, was developed with the issue of energy efficiency as a primary design goal.

15.3.3.1 Operation

The EC-MAC protocol is defined for an infrastructure network with a single base station serving mobiles in its coverage area. This definition can be extended to an ad hoc network by allowing the mobiles to elect a coordinator to perform the functions of a base station. Transmission in EC-MAC is organized by the base station into frames and each frame equals the basic unit of wireless data transmission.

The frame structure of EC-MAC protocol is shown in Figure 15.4. At the start of each frame, the base station transmits the frame synchronization message (FSM), which contains synchronization information and the uplink transmission order for the subsequent reservation phase. During the request and update phase, each registered mobile transmits a new connection request according to the transmission order received in the FSM. Collisions are avoided during this phase by having the base station send the explicit order of transmission using the FSM.

New mobiles that have entered the cell coverage area register with the base station during the new-user phase. Collisions during this phase are unavoidable and thus it may be operated using a variant of ALOHA. This phase also provides time for the base station to compute the data transmission schedule. The base station broadcasts a schedule message that contains the slot assignments for the subsequent uplink and downlink data transmissions (see Figure 15.4). Downlink transmission from the base station

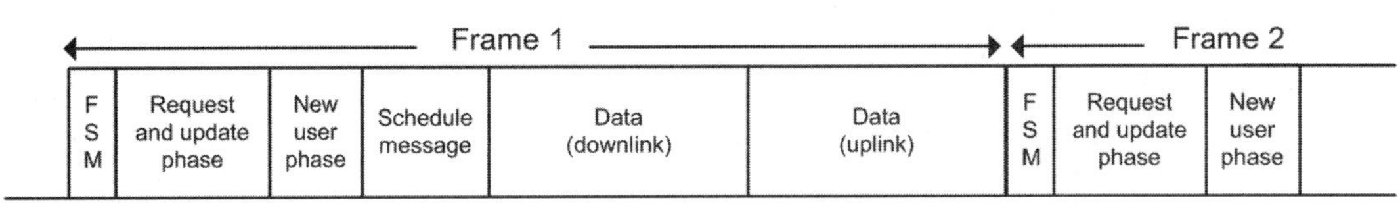

FIGURE 15.4 Frame structure in EC-MAC.

to the mobile is scheduled considering the QoS requirements; similarly, the uplink slots are allocated using a suitable scheduling algorithm.

15.3.3.2 Merits, Drawbacks, and Implications for WSNs

Energy consumption is reduced in EC-MAC due to the use of a centralized scheduler, as in Bluetooth. Therefore, collisions over the wireless channel are avoided, thus reducing the number of retransmissions. Additionally, mobile receivers are not required to monitor the transmission channel as a result of communication schedules. The centralized scheduler may also optimize the transmission schedule so that individual mobiles transmit and receive within contiguous transmission slots. This scheme highlights the fact that scheduling algorithms that consider mobile battery power level in addition to packet priority may improve performance for low-power mobiles. Techniques used to minimize the energy consumption and performance of EC-MAC in this regard are discussed in detail in Sivalingam et al. [6].

In contrast to Bluetooth, this scheme allows new mobile nodes to join the cluster without completely disassembling it. In certain WSN applications, the network may consist of a significant portion of mobile nodes among the stationary nodes. Certain stationary nodes may act as sink nodes analogous to the base stations or cluster heads discussed here. When mobile nodes roam around these clusters, a concept similar to EC-MAC can be used to attach new mobile nodes to an existing group or cluster. After such an attachment and schedule update, the mobile nodes can communicate with the cluster head, using the set schedule, at a minimal expense of its energy.

15.3.4 Power-Aware Multiple Access (PAMAS) Protocol

PAMAS (power-aware multiple access) is a contention-based protocol [7] designed for ad hoc networks with energy efficiency as the primary design goal. It modifies the MACA protocol [8] by providing separate channels for RTS/CTS control packets and data packets (out-of-band signaling), thereby avoiding overhearing among neighboring nodes.

15.3.4.1 Operation

In PAMAS, a mobile with a packet to transmit sends an RTS message over the control channel and awaits the CTS reply message from the receiving mobile. If CTS is received, then the node transmits the packet over the data channel. This procedure is shown with nodes N3 and N4 in Figure 15.5. With the start of receipt of the data packet, the receiving mobile transmits a busy tone (BT) over the control channel with more than twice the duration of RTS/CTS packets, thus enabling users tuned to the control channel to know that the data channel is busy. Also, if it hears any other RTS packets (from node N6 in Figure 15.5), it transmits a busy tone.

If an idle node receives an RTS, it will check whether any of its neighbors is transmitting (by sensing the data channel) or receiving (by sensing BT). In either case, it will not reply with CTS (shown with nodes N2 and N5), thus causing the sender of RTS (nodes N1 and N6) to back off using a binary exponential backoff (BEB) scheme. Power conservation is achieved by requiring mobiles that are not able to receive or send packets to turn off the wireless interface. The use of a separate signaling channel allows nodes to determine when and for how long to power off. A mobile should power off when: (1) it has no packets to transmit and a neighbor begins transmitting a packet not destined for it, or (2) it has packets to transmit but at least one neighbor pair is communicating.

Once a node is powered off, two main issues arise. First, the latency due to sleeping is an issue because, if some other node wants to transmit data to a sleeping node, it must wait until this node powers up again. However, it should be noted that, even if the node was awake in this scenario, the sender must wait until the other transmissions are finished. This is a valid argument as long as the node will wake up as soon as the neighboring transmissions and receptions are complete. Therefore, the mechanisms that a node will use to decide exactly when to wake up are crucial.

The second issue is determining the length of sleep duration. It is addressed using a special probe packet. When a node wakes up after some time, it will send out a probe packet over the signaling channel asking any receiving nodes how much time it will take for the current transmission to end. If no collision

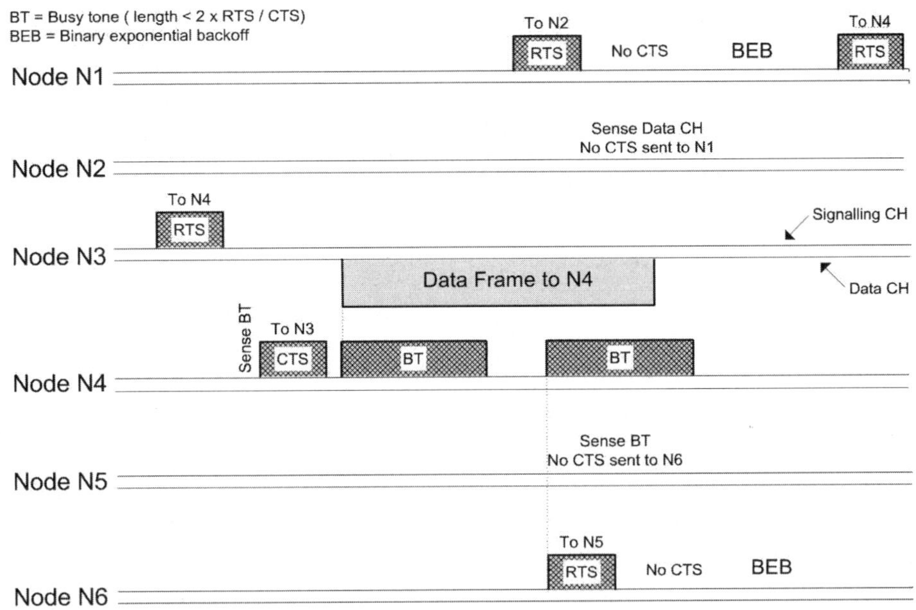

FIGURE 15.5 Operation of PAMAS.

occurs, the querying node will receive the exact time from the receiving node and will go back to sleep until this time. If the probe packet was destroyed due to a collision, the node will continue a binary search, sending more probe packets. Thus, the use of these probe packets ensures that the node sleeps no longer than necessary, thereby leaving the latency and throughput unchanged.

15.3.4.2 Merits, Drawbacks, and Implications for WSNs

Simulation results have established that this method reduces the power consumption by more than 50% in fully connected networks and at least 10% in highly loaded sparse networks. The essence of this scheme lies in the introduction of the additional signaling channel. This is also a major drawback of the scheme because employing an additional signaling channel requires additional hardware to be built into the nodes. It poses additional challenges, especially in WSNs in which node hardware is highly miniaturized.

Nevertheless, the significance of this protocol is that it can achieve high energy savings without compromising the network throughput and delay. Perhaps the concept of out-of-band signaling used here may be adaptable for intercluster communication in WSNs in which lack of synchronization does not permit using scheduling-based schemes. Such an out-of-band signaling channel can be used to set up data transfer directly between cluster heads or via gateway nodes. This will be remarkably effective for event-driven sensor networks (discussed in Section 15.4.2) in which intercluster communication occurs mostly in cases of a detected event and the regular communication is restricted primarily to nodes within the clusters.

A major contribution of the PAMAS protocol is the power savings achieved without sacrificing network throughput and latency. However, a major drawback observed here is that the power consumption of the nodes during excessive switching between the sleep and wake-up states is not given due attention. With the present WSN hardware designs, power consumption during state switching is significant. At the face of this, the PAMAS method may not perform satisfactorily without appropriate modifications for WSNs. Table 15.2 provides a brief comparison of these four medium access methods for WAHNs.

TABLE 15.2 Comparison of Media Access Protocols for Wireless Ad Hoc Networks

Protocol	Applications	Features	Implications to WSNs
IEEE 802.11 DCF	Wireless LAN	Optional PS mode	Redundant nodes can be sent to sleep and wake up as need arises
Bluetooth	Wireless networking for personal consumer devices	Piconets; centralized scheduling	Node clustering; local synchronization among nodes in clusters
EC-MAC	Wireless and mobile ATM networks	Scheduling for mobile nodes	Attaching mobile nodes to clusters without disassembling clusters
PAMAS	MAC protocol designed for WAHNs	Out-of-band signaling	Use of out-of-band signaling for intercluster communication

15.4 Design Challenges for Wireless Sensor Networks

As discussed in the first section, nodes in a WSN possess unique characteristics, especially the energy constraints, compact hardware, low transmission ranges, event- or task-based network behavior, and high redundancy. For a WSN, the extension of its lifetime is the most important issue. Therefore, power awareness is prominent in almost every aspect of the operation of WSNs. Currently, the research related to hardware of WSNs is focused on developing ultra low-power sensors, processors, and radio transceivers. Other drives are to reduce the form factor of batteries and improve technologies for power sources to keep nodes alive in active operation for many years. Meanwhile, software and middleware development is focused on minimizing power consumption during network operation. As highlighted in Section 15.2.1, it is extremely critical that the medium access control scheme be power optimal.

Energy consumption of a WSN occurs in three domains: sensing; data processing; and communications; among these, radio communication is the major consumer of energy. As highlighted in Pottie and Kaiser [12], energy for transmitting 1 kb over a distance of 100 m is estimated as 3 J. With the same amount of energy, a general purpose processor with 100 MIPS/W power could execute 3 million instructions. The sensing circuitry consumes less power than the processor board in a typical WSN platform such as MICA [13, 14]. However, the radio consumes two to three times the power of the processor during packet transmission. Power consumption of the radio during listening to the channel for reception is also higher than the processor at full operation, but relatively lower than the transmitting power. The MICA sensor network platform defines four modes of operation, and Table 15.3 shows the typical current draw and power consumption of each node.

Thus, it is clear that the research focus should be on optimizing the medium access method in order to extend the lifetime of the network. In addition to energy conservation, the ability of the MAC scheme to adapt to network size, node density, and topology is also important. To be used in sensor networks aimed for dynamic applications, a MAC scheme should be highly scalable. Other important attributes include fairness, latency, throughput, and bandwidth utilization. However, these issues are considered secondary compared to energy considerations because they determine the entire lifetime of the network.

TABLE 15.3 Modes of Operation in MICA

Mode	Typical Current Draw	Power Consumption
Transmit (peak power)	32 mA	95 mW
Receive	18 mA	55 mW
Idle/sense	8 mA	25 mW
Sleep	20 μA	60 μW

Source: Crossbow Inc., Data sheet for MICA2 wireless measurement system, 2003.

Similar to the schemes described in Section 15.3, MAC schemes for sensor networks can be fundamentally categorized into *contention-based* or *scheduling-based* schemes. The inherent advantages of contention-based schemes in the context of WSNs include:

- No synchronization requirements
- No central scheduler required
- More robust to network dynamics
- No clustering necessary
- More suitable for event-driven WSNs

However, in terms of energy savings, contention-based schemes are not very attractive. Several sources of energy wastage in contention-based schemes during communication [15] can be identified:

- *Collision*. Usually data gathered by a node are exchanged with others using the radio. Two nodes may transfer data to each other at the same time or several nodes transfer data to the same node at the same time. When a transmitted packet is corrupted, it must be discarded and, thus, the follow-on retransmissions increase energy consumption. Collision increases latency as well.
- *Overhearing*. When a node picks up packets destined to other nodes, overhearing occurs. In an ad-hoc fashion, a transmission from one node to another is potentially overheard by all the neighbors of the transmitting node; thus, all of these nodes consume power even though the packet transmission was not directed to them.
- *Control packet overhead*. Sending and receiving control packets such as routing updates consumes energy and effectively reduces the network bandwidth for data packets.
- *Idle listening*. Nodes must listen to the channel often in order to receive possible traffic that is not sent. This is especially true in many sensor network applications because, if nothing is sensed, nodes are in idle mode for most of the time. Actual measurements have shown that idle listening consumes 50 to 100% of the energy required for receiving in such networks.

Scheduling-based schemes attempt to determine network connectivity first (i.e., discover the neighbors of each node) and assign collision-free links to each node. Links may be assigned as time slots (TDMA), frequency bands (FDMA), or spread spectrum codes (CDMA). However, the miniature hardware design of nodes in a WSN may not permit employing complex radio transceivers required for FDMA or CDMA systems. Thus, TDMA schemes are preferred as scheduling methods for WSNs. Inherently, TDMA schemes have a distinct advantage over the other methods. Except for the transmission, receiving and sensing durations, nodes can be put to sleep in order to achieve the highest amount of energy savings possible.

Nevertheless, the task of assignment of channels (i.e., TDMA slots, frequency bands, or spread spectrum codes) to links between neighbors so that packets do not collide is difficult. To ease the assignment, often a hierarchical structure is formed in the network to localize groups of nodes and make the task of channel assignment more manageable. This requires formation of node clusters and elect leaders for each cluster.

For TDMA schemes, time synchronization is a crucial factor, also. In contrast to generic WAHNs, maintaining perfect synchronization over the whole network is almost impossible, mainly because of the wide range of deployment, lower transmission ranges, and less control packet transmissions permitted due to energy constraints. Under a hierarchical clustering scheme, synchronization within each cluster can be maintained, but intercluster communication poses problems because of lack of synchronization.

15.4.1 Why Existing Methods for Wireless Ad Hoc Networks Cannot Be Used

The main goals of WAHNs are to provide a high throughput and low delay at a high bandwidth. In such networks, all nodes are engaged in the same type of activity and each user deserves equal opportunity in accessing the media. Thus, per-node fairness is an important issue. Network lifetime is not considered significant because the energy sources can be recharged or replaced, although power-saving schemes are recommended.

In contrast, for a WSN, extending the network lifetime is one of the priorities. To this end, it is necessary to conserve energy at each node during network operation. Toward achieving this objective, one must be ready to compromise network throughput and latency to a certain extent. Moreover, based on the type and operation of the WSN, as described in Section 15.4.2, throughput and latency requirements will depend on the application.

Also in contrast to WAHNs, nodes in WSNs are highly redundant; thus, some nodes can afford to be in sleep mode until a need arises, while others are active. Certain nodes acting as cluster heads, gateways, or sink nodes would accumulate, process, and relay larger quantities of data than ordinary, leaf-level sensor nodes. Additionally, from time to time during certain applications, the importance of data sensed by a node may vary in its importance or relevance to the current network objective. These issues call for maintaining distinct node priorities in WSNs in contrast to per-node fairness desired in WAHNs. Thus, employing MAC schemes that are developed for WAHNs would not be satisfactory in sensor networks without proper modifications.

In summary, novel MAC protocols are needed for WSNs because:

- Extending network lifetime is the primary goal in WSNs.
- Throughput and delay performance become secondary goals.
- QoS requirements may vary from time to time (e.g., in an event-driven WSN).
- Per-node fairness is not desired; instead, distinct node priorities may need to be considered for resource allocation.

15.4.2 Communication and Application Types in Sensor Networks

This section attempts to categorize the application-led behavior and possible communication types in WSNs. This is a general categorization that may help to identify relevant issues in designing an optimum medium access scheme for WSNs. Depending on application characteristics, sensor networks will behave in one of the following ways. It is possible that the same network may adopt a different role due to changes in the system objectives or firing of certain events in the observation field.

- *Centralized* data gathering and decision making. These networks are hierarchically organized, thus easier to set up and manage. At the top of the hierarchy, one or more root (sink) nodes collect all data from leaf nodes. Local processing may be performed at the sensor nodes at the bottom of the hierarchy, but the root node is responsible for gathering final data and, for the most part, governing the operation of the whole network.
- *Distributed* data gathering and decision making. Tasks in these networks are highly distributed and it is difficult to identify a particular network architecture or a hierarchy. An example would be a set of sensor nodes dropped in a harsh environment with no central control. Individual nodes must perform whatever sensing operations they can, discover their own neighbors and perhaps collectively make decisions on discovered events, and relay the decisions to the outside world via any relay nodes in reach. Instead, they would control certain actuators within their reach to perform certain reactive actions individually or by temporarily appointed leader nodes.

In both data gathering schemes, the following types of communication can occur:

- Unicast messages:
  - *Local.* When a real-world event in the network occurs, nodes are expected to perform some in-network processing. This will generally involve local messages being exchanged between neighbors. In a cluster-based scheme, this will include the messages exchanged between two nodes in the cluster or between a member node and the cluster head.

  - *Multihop.* In centralized data gathering applications when a node requires sending data to the sink node (node-to-sink reporting), the sink node will not be in its direct reach most of the time. Thus, it will pass the message with the intended recipient address over multiple hops. This needs a proper addressing scheme as well as an efficient routing scheme.
- Multicast messages:
  - *Local.* These are messages originating from a node intended for several neighbors within its direct transmission range. For a clustering-based scheme, this is limited to multicast within the cluster.
  - *Multihop.* An example is a situation in which a sink node or a root node requires passing a control message to a set of nodes. In a clustering-based scheme, this may be communication from a root node to a set of cluster heads, or from one cluster head to several others. All such multihop messages need a proper addressing scheme as well as an efficient routing scheme, as mentioned earlier.
- Broadcast messages:
  - *Local.* Messages will be broadcast by a node to all the neighbors within its reach. Such messages will include anything of local importance to the neighborhood. In a clustering-based scheme, these will include messages broadcast among all nodes in a group.
  - *Multihop.* These are the messages that will impose the heaviest communication burden on the network. For instance, in a monitoring and surveillance application, a node may observe an alarming condition and may need to alert all others in the network. Unrestricted flooding may not be appropriate for such a situation, but a combination of multihop–multicast with local broadcast may be used.

Based on the application, optimal communication strategy for a sensor network would also be different. Two major categories of sensor networks dictated by their applications [16] have been identified:

- *Event driven sensor networks.* In an event-driven sensor network, sensor nodes do not send data (and are most likely asleep) until a certain event occurs. For example, in a fire-monitoring application, until a rise in temperature or smoke is detected, no data need be sent. In this way, node energy can be maximally saved. When an event occurs, how quickly the event can be reported to a central station, or how quickly other neighboring nodes can be alerted, become important issues. The main difficulty in an event-driven sensor network is to wake up the entire network, or at least the nodes along a path to the base station, when an event occurs. Moreover, the traffic pattern of the network may drastically change in case of an event.
- *Continuous monitoring sensor networks.* In a continuous monitoring sensor network, data are sampled and transmitted at regular intervals. For example, an ambient temperature monitoring station can take periodic readings and send it to a central monitoring station only at specific intervals. In these types of networks, the traffic patterns are more stationary and the routing tables (if any) remain unchanged most of the time. Scheduling-based MAC schemes can be used effectively in these networks for maximum energy savings.

The behavior and communication types identified in this section need to be considered for the optimal performance of energy-aware MAC schemes. Next, several MAC schemes proposed for WSNs will be discussed, along with their applicability to these different types of networks.

15.5 Medium Access Protocols for Wireless Sensor Networks

Recently, several authors have suggested energy-aware medium access schemes for WSNs, a number of which are modifications of existing protocols for WAHNs. This is still a growing area of research calling for attention to several open issues yet to be addressed. This section discusses four such recently proposed schemes, with their merits and drawbacks, in the context of WSNs. These include sensor MAC (SMAC) [15]; self-organizing MAC for sensor networks (SMACS) [17]; traffic adaptive medium access protocol

(TRAMA) [18]; and power-efficient and delay-aware medium access protocol for sensor networks (PEDAMACS) [19].

15.5.1 Sensor MAC (SMAC)

The main objective of SMAC [15] is to conserve energy in sensor networks; it takes into consideration that fairness and latency are less critical issues compared to energy savings. Thus, this scheme compromises fairness and latency to a certain degree. In order to save energy, SMAC establishes a low duty cycle operation in nodes. It reduces idle listening by periodically putting nodes into sleep in which the radio transceiver is completely turned off. As discussed in Section 15.2.1, a high bandwidth utilization is a goal in generic WAHNs, compelling nodes to operate in fully active mode all the time. In SMAC, the low duty cycle mode is the default operation of all nodes in the network. Nodes only become more active by changing the duty cycle when heavy traffic is present in the network, or once an event occurs in case of an event-driven WSN.

15.5.1.1 Operation

During the design of SMAC, the following assumptions have been considered:

- Short-range multihop communications will take place among a large number of nodes.
- Most communications will be between nodes as peers, rather than to a single base station.
- In-network data processing is used to reduce traffic.
- Collaborative signal processing is used to reduce traffic and improve sensing quality.
- Applications will have long idle periods and can tolerate some latency.
- Network lifetime is critical for the application.

All nodes in the network will be following a sleep-and-listen cycle called a frame, as shown in Figure 15.6. The duration of the listen period is normally fixed and the sleep interval may be changed according to application requirements, changing the duty cycle.

The same RTS/CTS/DATA/ACK procedure as that in IEEE 802.11 is adopted here for unicast packets in order to ensure collision avoidance and to avoid hidden terminal problem. Broadcast packets are sent without using RTS/CTS; NAV timer update information is included in all four types of packets. Thus, this scheme uses virtual and physical carrier sensing. After a successful exchange of RTS and CTS, the sender will start transmission and will extend it into the sleeping duration as well, if required. The nodes do not follow their sleep schedules until they finish the transmissions, thus increasing the performance.

15.5.1.2 Coordinated Sleeping

Although a node can freely choose its own active/sleep schedules in SMAC, it attempts to reduce the overhead by synchronizing schedules of neighboring nodes together. Nodes exchange their schedules by periodically broadcasting a SYNC packet to their immediate neighbors at the beginning of each listen interval. A set of nodes synchronized together will form a virtual cluster. Because the whole network cannot be synchronized together, neighboring nodes are allowed to have different schedules. However, neighboring nodes are free to talk to each other, no matter to which listen schedules they adhere. A considerable portion of the nodes will belong to more than one virtual cluster, enabling intercluster communication. Thus, this scheme is claimed to be adaptive to topology changes.

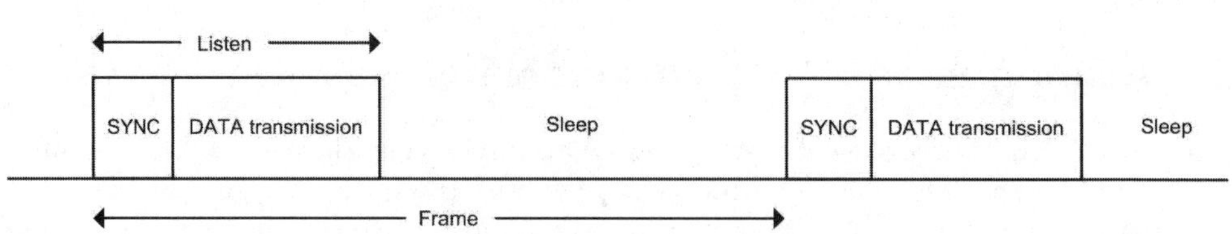

FIGURE 15.6 Periodic listen-and-sleep schedule in SMAC.

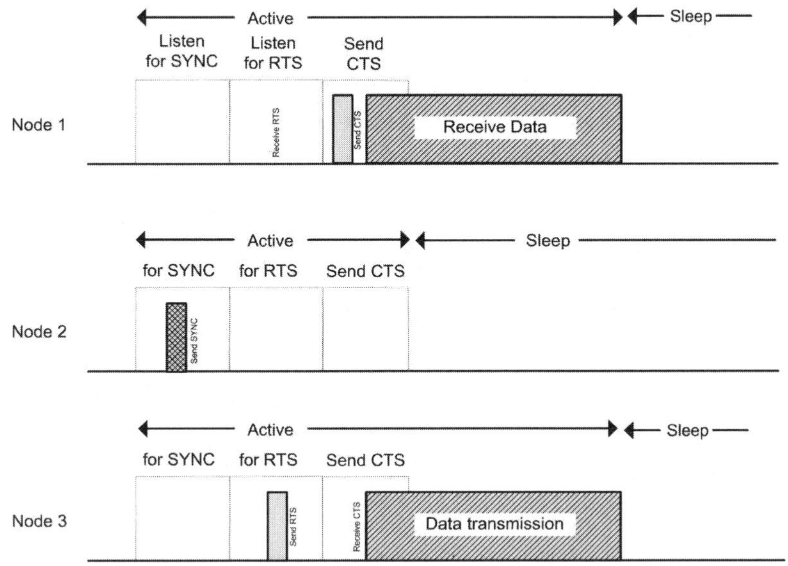

FIGURE 15.7 Timing schedules among different nodes in SMAC.

When the network is first deployed, each node tries to retrieve a sleep schedule from a neighbor first. In case of failure, it adopts one of its own and also tries to announce it to the neighbors by broadcasting a SYNC packet. Broadcasting SYNC packets must also follow the normal carrier sense and random backoff procedure. If a node receives a different schedule after it announces its own schedule, it must adopt one of the following:

- If the node detects no other neighbors, it can discard the current schedule and adopt the new.
- If it has one or more neighbors and is already a part of an existing virtual cluster, it can adopt both schedules by waking up at the listen intervals of both.

The active interval of a node is divided into three parts for SYNC, RTS, and CTS as shown in Figure 15.7. If CTS is received, data transmission will be immediately followed. Here nodes 1 and 3 are synchronized to the schedule of node 2 by receiving its SYNC packet, thus falling into the same virtual cluster. Node 3 initiates an RTS/CTS exchange with node 1 followed by a data transmission. While node 2 follows its normal sleep schedule, nodes 1 and 3 stay active until the completion of the data transfer, altering their usual schedule.

15.5.1.3 Neighbor Discovery in SMAC

When a new node powers on, it listens to the channel in anticipation of a SYNC packet. However, it is possible that a new node fails to discover an existing neighbor because of collisions or delays in sending SYNC packets by neighbor due to busy medium. To prevent a case in which two neighbors cannot find each other when they follow completely different schedules, SMAC protocol employs a simple periodic neighbor discovery procedure by requiring each node to listen periodically to the channel for the whole synchronization period. The frequency can be varied depending on the network conditions, etc.

15.5.1.4 Synchronization

Clock drift on each node can cause errors in the coordination of schedules among neighboring nodes. To minimize this problem, it uses relative timestamps in SYNC packets. Also, the listen period is made significantly longer than possible clock drift. Although this technique can tolerate relatively larger clock drifts, neighboring nodes are still required to update each other periodically with their schedules to prevent possible errors. Using experiments, authors claim that the clock drift between two nodes does not exceed 0.2 ms per second [15]; however, these figures may not be valid for certain applications of WSNs.

15.5.1.5 Adaptive Listening

An adaptive listening strategy that enables each node to adjust its schedule according to the network traffic is also used in SMAC to minimize latency. When a sensing event occurs, it is desirable that the sensing data can be passed through the network without much delay. When each node strictly follows its sleep schedule, potential delays are possible on each hop in the multihop path. The technique is that, if a node overhears its neighbor's RTS/CTS transmission during a listen period, it will receive the estimated length of that data transmission before going to sleep according to its normal schedule. However, the node will wake up for a short period of time at the end of that transmission to check whether it is the next hop in this multihop message. If so, the neighbor will immediately pass the data to it after RTS/CTS exchange, thus avoiding the neighbor's waiting for the next scheduled listen time of this node and minimizing latency.

15.5.1.6 Merits and Drawbacks

Compared to other schemes designed for the mobile ad hoc networks explained in Section 15.3, SMAC is designed particularly for use in wireless sensor networks. It attempts to combine the advantages of TDMA scheduling for power saving by periodically requiring sensing nodes to go to sleep. The sleeping patterns are coordinated in order to minimize the latency, as discussed before. Nevertheless, a solution with a fixed duty cycle does not give the optimal performance.

Authors claim that this scheme forms a flat topology and intercluster problems are absent; however, this may not be true in cases in which the application requires real clusters to be formed, at least temporarily. In such a case, the communication patterns will depend on the cluster formation and that these real clusters and the virtual clusters formed will coincide is not guaranteed. The adaptability of this scheme to such a situation should be investigated.

A significant portion of nodes will belong to two or more virtual clusters under this scheme. The energy consumption of such nodes would be higher compared to nodes within a single virtual cluster. Hence, the portion of such nodes and its effect on performance should be analyzed under real application scenarios. Also, the performance of this MAC scheme should be studied along with different routing schemes in order to assess its performance of intercluster communication, especially for multihop unicast and multicast messages. Data routing across virtual clusters needs to be studied further for its latency and throughput.

In WSN applications, it is possible for certain nodes to be exhausted with power and new nodes to be added. Performance of SMAC during times when a significant portion of nodes is discarded or added, or in cases with a higher portion of mobile nodes, should be studied. Another instance to be observed is what happens if the coordinated sleep schedules of two neighboring clusters are completely opposite. In such cases, it is not clear whether the bordering nodes could adopt both schedules.

15.5.2 Self-Organizing MAC for Sensor Networks (SMACS) and Eavesdrop and Register (EAR) Algorithms

Self-organizing MAC for sensor networks (SMACS) is designed for network startup and link layer organization in a static WSN [17]; it is one of the earliest attempts to develop MAC protocols for sensor networks. In this scheme, each node maintains a TDMA frame in which the node schedules different time slots to communicate with its known neighbors. During each time slot, it only talks to one neighbor. To avoid interference between adjacent links, the protocol uses different frequency channels (FDMA) or spread spectrum codes (CDMA). Although the frame structure is similar to a typical TDMA frame, it does not prevent two interfering nodes from accessing the medium at the same time. The actual multiple access is accomplished by FDMA or CDMA.

15.5.2.1 Operation

For the correct operation of the SMACS protocol, the following assumptions are made.

1. Nodes are able to tune the carrier frequency to different bands. It is assumed that the number of available bands is relatively large.

2. Nodes are randomly deployed. After deployment, each node wakes up at some random time according to a certain distribution.
3. The network is assumed to consist primarily of stationary nodes, with few mobile nodes.

Each node assigns links to its neighbors immediately after they are discovered. When all nodes hear all their neighbors, they have formed a connected, multihop network. Because each node is only partially aware of the radio connectivity in its vicinity, it is possible that collisions can occur if a simple TDMA scheme is used alone. To avoid this, frequency bands chosen at random from a large pool are assigned for each slot.

Length of the frame T_{frame} is fixed for all nodes in the network; however, these frames need not be synchronized and the time slots assigned inside the frame need not be aligned. This is possible because different frequency band or CDMA codes are used for communication during each slot. Such an ability to assign nonsynchronous slots is the key property that enables nodes to form links on the fly. This is illustrated in Figure 15.8, in which nodes A, B, C, and D are in the same neighborhood after deployment and they wake up at times T_1 to T_4, respectively.

Nodes A and B discover each other first and establish their own schedules for transmission and reception. Nodes C and D wake up at later times, discover each other, and establish their own schedules. However, note that all schedules are not aligned and also that the transmission slots of pair A/B and pair C/D overlap. This is made possible by using distinct frequencies f_x and f_y. After a schedule is established, a node will turn on its transceiver ahead of appropriate slots to communicate with others. Similarly, it will turn off the radio when no communication is scheduled, thereby enabling significant energy savings. In most WSN applications, mobile nodes will be present among other stationary nodes. In order to attach these mobile nodes in an energy conserving manner to the already formed network using SMACS, the eavesdrop and register algorithm (EAR) is introduced and discussed in the following section.

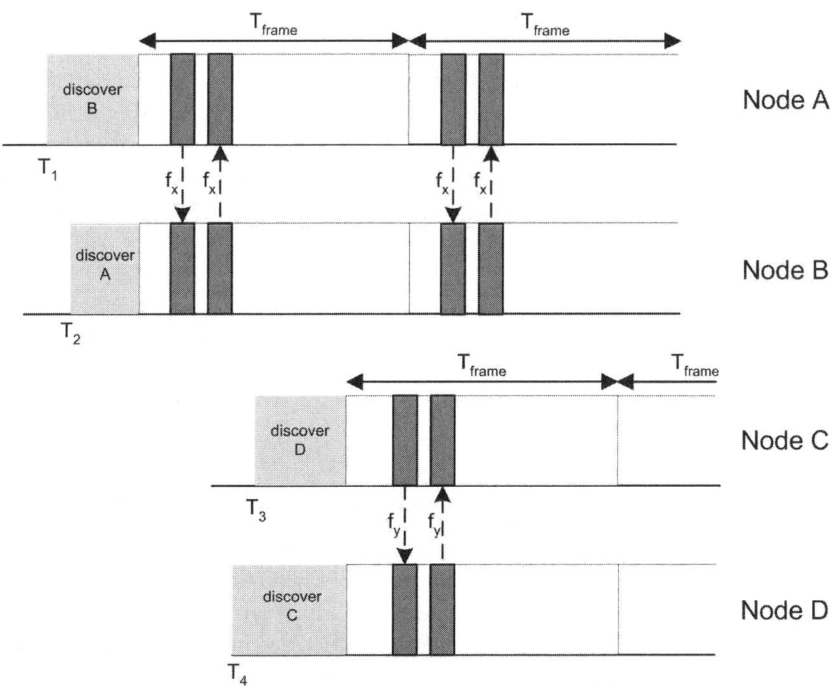

FIGURE 15.8 Nonsynchronous scheduled communication in SMACS.

15.5.2.2 EAR Algorithm

The EAR algorithm enables seamless interconnection of mobile nodes in the field of stationary wireless nodes. This protocol performs the mobility management of the network allowing mobile nodes to listen to the communication from the stationary nodes and establish connectivity with them. Because of energy limitations, the communication channels between the mobile and stationary sensors in the network must be established using as few messages as possible. This is accomplished by allowing the mobile node to decide when to invite the stationary node to establish a connection as well as when to drop a connection. In this manner, mobile nodes assume full control of the connection process to avoid the unnecessary use of power associated with lost messages.

According to the preceding third assumption, only a few stationary sensors will be within the reach of a mobile sensor at any given time. During some predetermined slot in the frame of each stationary node, it transmits an invitation message to the surrounding neighborhood with the intent of inviting new nodes to join the local network. Stationary nodes do not necessarily require a response to this message, but a mobile node with the intention of joining the network will be eavesdropping on such messages. These pilot messages will trigger the EAR algorithm in mobile nodes.

Each mobile node will maintain a list of neighbors according to the invitation messages received. It compares parameters, such as the received SNR, node ID, transmitted power, etc., and decides which node to connect. When the energy saving requirements are stringent, the decision will aim solely for minimal power connectivity. Accordingly, the mobile nodes will initiate a connection with a stationary node. Stationary nodes will also maintain a simple list of mobile nodes that have formed connections and remove the entries when the link is broken.

15.5.2.3 Merits and Drawbacks

Merits of this scheme include the ability to form links with any neighbor on the fly, with no restrictions on synchronization, which largely reduces the latency. Another advantage is that this scheme is applicable to WSNs with neither physical nor virtual clustering. This will allow the MAC scheme to function independently of any application-based clustering requirements of WSN.

However, a significant waste of resources is a trade-off to low latency. The main drawback is that the time slots are wasted if a node does not have data to send to the intended receiver. Also, the frames of a large number of nodes will mostly be vacant if the nodes are sparsely distributed in certain areas. Defining a smaller T_{frame} value will not be permitted in this case because some areas of the network may have a higher density of nodes. SMACS does not attempt to utilize these vacant time slots in order to maintain simplicity; rather, this protocol uses FDMA or CDMA and thus unnecessarily complicates the node hardware design. Bandwidth utilization would also be lower for the same reasons. Another major drawback is that the energy waste during the switching between sleep and active states is not considered. Because of assigning time slots on the fly without any synchronization, the nodes must switch between active and sleep states many times. This will drain the energy sources of nodes unnecessarily.

Apart from these drawbacks, this EAR protocol should be studied further in order to develop effective ways to manage WSNs with mobile as well as stationary nodes.

15.5.3 Traffic Adaptive Medium Access Protocol (TRAMA)

TRAMA is a recently introduced MAC protocol for energy-efficient and collision-free channel access in WSNs [18]. It uses traffic-based information to decide on schedules for individual nodes and thus is adaptive to network traffic. It is claimed that, because of this adaptability, it can deliver adequate performance and energy efficiency in both network types discussed in Section 15.4.2. TRAMA provides support for unicast, broadcast, and multicast traffic.

15.5.3.1 Operation

TRAMA assumes a single, time-slotted channel for data and signaling transmissions. The time schedule of each node is organized in two major sections, as shown in Figure 15.9. One consists of a collection of

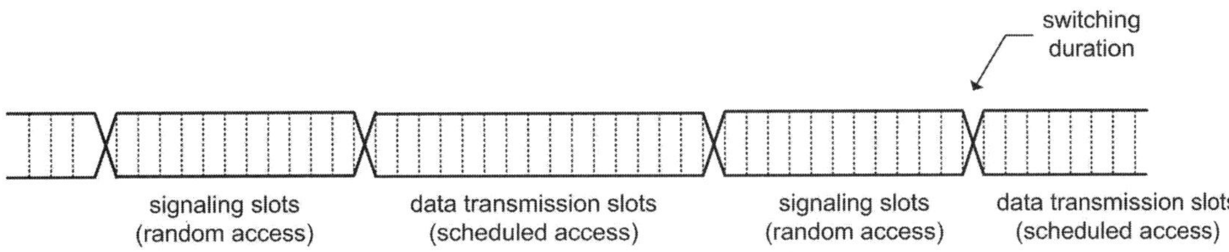

FIGURE 15.9 Time slot organization in TRAMA.

signaling slots using random access and the other of data transmission slots using schedules access. The duty cycle of switching between these states could be adjusted according to the application requirements and also according to the different network types described in Section 15.4.2. For stationary networks, the random access periods occur less frequently and vice versa for highly dynamic networks. Cycle duration is usually of the order of tens of milliseconds, making this scheme less prone to even significant clock drifts around 1 msec, which are highly unlikely in typical networks. Thus, the scheme assumes that adequate synchronization can be achieved using one of the synchronization schemes suggested for WSNs.

Communication in TRAMA consists of three major components: neighbor protocol (NP); the adaptive election algorithm (AEA); and the schedule exchange protocol (SEP). NP is used to exchange one-hop neighbor information among neighbors and to gather two-hop topology information for each node in the network. This is performed by exchanging small packets among neighbors during the random access period. Nodes always start in random access mode with NP. Also, the synchronization is performed during the random access period and the node should be in active state (transmit, receive, or listen) during this interval.

During the random access period, NP exchanges short signaling packets that include the information about connected neighbors of the sender; the goal is to provide the two-hop neighbor information to each node. Note that these include incremental information to keep the packet length small, i.e., it contains the node IDs of newly added neighbors and disconnected neighbors. These short packets are also used to maintain connectivity between neighbors.

During scheduled access, AEA selects transmitters and receivers so that collision-free transmission is achieved. AEA is based on the neighborhood-aware collision resolution protocol (NCR) proposed in Bao and Garcia–Luna–Aceves [20]. This technique claims to avoid data packet collisions among neighbors due to hidden terminals. AEA uses traffic-based information exchanged among nodes during SEP to make efficient use of the channel avoiding idle slots. The same traffic information is also used in AEA to perform receiver selection. During these selections, the node priorities in the network and two-hop neighbor information exchanged during NP are considered. By selecting the transmitter and receiver for each time slot, AEA enables nodes to switch into sleep mode whenever possible, thus achieving maximum energy savings.

SEP is used to exchange traffic schedules among neighbors during scheduled access mode. These schedules contain the set of receivers for the traffic currently originating at the node and its scheduled transmission slots. This information is periodically broadcast to the node's one-hop neighbors during scheduled access. Each node computes a schedule interval depending on how often it needs to transmit data according to its current application requirements. Following this, the node selects the highest priority slots it can acquire according to AEP; an example is illustrated in Figure 15.10. This information (schedule) is transmitted to its neighbors typically during the last slot of its schedule. Also, after emptying the current data buffers, it announces the release of its vacant time slots so that the other nodes can acquire them.

State of a node at a particular time slot is determined based on its two-hop neighborhood information and the schedules announced by its one-hop neighbors. Three possible states of a node are: transmit, receive, and sleep. At a given time slot, a node is in transmit state if it has the highest priority among its

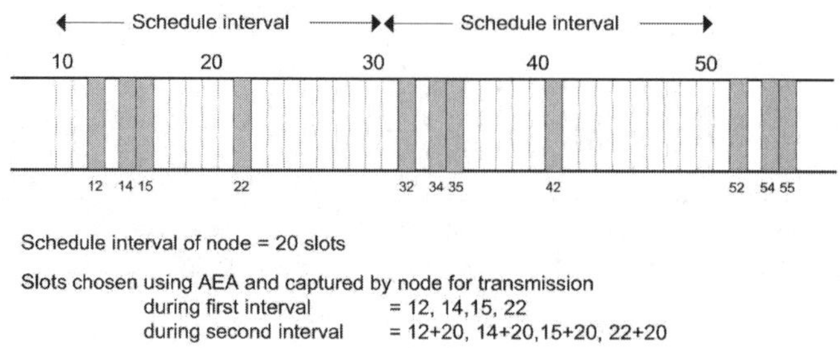

FIGURE 15.10 Example schedule of a node in TRAMA.

contending set and also if the node has data to send. A node is in the receive state when it is the intended receiver of current transmitter. If neither of these cases occurs, the node will be switched off to the sleep state in order to save energy. In other words, if a node is not the currently selected transmitter by AEP, it will consult the schedule sent by the current transmitter. If the transmitter does not have traffic destined for this node in the current slot, it can go to sleep. Under this scheme, a sleeping node is required to wake up at the schedule announcement slot (usually the last transmission slot in each schedule interval) to update itself on possible schedule changes. For this purpose, it should always be aware of the schedule of each of its one-hop neighbors.

15.5.3.2 Merits and Drawbacks

The authors provide extensive simulations to compare TRAMA with SMAC and several other comparable MAC schemes [18]. It is shown that the scheduled-based medium access protocol based on neighborhood-aware collision resolution protocol (NCR) achieves better data delivery than the contention-based protocols such as IEEE802.11, CSMA, and S-MAC. The main reason highlighted for the improvement in delivery is that the freedom from collision is guaranteed at all times during data transmission.

It is also shown that the scheduled-based medium access protocols incur higher average queuing delays. The average queuing delay for TRAMA is relatively large due to overhead involved in scheduling. Within every schedule interval, a transmission slot is used for announcing schedules in TRAMA. This decreases the effective channel access probability for data transmission and is not favorable for continuous data gathering type WSNs described in Section 15.4.2 because the traffic is homogenous across the network and all the nodes periodically generate traffic.

Simulation results also show that TRAMA exhibits high throughput compared to SMAC and IEEE 802.DCF because it avoids collisions due to hidden terminals using NCR protocol [20]. Energy savings of TRAMA depend mainly on the traffic pattern of the network as compared to duty cycle-dependent energy savings in SMAC. In TRAMA, the random access period duration plays a significant role in energy consumption. Significant features in TRAMA are the time slot reuse; using neighborhood information for collision avoidance; and use of a hybrid scheme of random and scheduled access for optimal performance.

15.5.4 Power-Efficient and Delay-Aware Medium-Access Protocol for Sensor Networks (PEDAMACS)

PEDAMACS [19] medium access protocol combines the characteristics of cellular networks with those of type 2 sensor networks for the continuous data gathering applications described in Section 15.4.2. It assumes that a single access point (AP) exists in the network and all nodes communicate with this AP. Also, it assumes that AP has no energy constraints and is capable of transmitting at higher power levels when needed so that it can reach any node in the network in a single hop. In contrast, the sensor nodes have limited transmission power and will reach the AP using multiple hops. Although it may not be possible always, in certain applications it may be possible to include a few nodes with higher energy resources to act as APs of each node cluster. The extra effort required may be compensated with optimal power savings in low-power sensor nodes.

15.5.4.1 Operation

The algorithm consists of three major phases: topology learning phase; topology collection phase; and scheduling phase. During the topology learning phase, each node identifies its interferers, neighbors, and parent node. This phase begins with a topology learning packet transmitted by AP over the longest range (highest power) in one hop to all sensors. This packet includes the current time so that each node updates its time and synchronizes with the other. Also, it includes the next anticipated incoming packet time so that every node will stop transmitting and listen for the next broadcast message of AP at this future time, as illustrated in Figure 15.11(a).

Following this, AP floods the network with a tree construction packet over a medium range (medium power). This packet contains a hop count field to avoid any retransmission loops and to facilitate choosing the parent node in the tree as shown in Figure 15.11(b). At the end of this phase, each sensor node decides the parent node to be the one with the smallest number of hops to AP, and the neighbors and interferers as the nodes with the received signal level above and below some interfering threshold, respectively. Because no prior topology information is available during this phase, the authors suggest a simple CSMA scheme with a random delay before carrier sensing.

The topology collection phase starts next with the AP transmitting a topology collection packet (with the same format as that shown in Figure 15.11(a) over the longest range [highest power]). The transmission time is announced in the incoming packet time field of the topology learning packet earlier. This packet also contains current time and next incoming packet time. Following this, each node transmits its topology packet containing its parent, neighbor, and interferer information to AP as shown in Figure 15.11(c). Here again, the CSMA scheme with some random delay before the transmission is used.

During the scheduling phase, each node is explicitly scheduled by AP based on the complete topology information obtained during the previous topology collection phase. The scheduling frame is divided into time slots. At the beginning of this phase, AP performs the scheduling of the sensor nodes in the network and announces the schedule of how all the traffic will be carried during the scheduling frame by broadcasting a schedule packet over the longest range. The schedule packet includes the transmitter information corresponding to each time slot in addition to current time and next incoming packet time fields as shown in Figure 15.11(d). At the beginning of the scheduling frame, each node samples the sensor and generates one packet, which is then carried to AP according to the schedule.

Header	Current time	Next packet transmission time	CRC

(a) Topology learning and topology collection packet from AP

Header	Number of hops	Parent transmission node ID	CRC

(b) Tree construction packet from AP

Header	Node ID	Node level	Parent ID	No. of neighbors	Neighbor IDs	No. of interferers	Interferer IDs	CRC

(c) Topology packet from nodes

Header	Slot seq. No.	Number of nodes scheduled for current slot	Scheduled node IDs	CRC

(d) Schedule coordination packet from AP

FIGURE 15.11 Packet formats in PEDAMACS.

15.5.4.2 Merits and Drawbacks

Although in certain specific applications such a scheme may be able to be used for sensor networks, characteristics of this scheme are not preferred for sensor networks in general. This is mainly due to the use of a central access point that can reach all sensor nodes using high transmitting power. This assumption is not realistic in most of the sensor network applications in which the nodes are distributed over large areas or in indoor environments. Authors argue that in such cases, several APs can be used, but fail to provide details on how to establish proper coordination and synchronization among all such APs. Furthermore, it has yet to be analyzed how the large overhead associated with such a scheme affects the network performance. This overhead also makes this scheme unusable for event-driven sensor systems or dynamic systems with frequent addition and removal of nodes from the network. Authors suggest using a CSMA scheme with implicit ACKs for the topology collection phase, but this may not be an appropriate solution to avoid the huge number of possible collisions.

15.5.5 Comparison

Table 15.4 summarizes and compares the previously discussed four MAC schemes for WSNs.

15.6 Open Issues

Having discussed application and communication categories of WSNs and compared several MAC schemes for WSNs, the open research issues yet to be addressed will be discussed in this section. Designing optimal, energy-aware MAC schemes for WSNs is still an open and fast growing research area. In this context, the following issues are highlighted for consideration in future research.

TABLE 15.4 Comparison of MAC Schemes for Sensor Networks

	SMAC	SMACS/EAR	TRAMA	PEDAMACS
Features	TDMA scheduling Coordinated sleeping schedules among neighbors Adaptive listening Virtual clustering	Hybrid TDMA/FDMA scheduling Mobile node attachment	Random access (CSMA) for neighbor discovery Scheduled access (TDMA) for data transmission	Access point (AP) with high-power transmitter Centralized TDMA scheduling by AP node Hierarchical organization
Applications	WSNs with more stationary nodes	Low traffic WSN with strict latency requirements	Event-driven WSNs	Centralized data gathering WSNs
Merits	Reduced latency for multihop messages Simple hardware for TDMA	Low latency Ability to create links on the fly No clustering requirements No synchronization requirements	TDMA slot reuse No collisions due to hidden nodes Traffic adaptable	Higher energy savings in centralized WSNs
Drawbacks	Synchronization required Virtual clusters may not coincide with physical clusters	Complex hardware for FDMA or CDMA Waste of time slots Low bandwidth utilization Frequent switching can cause heavy energy losses	Synchronization required Low bandwidth utilization in periodic data gathering WSNs	Centralized control necessary AP node requires high power High overhead for scheduling

Adaptability to network objectives. How much sensor node energy to spend on a particular task entrusted on a WSN depends on how critical current application objectives are. As explained in Section 15.2 and Section 15.4.2, the same network used for a low-frequency continuous monitoring application may be employed for mission-critical tracking or emergency threat alert in the next instance. In such a scenario, less critical goals of a sensor network become highly critical and the energy saving requirements become secondary as compared to latency and throughput. A challenging and open issue is to develop medium access schemes for WSNs that have changing missions. SMAC [15] and TRAMA [18] attempt to achieve this to a certain extent; nevertheless, more work must be done in this area.

Optimal schemes depending on WSN type. Certain applications such as habitat monitoring may have stationary traffic patterns mostly over the total lifespan of the WSN employed. For these types of applications, achieving energy savings to extend the network lifetime remains the primary objective throughout the monitoring period. Medium access schemes can be optimized for energy efficiency in WSNs used for such applications. TDMA scheduling-based schemes similar to SMAC [15] may be the ideal candidate for these applications. However, the synchronization requirements and virtual clustering need to be reconsidered in this respect.

Cross-layer design. Conventional WAHNs have neatly defined protocol stacks with independent operation of each layer. For example, medium access scheme would function independent of the node connectivity, routing requirements, and application context. However, the primary goal of energy saving is tightly coupled with all these factors in a WSN and thus medium access cannot be considered alone for optimal savings. It is increasingly clear that power efficiency cannot be addressed completely at a single layer in the networking stack [21, 22]. It will often be necessary to use parameters propagated from upper layers to adapt the medium access protocol, especially in situations in which network objectives change considerably from one time to another. Another issue that must be effectively coupled with medium access is the data aggregation. No significant research efforts have been observed so far in this regard and the next generation medium access schemes for WSNs beyond SMAC and TRAMA should take these aspects into thorough consideration.

Effects of time synchronization. It is observed that higher energy savings are mostly obtained using TDMA based-scheduling schemes in WSNs. Inherently, these schemes require time synchronization of participating nodes in a single schedule. Synchronization errors always tend to degrade the end-to-end throughput performance of the network. As highlighted in Section 15.3.2.2 and Section 15.4, it may be impractical to achieve network-wide synchronization in WSNs. As an alternative, it is better to have globally asynchronous and locally synchronous architectures in which local node clusters maintain synchronization for TDMA schemes aiming for maximum energy savings, while intercluster communication is mostly contention based and asynchronous. Coupling such schemes for optimal energy saving medium access has yet to be explored.

Cluster-based hierarchy. Most WSN applications may require hierarchical architectures. This favors clustering-based systems, which is assumed in most of the research on WSNs. Highly energy-efficient medium access methods could be developed for such networks, using the cluster head as a centralized scheduler, data sink, and relay for the whole cluster. Issues arising in such contexts include ways to achieve intercluster communication, minimizing intercluster interference, and the possibility of having same application-specific clusters in the MAC layer for optimal performance. All these issues need further investigation.

Scalability. Compared to generic WAHNs, WSNs have a larger node count as well as higher density. This should be a critical consideration in designing medium access schemes. Less scalable protocols may cause unbearable overheads when applied to large networks and may cause extreme energy drains in certain nodes, even causing network failure. On the other hand, quality of service degradations can be severe. In this context, the scalability of currently available MAC schemes should be further investigated.

Mobility management. In certain scenarios, several mobile nodes may be roaming the region of deployment of a WSN among stationary nodes already organized under a certain hierarchy for communication. Sometimes the mobile nodes might serve as gateways, sinks, or localization devices, requiring their proper attachment to certain stationary points of the network. This problem is addressed partially in the SMACS/EAR algorithm described in Section 15.5.2; however, the EAR algorithm does not ensure the optimal use of resources. Thus, further investigation is required in this regard.

Hardware constraints. It is argued that energy savings can often be improved using FDMA or CDMA scheduling schemes, for example, as in SMACS [17]. However, the complexity of the required radio interface poses challenges due to compact hardware of sensor nodes. Use of such schemes must be done in conjunction with a suitable TDMA scheme for energy savings because nodes must listen to the channel all the time under pure FDMA or CDMA schemes. Nevertheless, with possible future improvements in hardware fabrication, such hybrid schemes might have a potential to play a greater role in energy savings while giving superior throughput and delay performance. The main challenge in such schemes is to optimize use of resources, for example, time slots and frequency bands. Transmission ranges of nodes are relatively lower in WSNs, so frequency reuse may be possible in hierarchical, cluster-based WSNs as in traditional cellular networks. Moreover, frequency reuse might require nodes to listen only to a limited number of channels, making such hardware more feasible.

Comparison metrics. While novel medium access schemes for optimum energy savings are developed, due attention should be paid to the metrics used in comparing these schemes. Often a trade-off takes place between energy savings and network performance. Thus, unified metrics should be used during comparisons or this will lead to unfair conclusions and probable confusion. The total energy savings of a WSN depend on percentage sleep time and average length of sleep interval. If used alone, percentage sleep time does not account for the possible higher frequency of switching that may drain a significant amount of node energy. Average sleep length is a preferred metric because it can account for the node switching. Appropriate benchmarks should be developed to facilitate accurate comparison of metrics among different MAC schemes.

15.7 Conclusions

Unique features of WSNs in comparison with generic WAHNs were identified in this chapter. Four prominent ad hoc network medium access methods were briefly discussed, as well as their merits, drawbacks, and implications toward WSNs. Design challenges in MAC for WSNs were emphasized with a classification of application and communication types in WSNs. Four medium access schemes recently proposed toward energy savings in WSNs were discussed, comparing their merits and drawbacks. In addition to these four schemes, a few other medium access schemes that have been recently proposed for WSNs [23–26] were not discussed here to preserve brevity of this chapter.

Finally, several open issues related to energy aware MAC protocol design for WSNs were emphasized. Several research efforts are already underway in this area and are being tested on open source platforms like MICA/TinyOS [27]. Significant research efforts are still required to address these open issues in order to achieve the ultimate objective of energy optimal medium access in sensor networks. Also, it is anticipated that such research efforts will soon lead to open standards for WSNs similar to the currently available, commercially deployed standards for WAHNs.

References

1. I.F. Akyildiz, W. Su, Y. Sankarasubramaniam, and E. Cayirci, Wireless sensor networks: a survey, *Computer Networks*, 38(4), 393–422, 2002.
2. The Institute of Electrical and Electronics Engineers, Wireless LAN medium access control (MAC) and physical layer (PHY) specifications, IEEE Standard 802.11, June 1997.
3. J.C. Haartsen, The Bluetooth radio system, *IEEE Personal Commun. Mag.*, 28–36, Feb. 2000.
4. Bluetooth SIG Inc., Specification of the Bluetooth system: core, http://www.bluetooth.org, 2001.
5. C.E. Jones, K.M. Sivalingam, P. Agrawal, and J.C. Chen, A survey of energy efficient network protocols for wireless networks, *Wireless Networks*, 7(4), 343–358, 2001.

6. K.M. Sivalingam, J.-C. Chen, P. Agrawal, and M. Srivastava, Design and analysis of low-power access protocols for wireless and mobile ATM networks, *Wireless Networks*, 6(1), 73–87, 2000.
7. S. Singh and C.S. Raghavendra, PAMAS: power aware multi-access protocol with signaling for ad hoc networks, *ACM Computer Commun. Rev.*, 28(3), 5–26, July 1998.
8. P. Karn, MACA — a new channel access method for packet radio networks, in *Proc. ARRL/CRRL Amateur Radio 9th Computer Networking Conf.*, 1, 134–140, 1990.
9. V. Bhargawan et al. MACAW: a media access protocol for wireless LANs, *Proc. ACM Sigcomm '94*, 24(4), 212–225, 1994.
10. L. Kleinrock and F. Tobagi, Packet switching in radio channels: carrier sense multiple access modes and their throughput delay characteristics, *IEEE Trans. Commun.*, COM-23(12), 1400–1416, Dec. 1975.
11. Y. Tseng, C. Hsu, and T. Hsieh, Power-saving protocols for IEEE 802.11-based multi-hop ad hoc networks, in *Proc. IEEE Infocom*, 1, 200–209, New York, June 2002.
12. G.J. Pottie and W.J. Kaiser, Wireless integrated network sensors, *Commun. ACM*, 43(5), 51–58, May 2000.
13. Crossbow Inc., Expected battery life vs. system current usage and duty cycle URL: www.xbow.com/Support/Support_pdf_files/PowerManagement.xls, Energy specifications for MICA motes, 2003.
14. Crossbow Inc., Data sheet for MICA2 wireless measurement system, 2003.
15. W. Ye, J. Heidemann, and D. Estrin, An energy-efficient MAC protocol for wireless sensor networks, in *Proc. IEEE Infocomm*, 3, 1567–1576, New York, June 2002.
16. S.R. Madden, M.J. Franklin, J.M. Hellerstein, and W. Hong, The design of an acquisitional query processor for sensor networks, in *Proc. (SIGMOD'03)*, 1, 491–502, San Diego, CA, June 2003.
17. K. Sohrabi, J. Gao, V. Ailawadhi and G. Pottie, Protocols for self-organization of a wireless sensor network, *IEEE Personal Commun. Mag.*, 7(5), 16–27, Oct. 2000.
18. V. Rajendran, K. Obraczka, and J.J. Garcia–Luna–Aceves, Energy-efficient, collision-free medium access control for wireless sensor networks, in *Proc. ACM SIGMOBILE Int. Conf. Embedded Networked Sensor Systems (SenSys 2003)*, 1, 181–192, Los Angeles, CA, November 2003.
19. S. Coleri, PEDAMACS: power efficient and delay aware medium access protocol for sensor networks, M.S. Thesis, Department of Electrical Engineering and Computer Science, University of California, Berkeley, December 2002.
20. L. Bao and J.J. Garcia–Luna–Aceves, A new approach to channel access scheduling for ad hoc networks, in *Proc. IEEE MOBICOMM 2001*, 1, 210–221, Rome, 2001.
21. R. Min, M. Bhardwaj, S.-H. Cho, N. Ickes, E. Shih, A. Sinha, A. Wang, and A. Chandrakasan, Energy-centric enabling technologies for wireless sensor networks, *IEEE Wireless Communications*, 9(4), 28–39, August 2002.
22. E. Shih, S.-H. Cho, N. Ickes, R. Min, A. Sinha, A. Wang, and A. Chandrakasan, Physical layer driven protocol and algorithm design for energy-efficient wireless sensor networks, in *Proc. MOBICOMM 2001*, 1, 272–287, Rome, 2001.
23. A. Woo and D. Culler. Transmission control scheme for media access in sensor networks, in *Proc. MOBICOMM 2001*, 1, 221–235, Rome, 2001.
24. R. Kannan, R. Kalidindi, S.S. Iyengar, and V. Kumar. Energy and rate based MAC protocol for wireless sensor networks, *ACM SIGMOD Record*, Special section on sensor network technology and sensor data management, 32(4), 60–65, December 2003.
25. K. Arisha, M. Youssef, and M. Younis, Energy-aware TDMA-based MAC for sensor networks, in *Proc. IEEE Integrated Manage. Power Aware Commun., Computing Networking (IMPACCT 2002)*, New York City, May 2002.
26. J.M. Van-Dam, An adaptive energy-efficient MAC protocol for wireless sensor networks, M.S. thesis, Delft University of Technology, June 2003.
27. University of California, Berkeley, TinyOS homepage. URL http://webs.cs.berkeley.edu/tos/index.html, 2003.

16

Security and Privacy Protection in Wireless Sensor Networks

Sasha Slijepcevic
University of California at Los Angeles

Jennifer L. Wong
University of California at Los Angeles

Miodrag Potkonjak
University of California at Los Angeles

16.1 Introduction **16**-1
16.2 Unique Security Challenges in Sensor Networks and Enabling Mechanisms **16**-2
Security-Related Properties • System-Level Security • Mobile Code • Metering
16.3 Security Architectures **16**-4
Cell-Based WSNs • Ad Hoc Sensor Networks
16.4 Privacy Protection **16**-11
Principle of Minimal Generalization • Privacy of Location Information
16.5 Conclusion **16**-15

16.1 Introduction

Security and privacy protection are of extreme importance for many of the proposed applications of wireless sensor networks (WSNs). The list of potential applications that require protection mechanisms includes early target tracking and monitoring on a battlefield; law enforcement applications; automotive telemetric applications; room occupation monitoring in office buildings; measuring temperature and pressure in oil pipelines [1]; and forest fire detection. All these applications have unlimited benefits and potential; however, if the sensor information is not protected properly, possible compromises in user information, the environment, and even physical actuators could result.

The primary driving impetus for the development of sensor networks has been military applications, where security requirements are at their highest [2]. Although a WSN deployed on a battlefield can offer a reliable assessment of battlefield conditions without risking lives, an inadequately protected network could become a powerful weapon for an enemy. Strong security requirements for such applications are often combined with an inhospitable and physically unprotected environment. For commercial applications of WSNs, the issue of privacy protection is as important as secure and reliable functioning of a network. The protection of personal physiological and psychological information is expected by any user. As the applications of WSNs become more complex and widespread, the ability to protect such systems from any unauthorized access will become increasingly important.

Sensor networks operate in a variety of physical environments and under varieties of constraints. The limited resources of sensor nodes require the development of customized system architectures for each particular WSN application so that the sensor node resources are efficiently used. Because security and privacy protection mechanisms require a significant amount of computational and storage resources, such mechanisms must be tailored to the corresponding sensor system architectures and security threats specific to a given physical environment. Section 16.2 describes the unique properties of WSNs and the security challenges that they bring. Security implications and corresponding security solutions for two basic WSN system architectures, cell-based WSNs and ad hoc WSNs, are discussed in Section 16.3. The overview of the privacy protection solutions proposed for WSN, as well as the solutions originally developed for other environments but applicable in WSN, is given in Section 16.4. Section 16.5 summarizes and concludes the chapter.

16.2 Unique Security Challenges in Sensor Networks and Enabling Mechanisms

WSNs share several important properties with traditional wireless networks, most notably with mobile ad hoc networks. Both types of networks rely on wireless communication, ad hoc network deployment and setup, and constant changes in the network topology. Many security solutions proposed for wireless networks can be applied in WSNs; however, several unique characteristics of WSNs require new security mechanisms. In this section, four characteristics specific to WSNs and their resulting security challenges are discussed. Additionally, it presents work performed in the areas, system-level security, mobile code, and metering, which can be foundations for the development of security techniques in WSNs.

16.2.1 Security-Related Properties

Four properties that are specific for WSNs and require attention are hostile environment, limited resources, in-network processing, and application-specific architectures.

- *Hostile environment.* WSNs can be deployed in hostile environments such as battlefields. In these cases, the nodes cannot be protected from physical attacks. Security information potentially could be collected from compromised nodes. The development of tamper-proof nodes is one approach to security in hostile environments. However, as shown in Anderson and Kuhn [3], the development of such systems is far from simple and certainly not cheap in terms of computational and memory requirements. Because of the physical accessibility of sensor nodes, the security mechanisms for WSNs are specifically concerned with situations in which one or more nodes are compromised.
- *Limited resources.* Sensor network nodes are designed to be compact and therefore are limited by size, energy, computational power, and storage. The limited resources limit the types of security algorithms and protocols that can be implemented. Security solutions for WSNs operate in a solution space defined by the trade-off between resources spent on security and the achieved protection. Limited energy available to nodes allows for new types of attacks, such as a sleep deprivation torture attack [4].
- *In-network processing.* Communication between the nodes in a WSN consumes most of the available energy, much less than sensing and computation do. For that reason, WSNs perform localized processing [5] and data aggregation [6]. An optimal security architecture for this type of communication is one in which a group key is shared among the nodes in an immediate neighborhood. However, in an environment in which the nodes can be captured, the confidentiality offered by the shared symmetric keys is easily compromised.
- *Application-specific architectures.* As a result of the previously mentioned properties, WSN system architectures must be designed to be application specific. The flexibility of a general-purpose architecture is traded for the efficient utilization of the resources. Almost every aspect of a WSN can be adjusted to improve performance and optimize resource consumption in a network for a particular application. This allows a network designer to determine the importance of various security threats and adjust security mechanisms to these threats.

16.2.2 System-Level Security

Three types of cryptographic tools have been developed for practical security of real-life systems: firewalls, honeypots, and intrusion detection techniques. Each will be discussed briefly in order to illustrate the types of approaches that exist in the field. Although these techniques may not be well suited for WSNs as proposed, modifications of their notions may be excellent security approaches for WSN.

A firewall is a policy enforcement point (node) for a part of a network designed to restrict access from and to that subnetwork. Several classes of firewalls exist: packet filtering according to a particular set of rules; access to particular servers or ports; or application-level firewalls that protect by remembering the state of the network connection. Firewalls still face denial of service (DoS) attacks and they try to address them by filtering suspicious connections. Among the several limitations of firewalls is the fact that they do not protect the network from insider attacks and that filtering can only be done against already known attacks.

Honeypots are systems placed on networks specifically for the purpose of being attacked or compromised [7, 8]. Because they are not designed for true use, they exist only to detect and collect information about security attacks. Advantages of honeypots include low false positives; ability to capture unknown attacks; and ability to facilitate interaction with the attacker in order to gain better insights into actions and thinking. Intrusion detection techniques aim at recognizing statistical or pattern irregularities in the incoming or outgoing traffic. The most recent approach to detection of Internet attacks is probabilistic deduction of the IP traceback [9–12]. Finally, virtual private networks are logical extensions of private networks over insecure channels provided by the Internet.

16.2.3 Mobile Code

Once deployed, access to the nodes in a WSN for management and code updates poses security threats and drains resources. Despite difficulties, mechanisms that allow changes in application and system code on the nodes are necessary. One feasible solution for remote configuration and application code updates is network-wide deployment of mobile code. A legitimate mobile code is injected into the network through several nodes and then spread throughout the network [13]. This subsection surveys proposed code manipulation approaches (attacks) and techniques for secure execution of mobile code. Among mobile code intrusion techniques, four have been most popular: viruses, Trojan horses, buffer overflow, and covert communication channels.

A computer virus [14] can be defined as a small program that attaches to the host computer and co-opts its resources for the purpose of creating new copies of the virus. Detailed analysis of viruses and models of their proliferation can be found in Cohen [15] and Kephart and White [16]. Trojan horses [17, 18] disguise themselves as programs that appear to perform a function while actually performing another function. Buffer overflow has been by far the most common type of attack of computer security in the last decade [19]. These attacks use the functions of a privileged program in such a way that the attacker can take control of the program and corrupt the computer. This is commonly achieved by making suitable code available in the program address space and then inducing a program to jump to that space with suitable parameters. Recently, the first constraint-based analysis technique for automated detection of buffer overflow has been proposed [20].

Covert communication channels [21] arise from resource sharing in computer systems. For example, a process with high priority can pass information to a process with low priority by interfering or refraining from interfering with the timing of the process. The most popular and simplest is the timed Z-channel, in which the communication alphabet consists of time values [22]. Numerous generalization of the timed Z-channel have been proposed and analyzed [23–25].

Smaller mobile devices have created a strong impetus for the development of mobile code security techniques. At least three major approaches for mobile code security have emerged: code signing, sandboxes, and proof-carrying code. Code signing follows a typical client- and server-authenticated handshake protocol such as SSL or WTLS [26]. Recently, sandboxing has attracted a great deal of attention [27] as a security paradigm; Brigner [28] presented a 3-MB Java applet that implements a sandbox. In addition, Sekar and Uppuluri developed a security layer that includes a sandbox designed to protect the application against malicious users and the host from malicious applications [29]. Proof-carrying code is a mechanism that allows a host computer to determine if a program can be executed with certainty despite being provided by an untrusted source [30, 31].

16.2.4 Metering

One aspect of WSN security threat that is not often addressed as an attack is consumer access to the sensor data. As WSNs become more advanced and versatile, the notions of user access, application-specific sensor designs, and licensing of network usage will become an issue. Metering is one approach to handling these types of issues. Although many of these approaches are too computationally or memory intensive for WSNs, they provide a starting point for development of WSN techniques.

SiidTech Inc., an Oregon startup company, has proposed an approach for integrated circuit identification from random threshold mismatches in an array of addressable MOSFETs. The technique leverages on process discrepancies unavoidably formed during fabrication. This analog technique can be used in tracking semiconductor dies, authentication, and intellectual property (IP) tagging [32]. Sampling and auditing are the two main methods used for measuring the usage of media channels. Sampling conducted by Nielsen Media Research and NetRatings Inc. is based on surveys among a representative group of users [33]. Web page access metering has been addressed by a number of researchers and companies [34–36].

Licensing is the most common approach to protecting software. It provides a certain degree of control to the vendor in terms of software distribution and may prevent unauthorized duplication of software packages. The most common technique is based on the license key concept. A key is encrypted by using a string of data that contain software package ID and its usage constraints (e.g., expiration date) and the serial number of the computer where the key is installed. The invocation of the software package is done automatically when software is invoked by using one of the password schemes [37, 38]. A large number of patented licensing protocols have been proposed; for example, licenses can be used to authenticate the legal users, as well as to upgrade the products and other after-market information transmissions [39] or licensing using smart cards [40, 41].

16.3 Security Architectures

This section describes security protocols developed for two typical WSN system architectures: cell-based WSN and ad hoc WSN, with particular concentration on key establishment and distribution algorithms because they set up the necessary infrastructure for security protocols. The proposed WSN system architectures differ in many aspects, which is not surprising because WSNs operate in vastly different physical environments, supporting different applications and using different sensor nodes. The main benefit of the development of a specific architecture for each WSN application is the efficient utilization of scarce sensor node resources.

Many elements of WSN architecture, including hardware architectures of sensor nodes, routing protocols, and level of abstraction between the layers of the architecture, can be adjusted to improve performance and optimize resource consumption. One possible categorization of wireless ad hoc network systems, which includes WSN systems, is given in Law et al. [42]. Here, only the security architectures for WSN systems are discussed, while Papadimitratos and Haas [43] give an overview of security architectures for general wireless ad hoc network architectures. From the security point of view, the WSN system architectures can be broadly divided in two categories:

- Cell-based WSNs consisting of low-power low-cost sensor nodes and base stations, operating in relatively friendly environments of houses and office buildings, or in easily accessible outdoor areas
- Ad hoc WSNs consisting only of low-cost sensor nodes distributed in an ad hoc manner into remote and inhospitable environments without any wireless infrastructure

These two network architectures differ in terms of the security threats to which they are exposed and in terms of security requirements and abilities to support security architectures of various levels of complexity. The cell-based WSN allows for more sophisticated and resource-consuming protocols and algorithms because the additional computationally expensive workload can be assigned to the base stations.

16.3.1 Cell-Based WSNs

In cell-based WSN, the nodes are organized around one or more base stations that have significantly more computing and energy resources than the regular sensor nodes. These networks are most often used for user and object tracking systems in home and commercial building environments, as well as in outdoor perimeter-monitoring systems. The base stations collect information from the network and provide a link between the WSN and the outside world. Cell-based networks are often used in an environment in which it is easy to add new nodes, remove the ones that are not functioning, and even recharge the energy supplies for nodes. However, even in such an environment, the nodes can still be captured or damaged, and unauthorized nodes can be added.

The presence of base stations in a WSN offers at least two significant benefits:

- Base stations represent a trusted base that cannot be compromised. They can be used as a safe source of mobile code and configuration parameters, which enables safe bootstrapping and configuration of the network, as well as the addition of new nodes.
- Base stations offer computational resources that can be used in asymmetric security protocols in which they perform the majority of intensive computations. Such protocols allow stronger security, while not exhausting the limited resources of regular sensor nodes.

An example of a WSN organized around one or more base stations and SPINS, the security protocol suite for that network, is described by Perrig and colleagues [44]. The network consists of a trusted backbone of base stations with unlimited power supply and a large number of *motes* (low-cost, low-power sensor nodes described by Hill and colleagues [45]), distributed in the area covered by the base stations. The operation of the network is fully controlled from the base stations. A routing structure is formed as a set of routing trees; each base station is the root of one such tree. The traffic mainly consists of requests initiated at the base stations and sent down the trees to the nodes and the responses sent from the nodes back to the base stations. When the same request is sent to all nodes, the communication is most efficiently performed through broadcast messages. If a base station needs to send a unicast message to a particular node, source routing is used.

The SPINS protocol suite assumes that the base stations share a unique master key with each node in the network. The system architecture and security protocols require that the base station keep track of the route to each node and of the secret key. All other keys that the base station and a node use for communication are derived from the master key. Even though the base station is a single point of failure, it is trusted, implying no one can capture the station and recover all keys.

This security architecture efficiently uses the resources of the base stations. To keep a separate key for each node would not be possible in an architecture in which all nodes have limited resources. Also, this solution is not applicable to networks in which any two nodes are likely to communicate directly. However, because the bulk of traffic in the network is between the base station and the nodes, the inability of the nodes to communicate securely without involvement of the base station is of limited importance.

The SPINS protocols suite consists of two building blocks, sensor network encryption protocol (SNEP) and μTESLA. SNEP protects the unicast communication between the base stations and the nodes, while μTESLA provides secure broadcast communication. Each of these protocols will be discussed in more detail.

16.3.1.1 SNEP

The basic confidentiality of messages in any secure system is achieved through encryption. Encryption protects the network from adversaries who have the capability to listen to network traffic. SNEP uses

RC5 block cipher [46] for basic encryption. The original RC5 encryption algorithm is implemented with lowered functionality and generality in order to fit in the limited storage space of nodes. In addition to basic confidentiality, SNEP offers *semantic security*, which means that the encryption of the same plaintext produces a different encrypted message each time. This is achieved by keeping a shared counter on each of the two entities involved in the message exchange and incrementing the counter for each message.

Because the value of the counter is an initialization vector for the RC5 block cipher, it is guaranteed that the encrypted messages differ even if the content is the same. An additional benefit of the counters is that they ensure *freshness* of messages, i.e., a receiver can establish the partial message ordering of the messages from a particular sender. Finally, each node has its separate master key, so SNEP guarantees authentication of messages that the nodes receive from the base station.

An important property of such a solution is that it can be used in environments with relatively static forwarding structure, in which the nodes communicate with a limited number of other nodes or base stations, usually smaller than the number of neighbors. The number of the keys and counters can be estimated, and the efficiency of such a solution is known in advance. In a network with a dynamic forwarding structure in which any neighbor can be a previous or a next hop for any message, it may be prohibitively expensive to keep counters and separate keys for all possible sources and destinations. However, for a limited number of cases, two nodes that need to communicate directly can use their master keys to generate and exchange a session key through the base station.

16.3.1.2 μTESLA

The second element of the SPINS protocol suit is μTESLA. The master key shared between each node and the base station ensures confidentiality and authentication of unicast messages exchanged between the nodes and the base station. However, if the same message is sent from a base station to all nodes, it is much more efficient to broadcast the message. SNEP does not support secure broadcast because each master key is unique; allowing nodes to accept unencrypted, unauthenticated messages would allow an adversary to send arbitrary requests to nodes. Therefore, for secure broadcast communication, SPINS proposes μTESLA, the goal of which is to ensure authentication of broadcast messages sent from the base stations to the nodes.

In μTESLA, a base station generates a reverse key chain containing the keys $K_0, K_1, \ldots, K_n$. The key chain length and the key K_n are determined before the key chain is generated. Other keys are determined using one-way function F, $K_i = F(K_{i+1})$. The key K_0 is not used to authenticate any of the messages, but is distributed initially as a commitment to the key chain. The distribution of the commitment K_0 in μTESLA requires that each node and the corresponding base station share a secret key unique for that node. Then, the base station sends K_0 to all nodes as a sequence of unicast messages, before any broadcast message is transmitted.

The time is divided into the intervals $I_1, \ldots, I_n$, as shown in Figure 16.1, where each interval I_i corresponds to the key K_i. During the interval I_1, the base station sends broadcast messages with attached message authentication code (MAC) calculated using the key K_1. Because the key K_1 has not been disclosed yet, the messages could not have been forged by any of the nodes. The function F is a one-way function, so no one can determine K_1 from K_0. The nodes authenticate the messages received during the interval I_1 at the end of that interval, when the key K_1 is disclosed. At that time, the nodes compare K_0 with the value derived from $F(K_1)$. If the values match, then the messages authenticated with K_1 are sent from the base station, because only the base station could have known the value of K_1 before that key was disclosed. After K_1 is disclosed, the following broadcast messages are authenticated, using K_2, until K_2 is disclosed, and the process continues until the interval I_n expires.

Because the keys are disclosed in periodic intervals, the base stations and nodes must be at least loosely synchronized. If a node does not receive a message with a disclosed key and its clock is late, an adversary who received the disclosed key can forge and send messages with the MAC calculated using that key. A node with an unsynchronized clock would accept such messages for the interval equal to the delay of the node's clock.

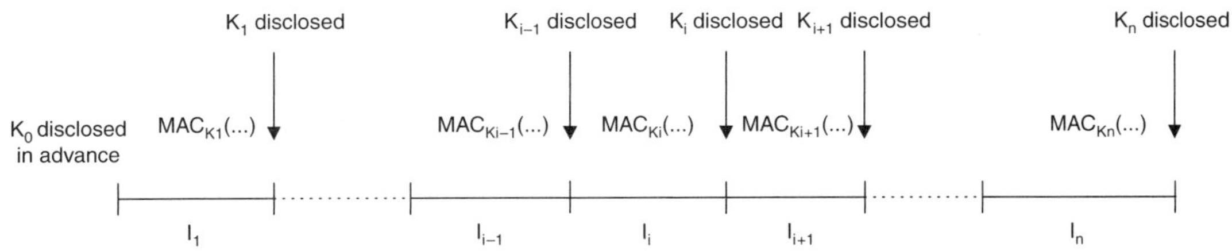

FIGURE 16.1 The key distribution timeline. The messages sent during the interval I_i have attached MAC calculated using the key K_i. The key K_i is disclosed at the end of the interval I_i.

Even if a node does not receive all the keys, it can still authenticate all messages. Once the node receives the key K_{i+1}, it can derive all previous keys $K_0,\ldots, K_i$ by successively applying the function F, and then use these keys to authenticate messages received in the corresponding intervals. However, buffering the messages for a prolonged time requires additional storage. Nodes have limited available memory, so additional mechanisms are needed to ensure that the keys are disclosed to all nodes in a timely manner.

If some nodes are deployed later and begin receiving messages during the interval I_i, they need to receive the key K_i at the end of the interval. The initialization process for those nodes is the same as for the nodes initialized when the whole network is bootstrapped; the only difference is that instead of the commitment K_0, the nodes receive the key K_{i-1} as a commitment. By doing so, new nodes save a certain amount of computation because they do not need to compute all keys from the chain $K_0,\ldots, K_{i-1}$ to compare the value of the key K_0 with the value of $F^i(K_i)$; they would need to do this if they received K_0 as a commitment. Instead, the nodes simply compare the commitment K_{i-1} with the value of $F(K_i)$.

The initial distribution of the key chain commitments requires that a base station send a separate unicast message to each node. In addition to the energy consumption of sending multiple unicast messages with the same information, the time required to initialize thousands of nodes is measured in tens of seconds, as calculated by Liu and Ning [47], who proposed that instead of costly initialization using broadcast messages, the commitment K_0 be embedded into the nodes during initialization, before deployment of the nodes. This solution brings significant savings for the majority of the nodes deployed together. For the nodes added later, there is a trade-off between computation expenses incurred when the new nodes authenticate a broadcast by comparing the embedded commitment K_0 and $F^i(K_i)$ and when the added nodes receive unicasts containing K_{i-1} and authenticating broadcast with only one calculation of the function $F(K_i)$.

The decision as to which mechanism for initialization of the added nodes is preferred could be potentially based on the number of new nodes that should be initialized. If the number is small, then the delay incurred by sending unicast messages is acceptable; however, if many nodes are added, it is more efficient to have the nodes use K_0 as a commitment. Unfortunately, the decision about the preferred commitment distribution mechanism cannot be made online because delivering the decision to the nodes would require sending authenticated unicast (which is the expense to avoid if the number of new nodes is large) or sending a nonauthenticated broadcast, which then could be a message forged by an adversary.

An implicit assumption in μTESLA is that the base stations have sufficient memory storage to hold a long key chain or that they have enough processing power to compute keys fast enough while keeping in memory only the last member of the chain. That assumption saves the nodes from buffering too many messages because the intervals can be arbitrarily short; the key chain is still long enough not to require frequent costly commitment distributions. Liu and Ning [47] propose a hierarchical organization of the keys that decreases the required memory storage at the base stations. The basic principle behind this solution is that the base stations keep only a high-level key chain in memory, while the elements of the low-level key chains are generated using the high-level keys. The keys $K_0,\ldots, K_n$ from the high-level chain are not used for message authentication. They only authenticate messages containing low-level key chain commitments $K_{<0,0>}, K_{<1,0>},\ldots, K_{<i,0>},\ldots, K_{<n,0>}$.

This extension of μTESLA needs to keep the property of the original scheme that even if some key disclosure messages are lost, the later key disclosures can be used to authenticate previous messages. Otherwise, the network would need to ensure that all messages are received at all nodes — an expensive proposition for WSNs. In order to enable authentication despite lost messages, the high-level key K_{i+1} is used to generate the last key for the low-level key chain for the interval I_i, $K_{<i,m>} = F_1(K_{i+1})$. Without this relation, if a message with the commitment $K_{<i,0>}$ is lost, the nodes could not authenticate the messages from the interval I_i. Even in this solution, if the message with the commitment $K_{<i,0>}$ is lost, the nodes must keep all messages from the interval I_i in a buffer until the key K_{i+1}, or some other later key, is disclosed in the interval I_{i+2}. Because the high-level intervals are intentionally kept long so that the number of keys stored in the memory of the base stations is small, the memory required to store the messages may be prohibitively large. One possible solution is to repeat messages frequently that contain key commitments.

Authentication of broadcast messages sent from the base stations to the nodes is supported by μTESLA. It may be possible to use the protocol in cell-based networks in which the nodes send broadcast messages too. Two possible solutions for this problem are: (1) a node sends a unicast to the base station using the key that the node and base station share, and then the base station broadcasts the message using the original μTESLA broadcast authentication mechanism; or (2) a node broadcasts the message and the base station handles the distribution of a key for that broadcast.

16.3.2 Ad Hoc Sensor Networks

Certain military, law enforcement, and disaster recovery WSNs are deployed in remote and inhospitable environments without any wireless infrastructure. Nodes must self-organize and bootstrap a network without any support from base stations. Such networks distributed in an ad hoc fashion must be capable of accepting requests from various points within the network because a user walking through the area may not have the capability to connect to the designated gateway. Any node in such an architecture can be a source of or a destination for messages.

Even more than in other networks, the nodes in such systems are exposed to a danger of capture or destruction. The most dangerous physical threat regarding security is physical possession of a sensor node by an adversary. Sensor nodes may contain keys that allow the adversary to decrypt the messages and even to inject false messages into the network. In circumstances in which long-term security of all nodes in a network cannot be guaranteed, the best solution is to extend the lifetime of the network as much as possible. There are two aspects of extending the lifetime of a network:

- The time period from when the network is deployed to the moment a node is compromised should be as long as possible. An adversary can determine the positions of nodes using various technologies. The easiest way is to listen to the messages exchanged between the nodes because they usually contain the locations of nodes that detected an event. In Slijepcevic et al. [48], messages are encrypted with a separate encryption algorithm for locations of nodes; this is stronger than the encryption for the rest of the message content so that the adversary has less encrypted text for cryptanalysis. If the information about the locations of nodes is adequately protected, the adversary is left using trilateration, which requires more equipment and effort than simply extracting the locations of nodes from the messages.
- Once some of the nodes are detected, the keys that these nodes contain can be extracted and used to decrypt previously exchanged messages as well as future ones. Key distribution mechanisms in WSN and secure protocols must be designed so that the security exposure is minimized when any of the cryptographic keys is compromised.

In a system architecture in which all nodes are potential senders or receivers, symmetric cryptography suits the low-power nodes better than public cryptography. Because symmetric cryptography assumes that the keys are shared, the design space between two extreme solutions remains: (1) all nodes share only one key embedded in them before the deployment; and (2) each pair of nodes shares a unique key.

The first solution is simple, does not require too much memory space, and has the broadcast primitive available. However, when one node is compromised, the adversary can decrypt all messages from the network. The second solution has a perfect security property: if a node is compromised, the recovered keys are useless because no other nodes use those keys. However, the memory space for all keys for networks of thousands of nodes is not available on most sensor node platforms. Even if only a handful out of thousands of keys is actually used when a network is deployed, the nodes must store them all because their exact physical locations are not known before the deployment, and they cannot know which nodes will be located close to each other.

Additionally, sending broadcast messages is not possible, so each broadcast message must be replaced with multiple unicast messages, and the energy consumption is multiplied accordingly. The key distribution algorithms proposed for WSNs try to find a trade-off among the various requirements. The important factors for the key distribution algorithms are:

- Impact of compromise of one or more nodes on security of the traffic in the network
- Ability of algorithm to include additionally deployed nodes into the security infrastructure
- No single point of failure
- Spatial and temporal variation in keys to reduce encrypted material for cryptanalysis
- Support for broadcast

16.3.2.1 Key Distribution Schemes

Most key distribution mechanisms shy away from key distribution after the nodes are deployed. Such schemes exist for various wired and wireless networks and they mainly include key distribution servers. They consider self-organized wireless network with no security infrastructure. Therefore, no central authority, no centralized trusted party, and no other centralized security service provider exist. The standard solutions for authenticated broadcast are not applicable. The solution used in wireless networks with more capable nodes [49, 50] employs public key cryptography to ensure authenticity of messages. However, in many WSN projects [44, 48, 51], the public key algorithms are considered too expensive in terms of memory and processing requirements to be used in WSNs, except as a one-time protocol for exchange of private keys. A thorough discussion about the energy requirements of the public key encryption algorithms and their performance on various processors is given by Yuan and Qu [52].

A possible solution for key distribution is to assign some nodes to be key distribution servers, delivering symmetric keys to nodes that need to communicate. The use of online key distribution servers in WSN has a disadvantage; because all nodes are physically exposed, key distribution servers would become single points of failure because of failures and especially because they would allow an adversary to get hold of all keys used in networks. Therefore, the key distribution algorithms presented here are based on key assignment before deployment. The addition of new nodes and the loss of previous nodes does not require an immediate key distribution process, as is the case in Internet multicast algorithms in which new nodes must not be able to read previously exchanged messages, and the old nodes must not be able to read future messages. In WSNs, new nodes are trusted and old nodes are most likely out of energy.

Eschenauer and Gligor [51] propose a probabilistic key distribution in which a node shares a key with a certain percentage of other nodes. Before the deployment, an initial pool of P keys is generated. For each node, k keys are selected from the initial pool for a key ring. After the deployment, the nodes announce and compare their key rings, looking for at least one key that belongs to both key rings. If such a key is found, those two nodes can communicate directly. When all such pairs are found, they represent the connectivity graph for the network. Now, even the pairs of neighboring nodes that do not have a direct connection can use an established secure path to generate a key, or pick a key from a set of unused keys from the key rings and exchange that key. Eventually, each pair of neighboring nodes will share the key, under the condition that the network graph was connected initially. If that was not the case, a certain number of nodes are permanently excluded from the network.

The main advantage of this scheme is its resiliency in the case of compromised keys. If a node is captured, all of its k keys are available to the adversary. The probability that a particular key is used for encryption of a link is the same for all keys, so a probability that the adversary can decrypt traffic on a particular link is k/P. As will be explained later, k is significantly smaller than P; thus, that probability is low. The scheme also achieves significant memory savings compared to a scheme in which each pair of nodes shares a key. Furthermore, when new nodes are added to the network, they announce their key rings in the same way as the nodes deployed during the initialization of the network.

Because different keys are used throughout the network, the amount of encrypted material for cryptanalysis is smaller in such a key distribution scheme than in the scheme with a shared key for all nodes. If the lifetime of the network is so long that the keys should be replaced, the nodes can revoke the keys with the expired lifetime. After some keys are revoked, the process of establishing secure connections must be run again, with fewer keys. The removal of keys decreases the probability that the network will be fully connected, so the key revocation has limited usage. Finally, the scheme does not support broadcast. However, after a node establishes secure paths to all its neighbors, it can distribute one of its keys as a broadcast key in the case of increased broadcast traffic. Obviously, the applications running on top of the WSN running this key distribution scheme need a certain amount of control over the deployment of certain mechanisms; such mechanisms are not always needed, but their deployment consumes energy.

The parameters of the scheme are determined as a trade-off between security in the case of compromised nodes and the probability that the network is connected. The parameter k, the size of a key ring, is determined by the size of the memory reserved for the keys. The size of the network and the average number of nodes within a communication range are determined by the network application and topology. The only parameter that can be changed over a large range of values is the size of the initial key pool, P. If P decreases, the probability that two key chains selected from the keys from P have one or more common keys increases. However, the value of the expression k/P, which represents the probability that an adversary can compromise a communication link when a node is compromised, also increases.

A result from graph theory, presented by Spencer [53], determines the probability, p, that an edge is between two vertices in a graph, for which the probability that the graph is connected rises from a small probability to "certainly true." Then, P is determined from the condition that the probability of two key chains with one or more common keys is equal to p. For a network of 10,000 nodes, with $d = 40$ neighbors per node on average and the size of a key ring $k = 15$, if the size of the initial pool is $P = 100{,}000$, the network is fully connected with the probability .99999.

Chan and colleagues [54] offer two improvements to the described scheme. The first change is that two nodes can establish a secure link only if they share q keys, instead of one as in the original scheme. The advantage of this approach is that, for a small number of captured nodes, the probability that any link in the network can be compromised is lower than if the nodes establish a link with only one shared key. However, with the increased number of captured nodes, the relation between these two probabilities changes, so with a sufficiently high number of captured nodes, an adversary has better chances of compromising the secure links than in the original scheme. This trade-off improves the protection of the network against small-scale attacks, which are easier to execute and therefore more likely, and decreases the protection against larger attacks, which are more expensive to perform.

The second improvement from Chan and colleagues [54] allows pairs of nodes that have a secure link between each other to establish new keys. That way, more keys are used, while the amount of the memory required to hold the keys is kept low, at the order of magnitude of the number of neighbors. The price paid for this improvement is increased traffic for key exchanges. It is also important to mention that the two schemes proposed here should not be used at the same time. The first scheme requires that the number of keys in the initial pool be kept low to ensure the connectivity of the network. At the same time, the second scheme tries to use different keys; however, during the exchange of the keys the probability of capture of these keys is increased.

16.4 Privacy Protection

The previous sections have examined the security architectures for two broadly defined types of WSN. The main goal of the presented architectures is to establish secure communication channels within a network in order to protect transmitted information from unauthorized access. In many WSNs, especially in military and law enforcement systems, sensor nodes and communication between them are the most exposed part of the network. In such networks, reliable and secure communication is the most important and best guarantee of uninterrupted functionality.

For another class of WSN systems — those intended for use in commercial settings — the privacy protection of individuals observed by a WSN whose living and working spaces are populated by sensor nodes is as important as the protection of applications' functionality. It is still necessary to ensure secure communication channels in commercial WSNs in order to prevent unauthorized access to the personalized information about the users of the system. However, even if the communication security architecture ensures that the personal information is well protected during transfer through the network, once such information is collected at a data collection point, the information is protected as much as the data collection hosting system is. Commercial systems tend to have lower standards for security protection of acquired information than military and law enforcement systems do. The news frequently reports about systems in which system security at data collection points was compromised and social security numbers, credit card numbers, and many other highly sensitive and personal data ended up publicly available to anyone on the Internet. These cases illustrate the need for additional mechanisms that will ensure a certain level of privacy protection without interfering with the functionality of commercial WSN systems.

16.4.1 Principle of Minimal Generalization

Sensor nodes' sensing capabilities, size, and low cost allow a large number of sensor nodes to be deployed in and a large amount of information to be acquired from physical surroundings. Except in rare cases, the larger the amount and the higher the precision of the sensing data available to a WSN, the better the performance of applications is. Although the applications may perform better if more data are available, privacy protection, by definition, strives for the minimum amount of data to be acquired about a single individual. Although the performance of applications and the need for privacy protection may seem to be two opposing goals, many applications can function effectively with their information precision at a lower level than the level of precision that WSNs are capable of delivering. That interval between the required and potential accuracy can be effectively used for privacy protection.

Samarati and Sweeney [55] have proposed a mechanism for generalization of data in databases in order to prevent matching individuals and their medical records. The medical records with the names removed, but with ZIP codes, dates of birth, and other information, are easily matched with the identities acquired from voter lists, city directories, and other publicly available sources. To prevent reidentification, nonessential information is removed, i.e., the year of the birth is kept, while the exact date is removed. The goal of the process is to depersonalize medical records so that multiple identities are equally likely to correspond to a particular medical record. This approach is called the principle of *minimal generalization*. The same principle can be applied in the context of the privacy protection in WSNs. Naturally, this may affect applications that operate on generalized data, so the principle can be applied only if the application can retain the required performance level. An informal description of the principle of minimal generalization is: accurate private information about the users of a system should be generalized so that the acquired data can be matched to no less than k identities, where k is the required level of anonymity.

Under the assumption that a data collection point and a WSN are under separate control, this principle is beneficial for both entities. The WSN offers higher privacy protection for its users, while the data collection system does not need to expend resources for privacy protection purposes and does not need to risk liability for possible breaches of privacy protection.

16.4.2 Privacy of Location Information

This principle is demonstrated on several WSN applications that rely on location information about users. Protection of the location information is highlighted for three main reasons:

- The most frequent tasks for WSN systems are concerned with detection of location of an event. Even if the goal of an application is to perform a more complex task, the location information is present as a part of the individual observations generated by sensor nodes.
- The privacy protection of location information for users observed by a WSN is a prime example of the importance of data protection because, with access to the location data for a user, an adversary can infer additional private information — for example, medical conditions, shopping habits, and patterns of social interactions between monitored users.
- Protection of location information allows the principle of minimal generalization to be demonstrated in an easily understandable case study. Generally, location discovery systems are often capable of locating users within meters indoors and within tens of meters outdoors. For many applications, that level of precision is more than required for basic functionality, so it is acceptable to reduce the precision of the information in order to achieve a required level of privacy protection.

The general system architecture for which privacy protection solutions are described is shown in Figure 16.2. The crucial part of the privacy protection framework for WSNs is the location server. The server is a part of the trusted zone, which in this context means that the server adheres to the same security policies and is controlled by the same entity as the accompanying network of sensor nodes. In fact, the location server is likely to be implemented as a service running on a gateway between the sensor network and the outside world. The assumption that sensor nodes are trusted and that they do not forward any information to an unauthorized party extends here to a location server. The responsibility of the location server is to transform the locations of users observed by the network into a representation that keeps the level of location privacy protection above a certain threshold. The main difference between various privacy protection algorithms is in the types of transformations performed by a location server. The transformed location information is then forwarded to any of the servers offering location-based services (LBS). The services are offered by various entities that do not share security trust with the WSN.

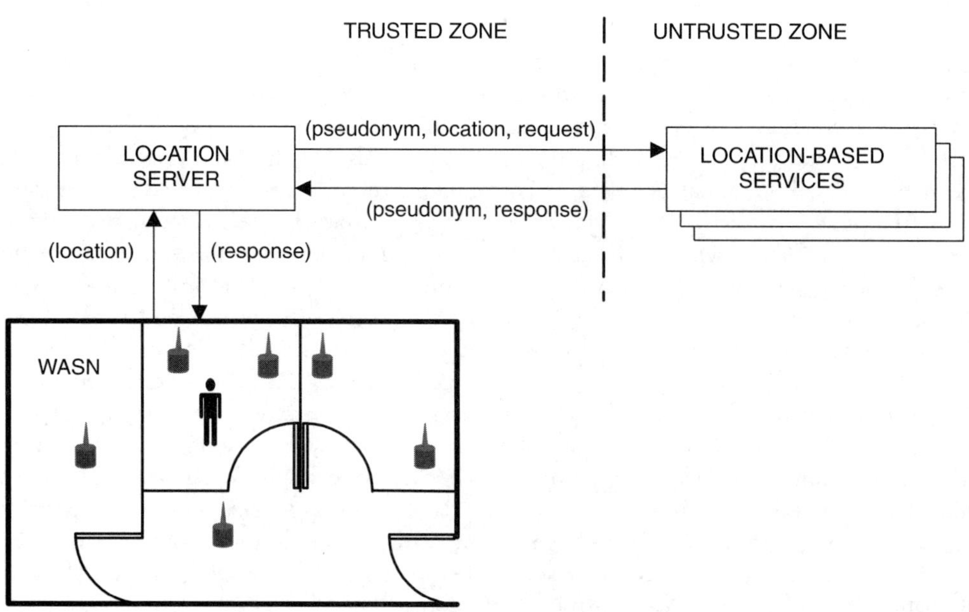

FIGURE 16.2 The architecture of the system connecting a wireless sensor network and a location-based service. The location server transforms location information, so the identities of users are hidden.

Many WSN and wireless network applications may run on top of a system architecture similar to the one shown in Figure 16.2. Two applications, which offer road maps and road condition information based on the location of a car, are proposed in several projects [56] and are commercially available [57–59]. In Beresford and Stajano [60], various proposed applications are implemented on the top of an indoor location system in an office building environment. Gruteser and coworkers examine the privacy concerns of applications that track the use of different building areas [61].

In all these applications, users send information about their locations to an LBS to update their locations or to request services offered in their vicinity. Without transformation performed in a location server, each user request or update would be accompanied with as precise location information as the location discovery technology used allows. In automotive applications, precision is defined by the precision of a GPS receiver, while in an indoor environment location precision depends on the density of the sensor network, usually precise enough to locate a room where a person is correctly located. If the information from these location discovery services is compromised, the location precision allows for easy recovery of users' movements by LBS.

The first step to protect users' privacy is to disconnect the location information from the explicit user identification. The location server performs this task by assigning an alternative identification or a *pseudonym* to each user. Some kind of identification is necessary because a response to each request among a possibly large number of requests handled by a location server must be forwarded to the original user. For many LBSs, it is not necessary for a service to be aware of a user's real identity. A road map can be generated for a user based only on the user's current location. However, two problems occur with privacy protection through anonymity of identifications. The first is the possibility that if a user uses the same pseudonym when connecting to various LBSs, the combined data from all LBSs can give a full overview of that user's activities. That problem can be solved simply by using different pseudonyms for various LBSs, similar to a solution for the same problem outside the context of WSNs, as proposed by Chaum [62].

The second problem is that a user can be easily identified despite different pseudonyms, if requests for LBSs are coming from specific locations that can be directly connected to the user. In the case of a request for a road map that an LBS issued from a location that can be identified as a private garage, the anonymous identification can be attached to the owner of the garage, and then all the movements of that ID can be personalized. In the same way, in an office building environment, an ID that spends most of the time in a particular office can be connected to the regular occupant of that office.

The solution for this problem is in the combination of the principle of minimal generalization and temporary anonymous identifications. Before details of the mechanism are described, it is necessary to include a more formal definition of privacy in order to be able to compare the benefits of different approaches to this problem. The measure of privacy in this context is the notion of *k-anonymity.* The meaning of the term *anonymity* in privacy protection research is formally defined by Pfitzmann and Koehntopp [63]; they define it as a quality of not being identifiable within an anonymity set containing a set of subjects. Then k-anonymity, as defined by Samarati and Sweeney [55], is anonymity within a set with the cardinality of k. This term is used to define an acceptable level of privacy protection in which a person cannot be distinguished from $k - 1$ other individuals. For the application using location information, k-anonymity means that the attached location information for that user comprises location information for $k - 1$ other users. If a stretch of a freeway of the length d contains k cars, a user whose location is defined with the resolution d is k-anonymous.

Gruteser and Grunwald [56] achieve k-anonymity by transforming precise, high-resolution GPS–originated location information available to a location server to low-resolution location information sent to the LBS. Additionally, the identity of a user is hidden in order to avoid continuous tracking of his location. If a user's location is defined by the intervals $[x_1,x_2]$ and $[y_1,y_2]$ in two-dimensional space and the time interval $[t_1,t_2]$ in time dimension, k-anonymity is achieved by extending and contracting the intervals until $k - 1$ or more objects share the resulting parallelepipe in the three-dimensional space-time coordinate system. Depending on the nature of the application, Gruteser and Grunwald have proposed two different algorithms [56] that transform resolution of location information:

- If the application requires a timely response, the spatial resolution is brought to a level at which k-anonymity is achieved. The implementation of this algorithm starts from the entire area covered by the LBS. The area is then divided into subareas, until the subarea containing the specified user also contains less than $k - 1$ other users. Examples of such applications are road map and road condition LBSs in which the delay must be on the level of minutes; otherwise, the information returned from the LBS cannot be effectively used.
- If a delay is acceptable, the application can set a threshold on the spatial resolution, requiring that the area confined with the intervals $[x_1,x_2]$ and $[y_1,y_2]$ is never above the given threshold. The property of k-anonymity for a user is then achieved by extending the temporal interval $[t_1,t_2]$ by simply waiting until k or more users pass through the space limited by the spatial intervals. Now, the LBS side of the application deals with a more precise location information, which is beneficial for the quality of service that LBSs offer.

In Gruteser and colleagues [61], the underlying application counts the number of people in various parts of a building to estimate the utilization of the rooms in the building. The nodes in the network are organized in a hierarchical structure, with sensor nodes at the lowest level detecting individuals in their vicinity, usually only in one part of a room. The nodes at the next level count the number of individuals in each room, using the sensing data from the nodes from the lowest level. The hierarchy structure assumes nodes at the floor level as the next level and then, finally, a location server that gathers the data from the floor level. Without privacy protection, the information about occupancy of the rooms would be simply transferred up to the location server. The application requirements are such that it can perform its task without identifying individuals occupying rooms. However, similar to the applications mentioned previously, if an adversary can access the data gathered at a data collection point, he can reidentify individuals from the rooms that each of them occupies most frequently. Now, identified individuals can be tracked by observing counterchanges in various parts of the building.

The solution proposed in this work leverages the hierarchical network architecture of the WSN [61]. A node determines a count of individuals in its area, by sensing, if at the lowest level of the system hierarchy, or by aggregating the counts received from the nodes one level below, if at one of the higher levels. The node then compares the count with a threshold value k. If the value is below k, that value is propagated to a higher level with decreased resolution of the location information. If the value is above k, the value sent to the upper level is the nearest multiple of k. In that case, the location information is accurate. Using this algorithm, even a location server does not need to belong to a trusted zone because the obfuscation of the location information is already performed in the network.

The work in Beresford and Stajano [60], in which anonymous users are tracked through a building, notes the same problem with permanent anonymous IDs that can be connected to a particular user based on location information from a private area. However, the applications from that work cannot allow for a location precision coarser than the level of a room, so the increased resolution is not an acceptable solution. On the other hand, at the room level, the location information can easily be used to find out which identifiers spend the most time in a particular room.

These authors propose a solution in which the concept of k-anonymity is used in special areas called *mix zones* where users change their temporary anonymous IDs [60]. The manipulation of identifiers is a responsibility of a location server. However, the authors also note an important weakness of the solution with *mix zones*. They demonstrate that the initial assumption that, if two individuals cannot be tracked when they enter a *mix zone* from opposite directions and then reappear from them with different pseudonyms does not hold well if an adversary uses a statistical analysis. The experimental results show that the probability that each individual will return to the same direction from which he came is 1%, so if an adversary assumes that a user who entered from one direction and the user who left the mix zone at the other end are one person, regardless of IDs, he will be correct in 99% of cases.

There are certainly applications that do not conform completely to the system architecture from Figure 16.2. The possible differences are the expansion of the trusted zone to include an LBS, in which case a location server is not needed. The example of this type of application is provided in Priyantha et al. [64]:

indoor user location detection system. The beacons embedded in the building transmit the information about their locations. A device carried by a user detects the signals from beacons, and then determines the location of a user from multilateration of distances acquired from the signal strengths of beacons' signals. In this case, the information about the location is kept on the user's personal device, so privacy is not a concern. However, this simple case has a downside because the user must perform additional work in order to match his location to the location of an interesting object or service. Such a solution is possible only for applications in which results are stored at a device controlled by a user.

Finally, in some applications, it is necessary to maintain relationships between an individual and his profile at a data collection point on an LBS. An example of such an application is Networkcar service [65]. A network of sensors in a car checks the state of the engine and other functioning units in a car. Each car has a built-in gateway that connects the car with a mobile telephony network. At the same time, the gateway uses a GPS client to determine the position of the car. All this information is stored on a Web server. A user of the service (an owner or an authorized car mechanic) can log on to the service through the Web and examine the current location of the car, the conditions of its engine, and other information. The owner of the car receives a message if it has been stolen and taken out of a certain area. This type of application cannot use the proposed techniques in which the precision of location information is reduced because the continuous connection between user and location information sent to the database must be maintained.

16.5 Conclusion

For many military and civilian applications of wireless sensor networks, security and privacy protection protocols and algorithms are an indispensable part of the system architecture. Because of their unique properties, most notably limited resources and physical exposure of sensor nodes, sensor networks require a new type of security protocols. These protocols are tailored to the underlying system architecture, patterns of network traffic, and specific security requirements so that security-related resource consumption is minimized. Physical exposure of nodes, as well as the threat that their cryptographic secrets are potentially available to an adversary, demands that security protocols in sensor networks protect the integrity of the network even if cryptographic secrets are compromised.

Privacy protection is especially important for certain commercial applications of sensor networks. Users who are monitored by sensor networks expect their private information not to be publicly available. However, sensor networks need services from other entities that may not have satisfactory privacy protection mechanisms. In cases in which applications require less precise data than are available, a certain level of privacy protection can be achieved by decreasing precision of the data; therefore, the data cannot be easily matched to any particular individual.

References

1. Industrial automation, Ember Corp. (2003). Retrieved August 20, 2003, from http://www.ember.com/products/solutions/industrialauto.html.
2. Dynamic sensor networks. Retrieved September 1, 2003, from http://dsn.east.isi.edu/.
3. Anderson, R. and Kuhn, M., Tamper resistance — a cautionary note, in *Proc. 2nd USENIX Workshop Electron. Commerce*, USENIX, Berkeley, CA, 1996, 1.
4. Stajano, F. and Anderson, R., The resurrecting duckling: security issues for ad-hoc wireless networks, in *Proc. 7th Int. Workshop Security Protocols*, Christianson, B., Crispo, B., and Roe, M., Eds., Springer-Verlag, Heidelberg, 1999, 172.
5. Meguerdichian, S. et al., Localized algorithms in wireless ad-hoc networks: location discovery and sensor exposure, in *Proc. 2nd ACM Symp. Mobile Ad Hoc Networking and Computing* (MobiHOC), ACM Press, New York, 2001, 106.

6. Krishnamachari, B., Estrin, D., and Wicker, S., The impact of data aggregation in wireless sensor networks, in *Proc. Int. Workshop Distributed Event-Based Systems*, IEEE Computer Society, Los Alamitos, CA, 575.
7. Cheswick, B., An evening with Berferd in which a cracker is lured, endured, and studied, in *Proc. of Winter USENIX Conf.*, USENIX, Berkeley, CA, 1992, 163.
8. Stoll, C., *The Cuckoo's Egg*, Doubleday, New York, 1989.
9. Bellovin, S., Leech, M., and Taylor, T., ICMP traceback messages, Internet draft (work-in-progress), IETF, 2003.
10. Savage, S. et al., Practical network support for IP traceback, in *Proc. ACM SIGCOMM Conf. Applications, Technol., Architectures, Protocols Computer Commun.*, ACM Press, New York, 2000, 295.
11. Snoeren, A. et al., Single packet IP traceback, *IEEE/ACM Trans. Networking*, 10(6), 1, 2002.
12. Song, D. and Perrig, A., Advanced and authenticated marking schemes for IP traceback, in *Proc. 20th Joint Conf. IEEE Computer Commun. Soc.* (INFOCOM), IEEE, 2001, 878.
13. Boulis, A. and Srivastava, M.B., A framework for efficient and programmable sensor networks, in *Proc. 5th IEEE Conf. Open Architectures Network Program.* (OPENARCH 2002), New York, June 2002.
14. Cohen, F., Computer viruses, theory and experiments, *Computers Security*, 6(1), 22, 1987.
15. Kephart, J.O. and White, S.R., Measuring and modeling computer virus prevalence, in *Proc. IEEE Computer Soc. Symp. Res. Security Privacy*, IEEE Computer Society, Los Alamitos, CA, 1993, 2.
16. Spafford, E.H., Computer viruses — a form of artificial life? in *Artificial Life II*, Langton, C.G. et al., Eds., Addison-Wesley, Redwood City, CA, 1992, 727.
17. Denning, D.E., *Cryptography and Data Security*, Addison-Wesley, Redwood City, CA, 1982.
18. Neumann, P., *Computer-Related Risks*, Addison-Wesley, Redwood City, CA, 1995.
19. Cowan, C. et al., Buffer overflows: attacks and defenses for the vulnerability of the decade, *DARPA Inf. Survivability Conf. Exposition*, IEEE Computer Society, Los Alamitos, CA, 1999, 1119.
20. Wagner, M.G., Robust watermarking of polygonal meshes, in *Geometric Modeling Process.*, IEEE Computer Society, Los Alamitos, CA, 2000, 201.
21. Lampson, B.W., A note on the confinement problem, *Commun. ACM*, 16(10), 613, 1973.
22. Gold, B.D. et al., A security retrofit of VM/370, in *AFIPS Conf. Proc.*, AFIPS Press, 1979, 335.
23. Golomb, S.W., The limiting behavior of the Z-channel, *Trans. Inf. Theory*, 26(3), 372, 1980.
24. Hu, W., Reducing timing channels with fuzzy time, in *Proc. IEEE Computer Soc. Symp. Res. Security Privacy*, IEEE Computer Society, Los Alamitos, CA, 1991, 8.
25. Simmons, G.J., The history of subliminal channels, *IEEE J. Selected Areas Commun.*, 16(4), 452, 1998.
26. Rubin, A.D. and Geer, D.E., Mobile code security, *IEEE J. Internet Computing*, 2, 30, 1998.
27. Gong, L. et al., Going beyond the sandbox: an overview of the new security architecture in the Java development kit 1.2, in *Proc. USENIX Symp. Internet Technol. Syst.*, USENIX, 1997, 103.
28. Brigner, P., Creating signed, persistent Java applets, *Dr. Dobb's J.*, 24, 82, 1999.
29. Bowen, R. et al., Building survivable systems: an integrated approach based on intrusion detection and confinement, in *Proc. DARPA Inf. Security Symp.*, 2000, 1084.
30. Colby, C. et al., A certifying compiler for Java, *SIGPLAN Notices*, 35, 95, 2000.
31. Necula, G.C., Proof-carrying code, in *Proc. 24th ACM SIGPLAN-SIGACT Symp. Principles Programming Languages*, ACM Press, 1997, 106.
32. Lofstrom, K., Daasch, W.R., and Taylor, D., IC identification circuits using device mismatch, in *Proc. Int. Solid-State Circuits Conf.*, IEEE, 2000, 372.
33. Pitkow, J., In search of reliable usage data on the WWW, *Comp. Networks ISDN Syst.*, 29, 1343, 1997.
34. Franklin, M.K. and Malkhi, D., Auditable metering with lightweight security, *J. Comp. Security*, 6(4), 236, 1998.

35. Naor, M. and Pinkas, B., Secure accounting and auditing on the Web, *Comp. Networks ISDN Syst.*, 30, 541, 1998.
36. Rivest, R.L., Electronic lottery tickets as micropayments, in *Proc. Int. Conf. Financial Cryptography*, Hirschfeld, R., Ed., Springer-Verlag, Heidelberg, 1997, 307.
37. Findley, R. and Dixon, R., Dual smart card access control electronic data storage and retrieval system and methods, U.S. patent 5629508, 1997.
38. Menezes, A.J., Oorschot, P.C., and Vanstone, S.A., *Handbook of Applied Cryptography*, CRC Press, Boca Raton, FL, 1997.
39. Ross, C.D. et al., Method and apparatus for electronic licensing, U.S. patent 5553143, 1996.
40. Aura, T. and Gollmann, D., Software license management with smart cards, in *Proc. USENIX Workshop Smartcard Technol.*, USENIX Association, Berkeley, CA, 1999, 75.
41. Thomas, D.C., Method and apparatus for providing security or computer software, U.S. patent 4446519, 1984.
42. Law, Y.W., Etalle, S., and Hartel, P.H., Assessing security-critical energy-efficient sensor networks, in *Proc. 18th IFIP TC11 Int. Conf. Inf. Security Privacy Age Uncertainty* (SEC), Gritzalis, D. et al., Eds., Kluwer Academic Publishers, Boston, MA, 2003, 459.
43. Papadimitratos, P. and Haas, Z.J., Securing mobile ad hoc networks, in *Handbook of Ad Hoc Wireless Networks*, Ilyas, M., Ed., CRC Press, Boca Raton, FL, 2002.
44. Perrig, A. et al., SPINS: security protocols for sensor networks, in *Proc. 7th Int. Conf. Mobile Computing Networking* (MOBICOM), ACM Press, New York, 2001, 189.
45. Hill, J. et al., System architecture directions for network sensors, in *Proc. 9th Int. Conf. Architectural Support Programming Languages Operating Syst.* (ASPLOS), ACM Press, New York, 2000, 93.
46. Rivest, R.L., The RC5 encryption algorithm, in *Proc. 2nd Workshop Fast Software Encryption*, Preneel, B., Ed., Springer-Verlag, Heidelberg, 1995, 86.
47. Liu, D. and Ning, P., Efficient distribution of key chain commitments for broadcast authentication in distributed sensor networks, in *Proc. 10th Symp. Network Distributed Syst. Security*, Internet Society, Reston, VA, 2003, 263.
48. Slijepcevic, S. et al., On communication security in wireless ad-hoc sensor networks, in *Proc. 11th IEEE Int. Workshops Enabling Technol.: Infrastructure Collaborative Enterprises* (WETICE), IEEE Computer Society, Los Alamitos, CA, 2002, 139.
49. Hubaux, J.P., Buttyan, L., and Capkun, S., The quest for security in mobile ad hoc networks, in *Proc. 2nd ACM Symp. Mobile Ad Hoc Networking Computing* (MobiHOC), ACM Press, New York, 2001, 146.
50. Carman, D.W., Matt, B.J., and Cirincione, G.H., Energy-efficient and low-latency key management for sensor networks, in *Network Assoc. Labs Advanced Security Res. J.*, 5(1), 2003, 31.
51. Eschenauer, L. and Gligor, V.D., A key management scheme for distributed sensor networks, in *Proc. 9th ACM Conf. Computer Commun. Security*, ACM, New York, 2002, 41.
52. Yuan, L. and Qu, G., Design space exploration for energy-efficient secure sensor networks, in *Proc. IEEE Int. Conf. Application Specific Syst., Architectures, Processors* (ASAP), IEEE Computer Society, Los Alamitos, CA, 2002, 88.
53. Spencer, J.H., *The Strange Logic of Random Graphs*, 1st ed., Springer-Verlag, Heidelberg, 2000.
54. Chan, H., Perrig, A., and Song, D., Random key predistribution schemes for sensor networks, in *Proc. 2003 IEEE Symp. Security Privacy*, IEEE Computer Society, Los Alamitos, CA, 2003, 197.
55. Samarati, P. and Sweeney, L. Protecting privacy when disclosing information: *k*-anonymity and its enforcement through generalization and suppression, technical report, SRI International, 1998.
56. Gruteser, M. and Grunwald, D., Anonymous usage of location-based services through spatial and temporal cloaking, in *Proc. ACM/USENIX Int. Conf. Mobile Syst., Applications, Services* (MOBISYS), USENIX, Berkeley, CA, 2003.
57. Navigation Technologies, Navtech. Retrieved September 1, 2003, from http://www.navtech.com.
58. Avis Assist, Avis, Inc. Retrieved September 1, 2003, from http://www.avis.com/AvisWeb/JSP/US/en/deals/us_assist.jsp.

59. Autodesk Location Services, Autodesk (2003). Retrieved August 9, 2003, from http://locationservices.autodesk.com/.
60. Beresford, A. and Stajano, F., Location privacy in pervasive computing, *IEEE Pervasive Computing*, 2(1), 46, 2003.
61. Gruteser, M. et al., Privacy-aware location sensor networks, in *Proc. 9th USENIX Workshop Hot Topics Operating Syst.* (HotOS), USENIX, Berkeley, CA, 2003.
62. Chaum, D., Security without identification: transaction systems to make Big Brother obsolete, *Commun. ACM*, 28(10), 1030, 1985.
63. Pfitzmann, A. and Koehntopp, M., Anonymity, unobservability, and pseudonymity — a proposal for terminology, in *Proc. Workshop Design Issues Anonymity Unobservability*, Federrath, H., Ed., Springer-Verlag, Heidelberg, 2001, 1.
64. Priyantha, N.B., Chakraborty, A., and Balakrishnan, H., The Cricket location support system, in *Proc. 6th Int. Conf. Mobile Computing Networking* (MOBICOM), ACM Press, New York, 2000, 32.
65. Networkcar technology, Networkcar, Inc. Retrieved August 21, 2003, from http://www.networkcar.com.

17

Low-Power Design for Smart Dust Networks

17.1 Introduction 17-1
17.2 Location 17-1
17.3 Sensing 17-2
17.4 Computation 17-2
Asynchronous Processors • Variable-Frequency Processors • Variable-Voltage Processors
17.5 Hardware–Software Interaction 17-5
17.6 Communication 17-7
Mote-to-Mote Communication • Mote-to-Central Station Communication
17.7 Orientation 17-10
17.8 Conclusion 17-10

Zdravko Karakehayov
Technical University of Sofia

17.1 Introduction

Distributed sensor networks (DSNs) are composed of numerous small, low-cost, randomly located nodes. The network can be scalable to thousands of nodes that cooperatively perform complex tasks such as intelligent measurement. The network must be able to self-organize, adapt to random node spacing, execute algorithms for signal processing, and operate as power efficiently as possible. The major applications of DSNs are for monitoring environmental conditions, tracking the movements of birds and small animals, monitoring product quality, and building automation and defense networks. Smart Dust is a term recently coined at the University of California, Berkeley, to describe massively distributed sensor networks consisting of cubic-millimeter sized motes [1, 2]. The small size and anticipated low cost of the motes will help to collect information cost-effectively and less intrusively.

Each mote depends on low-capacity batteries as energy sources. Practically, the chance for battery replacement is nonexistent. As a result, every aspect of the Smart Dust networks, from mote location through computing and communication, is viewed from the low-power perspective.

17.2 Location

A deployment may leave numerous motes located in different areas of a large geographical region. The location of the motes affects energy efficiency in a number of ways. Sensor readings are of interest if only bound to a known location. Interrogation of motes before a location procedure would be a loss of energy. The global positioning system (GPS) is able to locate network nodes in outdoor environments. However, cost, power consumption, and size of the currently available GPS receivers are prohibitive for Smart Dust motes. Optical communication emerges as the most efficient method if a central station may be harnessed to provide energy for location tasks. Because motes may move, some applications would demand updating the positions regularly. Also, radio-frequency (RF) communication can be used by motes to locate themselves via beacon signals from reference points [3, 4].

As soon as the location procedure has been completed, some nodes will be actively involved in sensing, while others will wait for events and can be turned off to save energy. An event tracking, such as following light shadow edges over a sensor field, can be organized in two ways:

- All motes deactivate all subsystems except sensors that can obtain relevant data. If the sensors provide binary readings, they can be used to awake the motes in case of events.
- A more sophisticated power reduction approach will turn off all motes, except motes in the close vicinity of the event, completely. However, in case of a dense deployment the distance alone is not sufficient as a criterion.

Liu et al. [5] have developed a method for event tracking. The method identifies motes that will not be immediately approached by the event and can be turned off to save energy. The method is based on dual space transformation [6]. Figure 17.1 shows an example for event tracking.

With no loss of generality, it can be assumed that the event is a moving light shadow edge. The edge is presented in the primal space as the E line and the motes' locations are indicated as points. The line is uniquely defined in the primal space by the p slope and the y-intercept q. The line is transformed into the e point in the dual space; in turn, the points from the primal space are transformed into lines in the dual space. As a result, the dual space is partitioned into cells. The e point is contained in the shaded cell. Because the e point cannot intersect the m2 line before it crosses one of the cell boundaries, the M2 mote can stay turned off as long as none of M1, M3, and M4 sense a transition.

17.3 Sensing

The mote's sensors vary from application to application: temperature, light, magnetic field, vibration, and acoustic. Recent advances in technology have made it possible for these sensors to be released in ultralow sizes and power versions [2, 7].

Sensors convert physical variables into electrical signals. Typically, the signals are in the microvolt or millivolt range. An input signal conditioner is used to filter and amplify the signals. Energy is consumed in the sensor, amplifier, and analog-to-digital converter (ADC). The power consumption can be reduced with appropriate power management. The ADC's resolution has a significant impact on the energy budget. For instance, if the ADC's resolution is increased from 15 to 16 bits while keeping the other parameters unchanged, the power consumption is increased from 100 to 400 mW [8].

A common method for analog-to-digital conversion is the successive approximation [9]. Because the ADC determines one bit of the result in each cycle, it would be possible to apply selective resolution. Consequently, different samples will have different numbers of bits and different energy costs. Finally, one may only want to test if the input value belongs to a certain range. In this case, a microcontroller with an on-chip analog comparator can be a power-efficient solution. Microcontrollers such as the Atmel ATmega161L are capable of turning off the comparator to reduce the power consumption [10].

17.4 Computation

Motes incorporate a processor to carry out computations locally. Functionality typically requires the processor to run in outbursts separated by idle periods. Within the idle period, the processor may enter a power reduction mode to save energy [9]. The battery lifetime is influenced by the power efficiency of a running processor and the balance between active and idle periods.

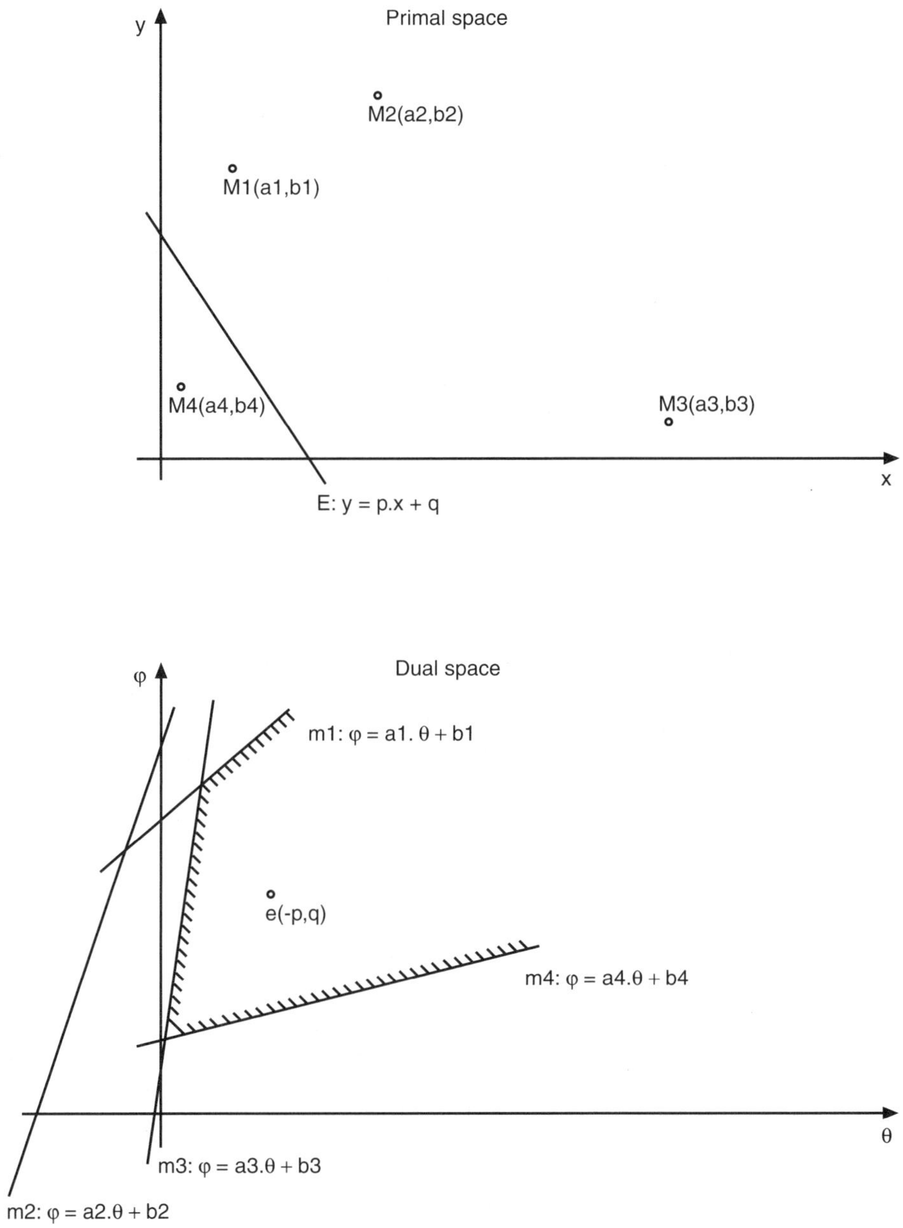

FIGURE 17.1 Primal-to-dual space transformation indicates the sequence of transitions.

17.4.1 Asynchronous Processors

The synchronous processor's clock distribution network is characterized by significant power consumption. Moreover, synchronous systems tend to maximize supply current transients. Smart Dust motes may have analog subsystems that are influenced by the electromagnetic radiation. Asynchronous designs promise to overcome the clock-related problems. In particular, a class of asynchronous implementations, termed self-timed systems, is capable of operating as fast as circumstances allow.

17.4.2 Variable-Frequency Processors

Using variable-frequency processors, power consumption can be gradually controlled by scaling the clock frequency. Typically, a phase lock loop (PLL) circuitry can multiply the oscillator frequency and an adjustable prescaler can divide the oscillator frequency. Based on the current task's deadline, the clock frequency may decline as much as possible. However, if the processor completes the task ahead of the deadline and enters a power-saving mode, the energy could be minimized [11]. In this case, the task's deadline period, T_{DL}, accommodates the active period, T_{ACT}, and the power-saving period, T_{PS}:

$$T_{DL} = T_{ACT} + T_{PS} \tag{17.1}$$

Assume that the power consumption scales linearly with the clock frequency:

$$P_{ACT} = k_{ACT} \times f_{CLK} + n_{ACT} \tag{17.2}$$

$$P_{PS} = k_{PS} \times f_{CLK} + n_{PS} \tag{17.3}$$

If the task's functionality requires NC processor clocks, the energy per task,

$$E_T = P \times T_{DL} = k_{PS} \times T_{DL} \times f_{CLK} + (n_{ACT} - n_{PS})\frac{\text{NC}}{f_{CLK}} + (k_{ACT} - k_{PS})\text{NC} + n_{PS} \times T_{DL} \tag{17.4}$$

Take the first derivative and calculate the critical numbers

$$f_{CLK} = \pm\sqrt{\frac{(n_{ACT} - n_{PS})\text{NC}}{k_{PS} \times T_{DL}}} \tag{17.5}$$

Consider two cases for the positive value:

- Let $n_{ACT} > n_{PS}$. Based on the second derivative test, the energy per task has a minimum for

$$f_{CLK,OPT} = \sqrt{\frac{n_{ACT} - n_{PS}}{k_{PS}}}\sqrt{\frac{\text{NC}}{T_{DL}}} \tag{17.6}$$

- If $n_{ACT} \le n_{PS}$, the clock frequency must be selected as low as possible. The power-saving mode is not used.

$$f_{CLK,OPT} = \frac{\text{NC}}{T_{DL}} \tag{17.7}$$

Equation 17.6 does not guarantee that the deadline will be met. In some cases, the calculated clock frequency must be increased to meet the deadline.

Figure 17.2 shows an example mesh plot for the clock frequency. Assume that the processor is characterized by n_{ACT} = 1 mW; n_{PS} = 0.1 mW; and k_{PS} = 1 mW/MHz. The example is based on 256 combinations of deadline periods and cycles per task. For two combinations, the optimal clock frequencies have been replaced by higher values.

Actual tasks, which require replacement of the optimal clock frequency, can be viewed as targets for further improvement. Optimization of the code or relaxing the timing constraints would be an appropriate course of action.

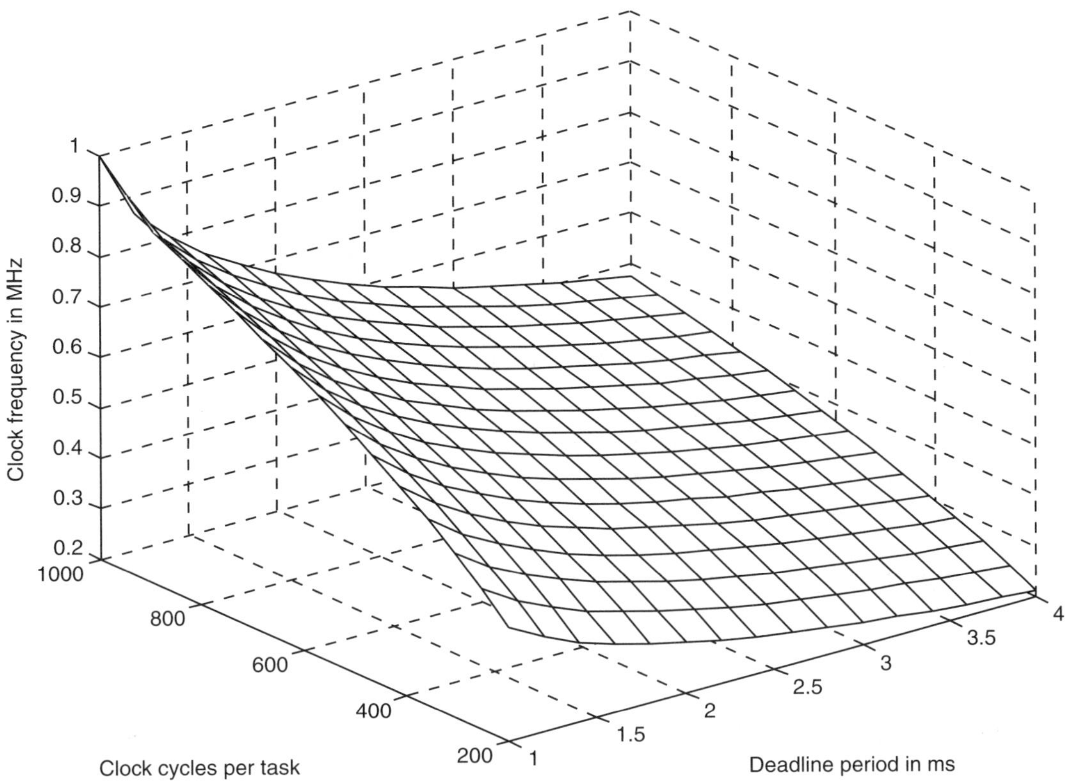

FIGURE 17.2 Mesh plot for the clock frequency.

17.4.3 Variable-Voltage Processors

Variable-voltage processors are capable of operating over a wide voltage range. Allocating such a processor for the network nodes allows power reduction by dynamically varying the supply voltage [12–15]. The method is often termed dynamic voltage scaling (DVS). DVS is an efficient method for power reduction; however, it imposes some limitations for the system:

- The system components must be capable of operating over a wide voltage range.
- A voltage converter loop hardware must be available.

Hong et al. [16] developed a design methodology for DVS. Figure 17.3 illustrates how to tune the voltages for extra power reduction. The tasks are specified by their arrival times, deadlines, and execution times at a nominal voltage. The schedule is viewed as a first iteration. It would be beneficial to extend the T2 task and reduce the V2 voltage. T1 is scheduled for V2 to shrink the execution time. The new border between T1 and T2 is placed just in the middle of the interval indicated by an arrow; no conflict takes place with the arrival time and the change is accepted.

Similarly, T3 is scheduled for V2 to allow extension for T2. The intention is to place the new border just in the middle of the interval marked by an arrow; however, T2 fails to meet the deadline and the border is aligned with the deadline. If the new schedule is more energy efficient, it is accepted.

17.5 Hardware–Software Interaction

A mote includes a CPU, memory, and peripherals. As a rule, peripherals possess three types of registers: data, control, and status. Data registers are employed as buffers between the CPU and peripherals, while control registers are used to adjust the I/O device functionality for a specific application. Status registers

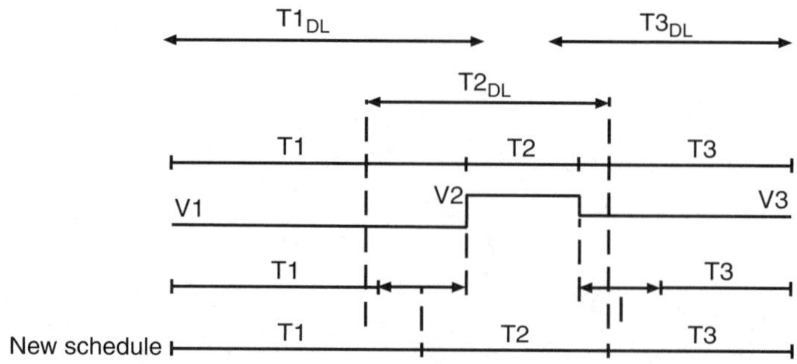

FIGURE 17.3 A schedule is modified for extra power reduction.

are read by the processor to check whether a specific operation is done. In spite of the huge variety of peripherals, the communication between the CPU and the I/O devices remains routine and easy to define.

Modifying one or more bits in a register, the CPU must keep the rest of the pattern unchanged. A common way to implement bit manipulation is to read a register, modify bits, and write the result back. The two memory accesses make the read–modify–write instructions power inefficient. In an attempt to improve the situation, Atmel has taken another approach with the AT91 microcontroller [17]. Instead of one control register, the microcontroller employs three registers mapped into three consecutive memory locations. The first register is used to set individual bits, the second to clear bits, and the third to obtain the current pattern. To set or reset a bit, a high bit is written to the corresponding position at the set or reset register.

In the AT91 microcontroller, a PLL circuitry and a programmable prescaler complement the ARM7TDMI core to a variable-frequency processor. The PLL circuitry multiplies the oscillator frequency; the highest multiplication factor is 64. As a result, the oscillator may run at a frequency 64 times lower than the actual clock and thus the oscillator saves energy. The programmable prescaler with a division factor of 64 allows the AT91 clock frequency to go down to 512 Hz. The CPU and embedded peripherals can be individually enabled and disabled. The ARM processor clock is enabled from the next interrupt or reset. The on-chip RAM reduces external memory accesses and allows further power reduction. Finally, the processor may switch to the 16-bit instruction set and benefit from a narrower memory.

Similarly to the analog-to-digital conversion, the measurement of time intervals also falls under the accuracy–power trade-off. Figure 17.4 shows how a counter/timer determines a time interval using different clock rates. The highest possible frequency provides the highest accuracy. If a Smart Dust application is based on the AT91 microcontroller, the number of counter transitions for a 50-ms period may vary with the frequency up to 50,000.

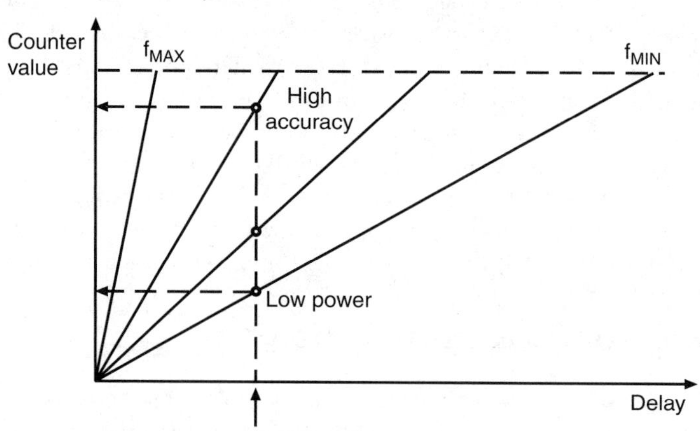

FIGURE 17.4 Using different frequencies to measure a time interval.

17.6 Communication

In a wireless sensor network, communication is the major consumer of energy. Smart Dust networks have two recognized communication styles: (1) RF is characterized with power consumption in the milliwatt range; (2) optical communication is associated with a lower energy cost but requires accurate pointing. Consequently, optical communication is more suitable for interaction between network motes and a central station. The RF approach is very common for communication between motes [8, 18, 19].

17.6.1 Mote-to-Mote Communication

The procedures for establishing and operating a network require the motes to communicate with one another. The task of routing packets from a source to a destination can be broken down into discovering the position of the destination and the actual forwarding of packets [20]. Furthermore, channel access can be implemented by two different methods: contention or explicit organization [21]. The contention-based approach is not suitable for DSNs because of its requirement to monitor the channel for a long span of time. Because the reception and transmission have almost the same energy cost, the organized channel access is characterized with better energy efficiency. At the same time, the process of establishing time division multiple access (TDMA) slots or frequency bands also consumes energy. In an attempt to alleviate this problem, some protocols employ a hierarchical structure that requires partitioning the network.

The two basic schemes to limit the mote's RF transmission power are: (1) a transmitter can vary its power to cover different distances under different environmental conditions; or (2) the link can be partitioned into several short intermediate hops and use constant transmission power. Any DSN with a sufficient density of nodes can benefit from multihop communication.

The energy used to send a bit over a distance d may be written as

$$E = A \times d^n \tag{17.8}$$

where A is a proportionality constant and n depends on the environment [18, 22]. The greater-than-linear relationship between energy and distance promises to reduce the energy cost when the link is partitioned.

Rewrite Equation 17.8 for NH number of hops. Also, include the energy for receiving E_R and energy for computation E_C:

$$E = A\left(\frac{d}{NH}D\right)^n NH + (E_R + E_C)NH \tag{17.9}$$

Assume equal distances for each hop. $D > 1$ is introduced to take into account the longer path inevitably associated with multihop communication. The energy has a minimum for

$$NH_{OPT} = (d \times D)\sqrt[n]{\frac{A(n-1)}{E_R + E_C}} \tag{17.10}$$

Figure 17.5 shows a plot for the energy per bit using different numbers of hops. The distance $d = 50$ m; $n = 4$; $A = 0.2$ fJ/m^4; $D = 1.2$; and $E_R + E_C = 30$ pJ. The energy per bit has a minimum for four hops.

A subtle effect of multihop communication is that energy consumption is distributed over the motes fairly. If the motes consume energy at about the same rate, the system lifetime is increased. Chen et al. [23] developed a coordination algorithm to increase the energy efficiency further. The algorithm is based on an assumption that when a wireless network has an ample density of nodes, only a small number of them need to be active to forward traffic. A distinctive feature of the method, named SPAN, is that the

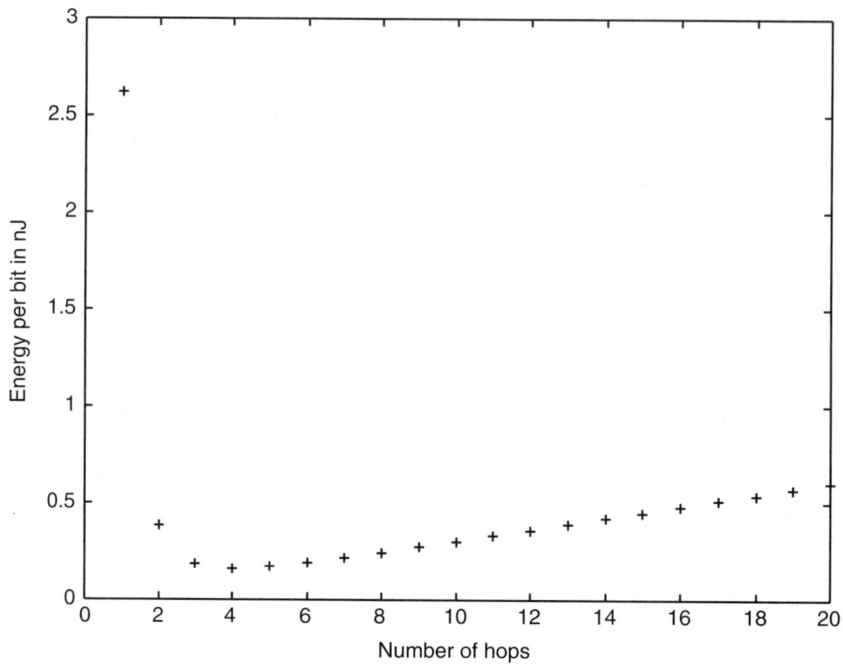

FIGURE 17.5 Energy-per-bit scales with the number of hops.

motes make a decision whether to sleep or be active based not only on the topology of the network, but also on the amount of energy available in the battery.

All motes of the network are dynamically split into two sets: motes that sleep and motes that stay awake to participate in the forwarding backbone topology. According to SPAN's terminology, the active motes are named coordinators. Each mote of the network makes periodic, local decisions on whether to sleep or become a coordinator. Coordinators are elected to achieve two goals: improved connectivity of the network and equal levels of energy remaining at each mote. All noncoordinator motes periodically participate in an election procedure to become coordinators; in parallel, all coordinators periodically pass through a withdrawal procedure to switch back to a sleep state. Figure 17.6 shows this election–withdrawal cycle. A mote becomes a coordinator to link two neighbor motes that cannot communicate directly or via one or two coordinators. Because several motes can run an election procedure simultaneously, there might be an overlap in the connectivity they introduce. The method attempts to minimize the number of coordinators to save energy.

To resolve contention, the election procedure is extended with a variable delay. As soon as the delay period is over, a coordinator announcement is sent out. If, at the end of the delay the mote receives other announcements for new coordinators, it reconsiders the need to become a coordinator. The election procedure distinguishes between two cases:

- All applicants for coordinators have equal energy left in their batteries. In this case, the more pairs of motes the applicant connects, the shorter is the delay. Also, to rotate coordinators with time, a random value influences the delay.
- The participating motes have unequal energy available in their batteries. In this case, the delay period is calculated on the base of the connection improvement and the amount of energy scaled to the maximum amount of energy the mote can have. The random factor is still included.

Each coordinator periodically runs a withdrawal procedure. A coordinator can go back to sleep if every pair of its neighbors can reach each other directly or via some other coordinators. Initially, the mote will stay as a coordinator if its withdrawal affects the network connectivity. However, after some time it will

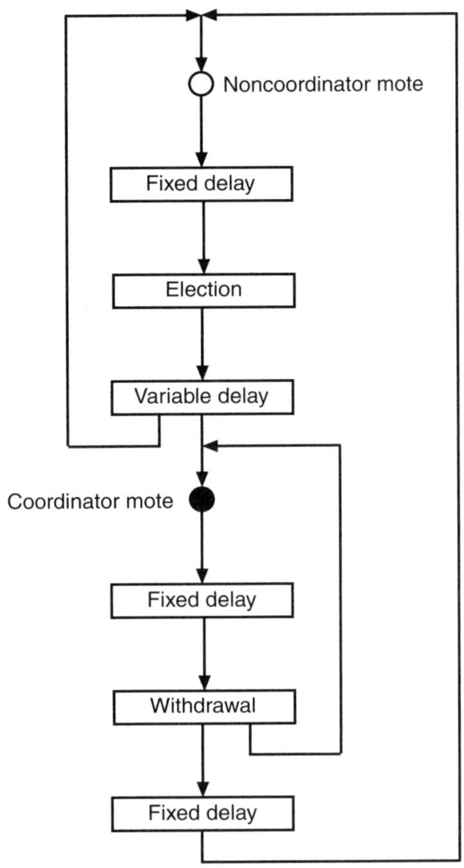

FIGURE 17.6 Span's election withdrawal cycle.

switch to noncoordinator state to give other neighbors a chance to become coordinators. As shown in Figure 17.6, a mote continues to serve as a coordinator for a fixed period of time after its withdrawal announcement is sent out. Thus, the routing protocol can use the old coordinator until a new coordinator is elected.

17.6.2 Mote-to-Central Station Communication

When one or more central stations communicate with a field of dust motes, optical systems are characterized by the lowest energy budget. Two methods can be used to apply optical communication for Smart Dust: passive reflective systems and active-steered laser systems [2]. Figure 17.7 shows an example of a passive reflective device, a corner-cube retroreflector (CCR). A CCR reflects the light via three mutually orthogonal mirrors. When a light beam enters the CCR, it bounces off the mirrors and is reflected back parallel to the direction from which it entered. Because one of the mirrors is mounted on a spring at an angle slightly askew from perpendicularity to the other mirrors, in this state little light returns to the remote receiver. No reflection of the light is considered a low logic level. To return the light to its source, high logic level, the mirror is shifted to a position perpendicular to other mirrors. The low-to-high transition consumes less than a nanojoule [2]. The high-to-low transition requires almost no energy.

Active-steered laser systems are suitable for mote-to-mote and mote-to-central station communications. The device consists of a semiconductor diode laser, collimating lens, and a two-degree-of-freedom micromirror [2]. Central stations can use imaging receivers to process transmissions from different angles.

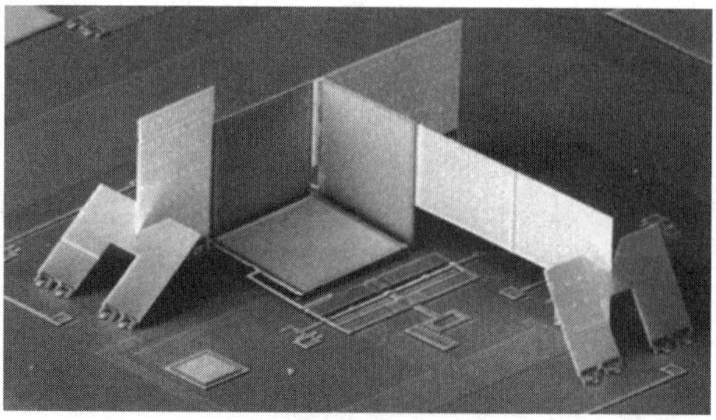

FIGURE 17.7 Microfabricated corner-cube retroreflector. (From Hsu, V., Kahn, J.M., and Pister, K.S.J., www-ee.stanford.edu/~jmk/pubs/hsu.ms.11.99.pdf. With permission.)

This approach of separating transmissions according to their originating location is termed a space division multiple access (SDMA).

17.7 Orientation

Many applications will deploy motes in random orientation. Consequently, it will not be possible for all CCRs to return light to the central station. A CCR quadruplet is a solution that improves the accessibility of the motes. At the same time, some directions may be characterized with noise emissions and should be avoided. Furthermore, applications may require the motes to be invisible from a certain area.

It is proposed that the motes be magnetized and the CCR oriented to a predefined direction. When the motes fall through the air after being deployed, they will orient themselves. If the network has a sufficient density of motes, it may not need the motes, which change orientation upon landing. This approach for zero-power orientation is even more efficient for motes floating on the water. They could freely rotate to orient themselves. Figure 17.8 shows a deployment of two types of motes that differ in their CCR orientation; two central stations interrogate the motes. The DSNs community is growing and projects that simultaneously employ a single field can benefit from SDMA.

17.8 Conclusion

The low-power design of Smart Dust networks has a lot in common with many other computer applications. By allocating variable-frequency processors for the Smart Dust motes, clock frequency scaling can be applied to decrease power consumption. It is necessary to distinguish between two types of processors in order to decide whether it is more power efficient to operate quickly and then wait quietly, or just operate at the minimum speed possible. For the first case, the optimal clock frequency is calculated based on the required number of clock cycles and a deadline period. This approach also allows identifying tasks that require replacement of the optimal clock frequency. Thus, a set of tasks emerges as a target for further improvement. Variable-voltage processors could combine voltage scaling with frequency scaling if the hardware overhead is not prohibitive for a cubic-millimeter sized mote.

Hardware–software interaction also provides ample reserve for power reduction. Scaling down the theme of variable frequency from processors to counters, motes could measure time intervals, trading accuracy against number of transitions. The hardware–software interaction and the sensing show that redundant accuracy wastes energy in the same way as redundant computation speed.

The energy spent for communication is crucial for the success of wireless networks such as Smart Dust. Multihop communication can help power consumption to decline significantly and avoid obstacles for RF and optical systems. As an additional benefit, multihop transmissions distribute power consumption over the motes fairly and increase system lifetime.

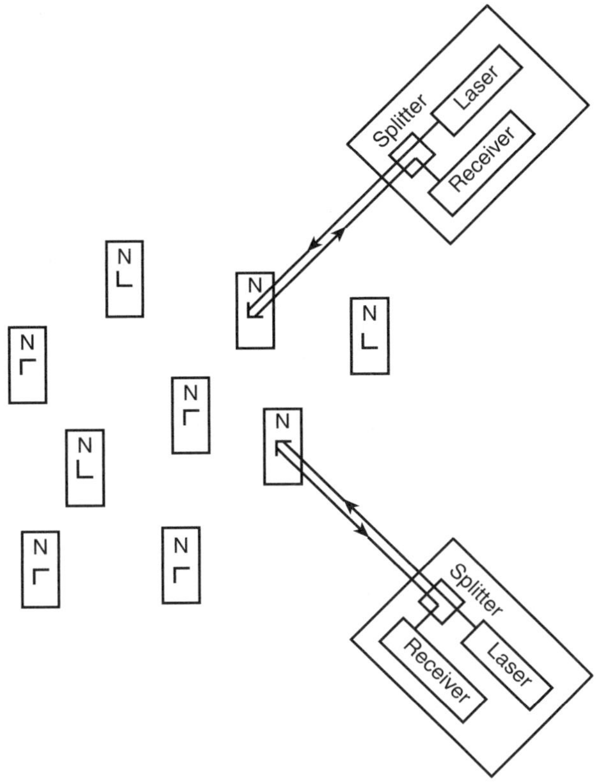

FIGURE 17.8 Each central station interrogates its own motes.

The location of the motes is a process specific to the network operation. Some applications may require only relative positions. Relative positions can be used to turn off motes, especially in case of event tracking. Finally, optical communication is associated with pointing and orientation. By using the Earth's magnetic field, zero-power orientation of the motes can be implemented for SDMA.

Acknowledgment

The author is grateful to Brett Warneke for important advice and to Victor Hsu for the CCR photograph.

References

1. Kahn, J.M., Katz, R.H., and Pister, K.S.J. Next century challenges: mobile networking for "Smart Dust," *J. Commun. Networks*, 2(3), 188, September 2000.
2. Warneke, B., Last, M., Liebowitz, B., and Pister, K.S.J. Smart Dust: communicating with a cubic-millimeter computer, *Computer*, 34, 44, January 2001.
3. Bulusu, N., Heidemann, J., and Estrin, D. GPS-less low-cost outdoor localization for very small devices, *IEEE Personal Commun.*, 7, 28, 2000.
4. Hightower, J. and Borriello, G. Location systems for ubiquitous computing, *IEEE Computer*, 34, 57, 2001.
5. Liu, J., Cheung, P., Guibas, L., and Zhao, F. A dual-space approach to tracking and sensor management in wireless sensor networks, Palo Alto Research Center technical report P2002-10077, 2002, available at www2. parc.com/spl/projects/cosense/pub/dualspace.pdf.
6. Berg, M., Kreveld, M., Overmars, M., and Schwarzkopf, O. *Computational Geometry: Algorithms and Applications*, Springer-Verlag, Berlin, 1997.

7. Gardner, J.W. *Microsensors: Principles and Applications*, John Wiley & Sons, New York, 1994.
8. Doherty, L., Warneke, B.A., Boser, B.E., and Pister, K.S.J. Energy and performance considerations for smart dust, *Int. J. Parallel Distributed Syst. Networks*, 4, 121, 2001.
9. Karakehayov, Z., Christensen K.S., and Winther, O. *Embedded Systems Design with 8051 Microcontrollers*, Marcel Dekker, New York, 1999.
10. Atmel Corporation, *AVR RISC Microcontroller, Data Book*, 1999.
11. Karakehayov, Z. Zero-power design for Smart Dust networks, *Proc. 1st IEEE Int. Conf. Intelligent Syst.*, Varna, 1, 302, 2002.
12. Macken, P., Degrauwe M., Paemel V., and Oguey H. A voltage reduction technique for digital systems, *Digest of Technical Papers, 37th IEEE Int. Solid-State Circuits Conf.*, 238, 1990.
13. Mudge, T., Power: a first-class architectural design constraint, *IEEE Computer*, 34, 52, 2001.
14. Burd, T. *Energy-Efficient Processor System Design*, Ph.D. thesis, University of California, Berkeley, 2001.
15. Sinha, A. and Chandrakasan, A. Dynamic power management in wireless sensor networks, *IEEE Design Test Computers*, 18, 62, 2001.
16. Hong, I. et al. Power optimization of variable-voltage core-based systems, *IEEE Trans. Computer-Aided Design Integrated Circuits Syst.*, 18(12), 1702, 1999.
17. Atmel Corporation, *AT91 ARM Thumb Microcontrollers, AT91M55800A*, 2002. Available at www.atmel.com.
18. Rabaey, J.M. et al. PicoRadio supports ad hoc ultra-low power wireless networking, *Computer*, 33, 42, July 2000.
19. Hill, J.L. and Culler, D.E. Mica: a wireless platform for deeply embedded networks, *IEEE MICRO*, 22, 12, November–December 2002.
20. Mauve, M. and Widmer, J. A survey on position-based routing in mobile ad hoc networks, *IEEE Network*, 15, 30, 2001.
21. Sohrabi, K. et al. Protocols for self-organization of a wireless sensor network, *IEEE Personal Commun.*, 7, 16, 2000.
22. Akyildiz, I.F. et al. A survey on sensor networks, *IEEE Commun. Mag.*, 40, 102, 2002.
23. Chen, B. et al. Span: An energy-efficient coordination algorithm for topology maintenance in ad hoc wireless networks, *ACM Wireless Networks J.*, 8, 481, 2002.

Index

A

Access control lists, 2:7
Accounting management, 3:14–15
Accuracy of sensor networks, 1:5
Acquisition tasks, energy consumption, 1:9
Active Badge system, 11:10
Ad hoc networks, 15:4–12, 16:8–10
  architecture, 7:9
  layered protocol stack, 7:21
  management protocol, 3:24
  medium access control protocols, 15:4–10
  multihop wireless architecture, 8:15–16
  sensor networks, contrasted, 10:1–2
  wireless sensor networks, contrasted, 1:4, 7:8, 10:1–2
Ad hoc on-demand distance vector (AODV), 1:7
Adaptive listening, 15:16
Address-centric data forwarding, 7:20
Agriculture, wireless sensor networks in, 1:3
Agriculture monitoring, sensor networks in, 1:3
Air-borne dispersal, 5:1
Aircraft drug reduction, 1:3
Alternative identification of users, 16:13
Ambient light sources, photocurrents from, charging capacitors with, 1:6
Analog-to-distal conversion, power savings, 17:2
Anonymity in privacy protection, defined, 16:13–15
Antennas, 1:6, 2:8
AODV. *See* Ad hoc on-demand distance vector
Application layer protocols, 10:2–4
Application-specific design, 7:10
Artificial retina, sensor networks for, 7:5
Asset monitoring, sensor networks in, 1:3
Associative broadcast, mechanism of, 6:5
Asynchronous processors, power savings, 17:3–4
ATM networks, energy-conserving, 15:7–8
Attenuation requirements, 1:6
Attribute-based naming, 6:3–4
Automatic warehouse inventory tracking, sensor networks for, 7:6
Automotive telematics, sensor networks in, 1:2

B

Bandwidth efficient radios, 8:6
Bargaining, communication mode, 3:11
Base station, sensor network with, 1:5
Bat system, 11:10
Batteries, 14:13–14:14. *See also* Power conservation
  classification, 3:7
  in development, 5:11
Battle-space monitoring, wireless sensor networks in, 1:4
Berkeley Mote node, 8:8–9
Biological attack, remote sensing, 7:3
Blackboard-based model, system-level programmability, 4:6
Bluetooth, 1:6, 15:6–7
Boxes, tracking of, 1:3
Broadcast messages, 15:13
Broadcast trees, 1:6
Bulk micromachining, 5:2
Burstiness, 7:14–15
Business management, 3:5
BWRC PicoNode, 8:11

C

Cancer detection, sensor networks for, 7:5
CAOC. *See* Combined air operations center
Capacity of wireless networks, research on, 1:7
CCLK. *See* Core clock
CCR. *See* Corner-cube retroreflector
Cell-based wireless sensor networks, security, 16:5–8
CERN, European Laboratory for Particle Physics, 12:18
Channel access, 1:7–8
Chemical process control, sensor networks for, 7:6
Chemical weapon, remote sensing, 7:3
Civil engineering, sensor networks in, 1:3
Classification of wireless sensor networks, 7:6–7
Climate control, 7:3–4
Clock frequency, mesh plot, 17:5
Clock synchronization, 7:22
Clustering
  hierarchical architectures, 7:16–19
  for static sensor networks, 1:6
CMIP. *See* Common management information protocol
CMOS process, 5:2–3
CMOS transistors, with micromachining capabilities, 5:3–4
COBRA. *See* Common object request broker architecture
Collaborative signal, information processing, 7:19–20
Combined air operations center (CAOC), 2:3–5
Commercial applications, 10:2
Commercial interest in sensor networks, 1:2
Common management information protocol (CMIP), 3:24
Common object request broker architecture (COBRA), 2:11
Communication approach, classification of, 3:11
Communication coverage area map, 3:24
Communication protocols, 10:1–16
Compatibility issues, 3:4
Components of nodes, 5:1
Computational module, 3:7
Computational resources guarantees, 2:8–10
Conference sessions on sensor networks, 1:2
Configuration
  functionality, factors related to, 3:8

management, 3:12–13
under varying conditions, 3:2
Connecting with communication network, technical challenges, 1:4–11
Containers, tracking of, 1:3
Contaminant transport, sensor networks in, 1:3
Contention-based schemes, 15:11
collision, 15:11
control packet overhead, 15:11
energy wastage, 15:11
idle listening, 15:11
overhearing, 15:11
Continuous monitoring sensor networks, 15:13
Cooperative computing, 13:1–20
data brick serialization, 13:9
distributed object tracking, 13:3
Kilobyte Virtual Machine, 13:8
multiple smart messages, 13:16
node architecture, 13:4–5
programming interface, 13:7–8
prototype implementation, 13:8–12
readTag primitive, 13:11
smart messages, 13:5–7
directed diffusion, 13:15
life cycle, 13:6
self-routing, 13:7
tag space operations, 13:11–12
Cooperative engagement capability systems, 2:3
Coordinated sleeping, 15:14–15
Copper electrodes, 5:12
Core clock (CCLK), 14:7–8
configuration register, 14:7–8
Corner-cube retroreflector (CCR), 5:13, 17:9–17:10
Cost map, 3:24
Costs of implementation, 1:11, 2:1, 3:24, 7:2, 13:12
COTS processor systems, 2:9
Cougar system, 4:7
CreateSM, 13:7
Cricket, 11:11
Cross-layer designs, 7:20, 15:23
Cross-layer protocol stack, for wireless sensor networks, 7:21
Cubic millimeter sensor nodes, 5:13

D

Data advertisement protocol, 10:3
Data aggregation algorithms, 7:19–20
Data brick serialization, 13:9
Data-centric data, 7:20
Data clustering, 12:16–18
Data dissemination protocol, 10:3–4
Data-link layer, 3:25
Data link layer protocols, 10:11–13
error control, 10:13
medium access control, 10:12–13
Data processing techniques, 12:1–20
sensor processing systems, 12:3
distributed sensor processing, 12:3
hierarchical systems, 12:3
multisensor integration, 12:3
sensor collection, 12:3
tree systems, 12:3
types of, 12:3
Data redundancy, 7:8
Decision-making networks, 15:12
Delay-aware-medium-access protocol, 15:20–22
Delay balancing, 1:8
DeleteTag, 13:7
Denial of service (DoS), 3:5–3:6, 3:14
Dependent model battery, 3:7
Deployment strategies, 7:13–14
Destination-sequenced distance vectors, 1:7
DFuse, 4:11
Diabetes glucose monitoring, 7:5
Diagnosis of sensor networks, 6:9
Diffused computation operation, 6:5
Directed diffusion, 10:11
with in-network processing, 4:6–7
using smart messages, 13:15
Directional eight-neighbor node, 9:14
Directional source-aware routing protocol (DSAP), 9:13–19
three-dimension analysis, 9:18–19
two-dimension analysis, 9:15–18
Directly attached storage, 2:8
Disaster detection, sensor networks in, 1:3
Disaster prevention, relief, 7:4
Disaster recovery, sensor networks in, 1:3
Disk model, wireless link, 1:8–1:9
Distributed data gathering, decision making networks, 15:12
Distributed object tracking, 13:3
Distributed programming, complications, 4:4
Distributed signal processing, 1:11
DoS. *See* Denial of service
DPNs. *See* Dynamic private networks
Drug administration, sensor networks for, 7:5
DSAP. *See* Directional source-aware routing protocol
DSR. *See* Dynamic source routing
Dual space transformation, 17:2
Duty cycle assignment, 6:12
DVS. *See* Dynamic voltage scaling
Dynamic information, 3:23–24
communication coverage area map, 3:24
cooperational model, 3:24
cost map, 3:24
dependence model, 3:24
network topology, 3:24
production map, 3:24
residual energy, 3:24
sensing coverage, 3:24
structural model, 3:24
usage standard map, 3:24
Dynamic power management, 14:1–16
active power management, 14:5–6
battery life, 14:13–14:14
circuit schematic, dynamic voltage scaling, 14:7
component description language, 14:11
core clock, 14:7–8
configuration register, 14:7–8
dynamic voltage, 14:4
scaling, 14:9
scaling circuit, 14:6–8

exception handling, 14:11
hooks, idle power management, 14:8–9
idle mode, 14:10
idle power management, 14:2–5
MAX1717 step-down controller, 14:6
memory clock, 14:7–8
microelectricalmechanical systems technology, advances in, 14:1
multiple shutdown states, 14:2–3
node architecture, 14:3–4
operating system architecture, 14:11–12
operating system-directed dynamic power management, 14:2
power consumption, 14:9–10
power management functions, 14:12
power savings distribution, 14:13–14:14
processor power modes, 14:9–11
Redhat eCos, 14:11
run mode, 14:10
SA-1110 processor
core clock configuration, 14:8
power consumption, 14:11
scheduler, 14:11
sensor-specific application programming interface extensions, 14:12
sleep mode, 14:10
sleep state
power, 14:5
sensor nodes, 14:3
transition policy, 14:4–5
system implementation, 14:6–12
variable voltage processing, 14:5–6
Dynamic private networks (DPNs), 2:7
Dynamic sensor network programming, 4:3
Dynamic source routing (DSR), 1:7
Dynamic voltage, 14:4
Dynamic voltage scaling (DVS), 14:9, 17:5–6
circuit, 14:6–8

E

Earliest deadline first (EDF), 1:7
Eavesdrop algorithm, 15:18
Ecosystem monitoring, 7:3
EDF. *See* Earliest deadline first
Edge routing, 9:10–11
vs. interior routing, 9:11
EDP. *See* Ethylene diamine pyrocatechol
Efficient sensor network applications, aspects of, 4:2–3
Eight-neighbors wireless sensor network, 9:6–9:7
Election withdrawal cycle, 17:9
Electricity meters, sensor use for, 7:5
Electro-optical or infrared (EO/IR), 2:2–3
Element management, 3:7
Embedded programming, complications, 4:4
Embedded signal processor evolution, 2:9
Embedded systems, 8:13–16
Encryption protocols, 16:5–6
Energy efficiency, 1:4–10, 3:7, 5:4, 5:12, 7:14–16, 8:5, 9:9–13, 14:1–16, 15:1–26. *See also* Dynamic power management; Medium access control
active power management, 14:5–6
analog-to-distal conversion, 17:2
asynchronous processors, 17:3–4
battery classified into following types, 3:7
clock frequency, mesh plot, 17:5
communication, 17:7–10
computation, 17:2–5
corner-cube retroreflector, 17:9–17:10
distribution, 14:13–14:14
dual space transformation, 17:2
dynamic voltage scaling, 17:5–6
election withdrawal cycle, 17:9
gathering efficiency, 7:18
global positioning system, 17:1
hardware-software interaction, 17:5–6
idle power management, 14:2–5
low-power designs, 17:1–12
management functions, 14:12
mote-to-central station communication, 17:9–10
mote-to-mote communication, 17:7–9
node location, 17:1–2
optical communication, 17:1–2
orientation, 17:10
phase lock loop circuitry, 17:4
power-aware multiple access protocol, 15:8–10
protocols for, 1:6
radio frequency communication, 17:2
resources, 3:2
schedule modification, 17:6
sensing, 17:2
sleep modes, 1:6
storage, 5:10–12
time division multiple access, 17:7
variable-frequency processors, 17:4–5
variable-voltage processors, 17:5
wireless sensor networks, 1:4
Engineering, wireless sensor networks in, 1:2–3
Environment, sensor execution, 6:7–8
Environment detection, monitoring, 7:3–4
Environmental applications, 10:2
monitoring, 1:3
EO/IR. *See* Electro-optical or infrared
Error-prone wireless medium, 7:9
Ethylene diamine pyrocatechol (EDP), 5:2
European Laboratory for Particle Physics, CERN, 12:18
Event driven sensor networks, 15:13
Exception handling, 14:11
Experimental location systems, 11:10–12
Extraction of information, 2:1

F

Factory instrumentation, sensor networks for, 7:6
Fault management, 3:13
Fault tolerance, 1:5, 7:2, 7:10, 8:5
Fingertip accelerometer virtual keyboards, 1:2
Fire detection, with remote sensing, 7:4
Five-neighbors wireless sensor network, 9:4–9:5
Flat wireless sensor networks, agents in, 3:17–18
Flooding, 10:11
communication mode, 3:11

Flynn taxonomy for computer architectures, 12:3–4
Four-neighbors wireless sensor network, 9:4–9:5
Free-space optical transmission, 1:6, 5:9–10
Freshwater quality monitoring, sensor networks in, 1:3
Fuel leaks, tracking of, 7:4
Functional architecture, 3–4, 6:3–4
    components of, 6:3–4
Fusion center, sensor network with, 1:5

G

Gas leaks, tracking of, 7:4
Gas meters, sensor use for, 7:5
General engineering, sensor networks in, 1:2–3
General interORB protocol (GIOP), 2:11
Geocast, 11:2
Geolocation, mobile targets, 2:2–3
Geophysical monitoring, sensor networks in, 1:3
GFLOPS, 2:5, 2:9
GIOP. *See* General interORB protocol
Global positioning system, 17:1
Glucose level monitoring, sensor networks for, 7:5
GMTI. *See* Ground moving target indication
Gossiping, 3:11, 10:11
Graph generator, 2:12–13
Graph search, 2:14–15
Greedy aggregation approach, 7:20
Ground moving target indication (GMTI), 2:2–3
Guarantee for network resources, 2:6–8

H

Habitat monitoring, sensor networks in, 1:3
Hardware-software interaction, power savings in, 17:5–6
Health care applications, sensor networks for, 7:5
Health Improvements through Space Technologies and Resources (HI-STAR) Project, 7:4
Health monitoring, sensor networks for, 1:4
Heterogeneous wireless sensor networks, agents in, 3:18
HI-STAR project. *See* Health Improvements through Space Technologies and Resources project
Hierarchical homogeneous wireless sensor networks, agents in, 3:18
Hierarchical systems, 12:9–11
High-energy physics, architecture evaluation in, 18–19
High-frequency transmission, attenuation requirements, 1:6
High-range resolution (HRR) profiles, 2:3
Home intelligence, sensor networks for, 7:5
Homogeneous wireless sensor networks, agents in, 3:17
Hostile environments, 16:2
HRR profiles. *See* High-range resolution profiles

I

Identification of mobile targets, 2:2–3
Idle power management hooks, 14:8–9
IEEE 802.11 standard, 15:4–6
    power-saving mode, 15:5
Implementation architectures, 4–12
Implosion, 3:19–20
Industrial plants, sensing, maintenance in, 1:2
Information abstraction, 6:5
Information architecture, 3:21–24
Information gathering methods, 6:5–6
Integrated circuit sensors, 5:5
Integrated network sensors, wireless, 7:12
Integrating architecture, MANNA, 15–25
Interior routing, 9:10–11
Intruder identification, sensor networks for, 7:6

J

Journal publications devoted to sensor networks, 1:2

K

K-anonymity, 16:13–15
    in mix zones, 16:14
Key distribution schemes, 16:9–10
    online key distribution servers, 16:9
    probabilistic key distribution, 16:9–10
    public key cryptography, 16:9
Kilobyte Virtual Machine (KVM), 13:8
KVM. *See* Kilobyte Virtual Machine

L

LAN. *See* Local area network
Latency, sensor networks, 1:5
Layer design, 15:3
Layered design, protocol stack, ad hoc networks, 7:21
Lifetime of network, extending, 16:8
Light compass node, 8:11–13
Linda-like model, system-level programmability, 4:6
Linear model battery, 3:7
Listening, energy consumption, 1:9
LLA. *See* Logical layered architecture
Local area network (LAN)
    data storage centers in parallel, 2:8
Local messages, 15:12–15:13
Local network resource managers, wide-area network resource managers, interface, 2:16
Localization of communication, 1:11, 4:2–3
Localization protocols, 10:4–5
Location information, protection of, 16:12–15
Location tracking, 11:8–9
Logical layered architecture (LLA), 3:3
Low duty cycle electronics, 7:12–13
Low-power adaptive clustering hierarchy, 9:2, 10:11. *See also* Power conservation
Low-power designs. *See* Power conservation

M

MagnetOS, 4:11
Maintenance, 3:11–12
Management functional areas, 3:3–4
Management levels, 3:3–4
Management of assets, wireless sensor networks in, 1:3

MANNA integrating architecture, 3:15–25
Mapping database, 2:15
Mars measurement collection, sensor networks for, 7:5
Maté, 4:9
MAV. *See* Microaerial vehicles
MAX1717 step-down controller, 14:6
Maximum lifetime data aggregation (MLDA)
  clustering-based heuristic for, 7:20
MCLK. *See* Memory clock
Medical care, use of sensor networks in, 7:4–5
Medium access control
  ad hoc networks, 15:4–10
  adaptive listening, 15:16
  application types, 15:12–13
  ATM networks, energy-conservation, 15:7–8
  Bluetooth, 15:6–7
  broadcast messages, 15:13
  centralized data gathering, decision making networks, 15:12
  cluster-based hierarchy, 15:23
  communication types, 15:12–13
  comparison metrics in, 15:24
  comparison of schemes, 15:22
  contention-based scheme, 15:11
  continuous monitoring sensor networks, 15:13
  coordinated sleeping, 15:14–15
  cross-layer design, 15:23
  delay-aware-medium-access protocol, 15:20–22
  design challenges, 15:10–13
  distributed data gathering, decision making networks, 15:12
  eavesdrop algorithm, 15:18
  energy-efficiency, 15:1–26
  event driven sensor networks, 15:13
  frame structure, 15:7
  hardware constraints, 15:24
  IEEE 802.11 standard, 15:4–6
    power-saving mode, 15:5
  layer design, 15:3
  local messages, 15:12–15:13
  mobility management, 15:23
  multicast messages, 15:13
  multihop messages, 15:13
  neighbor discovery, 15:15
  operation, 15:4–5
  Piconet configurations in Bluetooth, 15:6
  power-aware multiple access protocol, 15:8–10
  protocols, 15:13–22
  register algorithm, 15:18
  research issues, 15:22–24
  scalability, 15:23
  scheduling-based scheme, 15:11
  self-organizing medium access control, 15:16–18
  sensor medium access control, 15:14–16
  sensor network architectures, 15:2
  synchronization, 15:15
  time synchronization, effects of, 15:23
  timing schedules, 15:15
  traffic adaptive medium access protocol, 15:18–20
  unicast messages, 15:12
Medusa MK-2 node, 8:9–11
Memory clock (MCLK), 14:7–8
Mesh networking, 1:1
Messages received by node, dispatching of, 6:8
Metering, with security threats, 16:4
Metrics object, 2:13–14
Microaerial vehicles, 5:1
Microbattery, features of, 5:10
Microelectromechanical systems, 1:5, 3:1, 5:1–20
  advances in, 7:12, 14:1
Micromachining techniques, 5:2–3, 5:13
Micromechanical relays, 5:8
Micropower sources, 5:10–12
Microrocket engine combustion, utilizing heat from, 5:12
Microsurgery, sensor networks for, 1:4
Middleware, 2:6, 2:11, 7:21
  sensor information networking architecture, 6:5
Military applications, 1:3–4, 7:2–3, 10:2
MIMO antenna systems. *See* Multiple input, multiple output antenna systems
Minefields, self-healing, wireless sensor networks in, 1:4
Miniaturizing sensor networks, 5:1–20
MLDA. *See* Maximum lifetime data aggregation
Mobile agent model, benefits of, 3:24–25
Mobile code, 16:3–4
Mobile targets, 2:2–3
Mobility, 1:10, 15:23
Molybdenum oxide batteries, 5:11
Monitoring of structures, sensor networks in, 1:3
Mote-to-central station communication, power savings, 17:9–10
Mote-to-mote communication, power savings, 17:7–9
Multicast, communication mode, 3:11
Multicast messages, 15:13
Multicast trees, 1:6
Multihop ad hoc wireless architecture, 8:15–16
Multihop clustering architecture, 7:17
Multihop communication, 1:7
Multihop messages, 15:13
Multilateration, 11:5–6
Multilevel hierarchical clustering architecture, 7:18
Multiple input, multiple output (MIMO) antenna systems, 2:8
Multiple shutdown states, 14:2–3
Multiple smart messages, 13:16

N

Nanosensors, 5:5
National Chiao Tung University, 11:11–12
Nearest neighbor algorithms, 11:7
Neighbor discovery, 15:15
Neighbors in wireless sensor network, 9:1–20
Network applications, 7:2–6
Network-attached storage, 2:8
Network-based tracking, 11:9–10
Network connectivity, 1:10
Network functionalities, 3:8–12
Network layer, 3:25
  design principles, 10:9
  protocols, 10:9–11
Network management, 3:1–28

Network modeling, 1:8–10
Network resource management, 2:11–16
Network resource manager (NRM)
  object model for, 2:12
Network resource manager agents, 2:15
Network resource manager fault tolerance, 2:16
Network resource manager federation, 2:16
Network resource manager system, 2:6
Network topologies, 9:3–8
  eight-neighbors wireless sensor network, 9:6–9:7
  five-neighbors wireless sensor network, 9:4–9:5
  four-neighbors wireless sensor network, 9:4–9:5
  seven-neighbors wireless sensor network, 9:6–9:7
  six-neighbors for three dimensions, 9:6, 9:8
  six-neighbors wireless sensor network, 9:5–9:6
  three-neighbors wireless sensor network, 9:4
Networking, real-time computing convergence, 2:6–10
Networking architecture, sensor information, 6:4–9
Next-generation technologies, 2:1–22
Nibble system, 11:11
Nodes, 8:7–13
  admission manager, 13:4
  architecture, 4–5, 13:4–5, 14:3–4
  battery life, 14:14
  Berkeley Mote Node, 8:7–8
  code cache, 13:4
  components, 5:1, 8:4–7
  cubic millimeter, 5:13
  current drawn, 8:7
  density, 7:6
  directional eight-neighbor, 9:14
  dispatching of messages, 6:8
  distance to base station, 7:6
  distribution, 1:10
  eavesdrop, register algorithms, 15:18
  heterogeneous, 7:22
  identification, 3:2
  internetworking, 10:10
  light compass, 8:11–13
  location of, 17:1–2
  low-power, 7:11–12
  Medusa MK-2 node, 8:9–11
  managing, technical challenges, 1:4–11
  PicoNode, 8:11
  processor, 8:4
  single level, 7:21
  sleep states for, 14:3
  smoke sensor, 7:6
  spurring, 7:12
  state transition diagram, 7:15
  storage, 8:4–5
  system architecture, 7:12
  tags, 13:5, 13:13
  timing schedules, 15:15
  UCLA Medusa MK-2 node, 8:8–11
  with video camera, 13:12
  virtual machine, 13:4–5
NRM. *See* Network resource manager
Nuclear attack, remote sensing, 7:3

O

OASIS. *See* Operator assisted integrated systems
On board super-resolution processing, 2:5
On-chip high-Q inductors, 5:6–7
One hundred billion floating-point operations per second. *See* GFLOPS
Online key distribution servers, 16:9
Operating system-directed dynamic power management, 14:2
Operator assisted integrated systems (OASIS), 2:17–19
Optical communication, 5:8–10, 17:1–2
Optical transmission techniques, 1:6
Organ monitoring, sensor networks for, 7:5
Orientation, power conservation, 17:10
  random orientation, 17:10
  zero-power orientation, 17:10

P

Packaging, 5:5, 5:12–13
Parallel processing, 2:11
Parallel system, 12:8–9
Parameterization, parallel, hierarchical architectures, 12:6–8
Patient tracking, monitoring, use of sensor networks, 7:5
Pattern matching, 11:6–8
PBNM. *See* Policy-based network management
Performance management, 3:13–14
  performance metrics, 3:14
  qualitative features, 3:14
Phase lock loop (PLL) circuitry, 17:4
Photocurrents, charging capacitors with, 1:6
Piconet configurations, Bluetooth, 15:6
PicoNodes, 7:12, 8:11
PicoRadio project, 7:12
Planar micromachining, 5:2
Planetary exploration, sensor networks in, 1:3
PLL circuitry. *See* Phase lock loop circuitry
Policy-based network management (PBNM), 3:4
Polymer encapsulation process, for packaging, 5:14
Polynomial-time algorithm, maximum lifetime data aggregation, 7:20
Positioning algorithms, 11:4–10
Potential hazards, detection of, 7:4
Potential terrorist attack, remote sensing, 7:3
Power conservation, 1:4–10, 3:7, 5:4, 5:12, 7:14–16, 8:5, 9:9–13, 14:1–16, 15:1–26. *See also* Dynamic power management; Medium access control
  active power management, 14:5–6
  analog-to-distal conversion, 17:2
  asynchronous processors, 17:3–4
  battery classified into following types, 3:7
  clock frequency, mesh plot, 17:5
  communication, 17:7–10
  computation, 17:2–5
  corner-cube retroreflector, 17:9–17:10
  distribution, 14:13–14:14
  dual space transformation, 17:2
  dynamic voltage scaling, 17:5–6
  election withdrawal cycle, 17:9

gathering efficiency, 7:18
global positioning system, 17:1
hardware-software interaction, 17:5–6
idle power management, 14:2–5
low-power designs, 17:1–12
management functions, 14:12
mote-to-central station communication, 17:9–10
mote-to-mote communication, 17:7–9
node location, 17:1–2
optical communication, 17:1–2
orientation, 17:10
phase lock loop circuitry, 17:4
power-aware multiple access protocol, 15:8–10
protocols for, 1:6
radio frequency communication, 17:2
resources, 3:2
schedule modification, 17:6
sensing, 17:2
sleep modes, 1:6
storage, 5:10–12
time division multiple access, 17:7
variable-frequency processors, 17:4–5
variable-voltage processors, 17:5
wireless sensor networks, 1:4
Power cycling, 5:4–5
Precision agriculture, sensor networks in, 1:3
Preprocessing elements, 12:12–18
Principle of minimal generalization, 16:11
Prioritization, energy balancing, 1:7
Priority scheduling, 1:7
Privacy protection, 16:11–15
algorithms, 16:13–14
alternative identification or pseudonym for user, 16:13
anonymity in privacy protection, defined, 16:13–15
architecture, 16:12
k-anonymity, 16:13–15
in mix zones, 16:14
location information, 16:12–15
principle of minimal generalization, 16:11
users, 16:12
Probabilistic key distribution, 16:9–10
Probability-based algorithms, 11:7–8
histogram method, 11:8
kernel method, 11:8
Processor interruption, 2:9–10
Production map, 3:24
Productivity of network resources, 3:2
Programmability models, 4:1–14
Programming languages, 6:6–7
Propagation of information, speed of, 1:10
Pseudonym for users, for privacy, 16:13
Public key cryptography, 16:9

Q

QoS. *See* Quality of service
Quality of service (QoS), 1:10–11, 3:2, 3:5, 7:9
dissemination, 3:6
model, 3:5
processing, 3:6
routing, 1:7
sensing, 3:5–6
Query, tasking language, 7:21
Querying applications, 6:2

R

RADAR system, 11:11
Radar technology evolution, 2:2
Radio, 8:6–7
Radio characteristics, 9:9
Radio frequency communication, 1:6, 5:5–8, 7:1, 17:2
transmission techniques, 1:6
Radioactive isotopes for energy storage, 5:12
Random deployment, 7:8
Rayleigh fading model, wireless link, 1:9
Reactive ion etch, 5:2
ReadTag primitive, 13:11
Real-time computing, networking convergence, 2:6–10
Real-time distributed network computing, 2:5–6
Real-time operating system (RTOS), 2:10
software, 8:7
Reconfiguring, under varying conditions, 3:2
Reconnaissance, 7:2–7:3
information extraction, 2:1
Redhat eCos, 14:11
Region of interest (RoI)
builder placement, 12:13
hierarchical systems with, 12:14
Register algorithm, 15:18
Relaxation model battery, 3:7
Remote virus monitoring, 7:4
Reprogramming, wireless, 1:11
Rescue operations, 7:4
Research in sensor networks, increasing amount of, 1:2
Research project funding for sensor networks, 1:2
Residential security, wireless sensor networks in, 1:3
Retail stores, tracking of goods in, 1:3
Retrieval of network state, 6:12
Revolutionary nature of sensor networks, 1:2
Robot control, 7:6
Robustness, network topology, 1:7
RoI. *See* Region of interest
Routing, 1:4, 9:12
protocols, 1:7–1:8
RTOS. *See* Real-time operating system
Run mode, 14:10

S

SA-1110 processor
core clock configuration, 14:8
power consumption, 14:11
Sampling operation, 6:5
SAR. *See* Synthetic aperture radar
Scalability of sensor networks, 1:5
Schedule modification, power savings, 17:6
Scheduler, 14:11
Scheduling, 1:8
algorithm, 1:8
channels, 1:7–8
Scheduling-based schemes, 15:11

Scientific exploration, sensor networks for, 7:5
Seawater, generation of power from, 5:12
Security management, 1:11, 3:14
- ad hoc sensor networks, 16:8–10
- application-specific architectures, 16:2
- architectures, 4–10, 16:4–10
- cell-based wireless sensor networks, 16:5–8
- challenges to, 16:2–4
- denial of service attack, 3:14
- design, 7:10
- encryption protocol, 16:5–6
- hostile environment, 16:2
- in-network processing, 16:2
- key distribution schemes, 16:9–10
- lifetime of network, extending, 16:8
- limited resources, 16:2
- metering, with security threats, 16:4
- mobile code, 16:3–4
- security-related properties, 16:2
- system-level security, 16:3
- wireless sensor networks in, 1:3

Self-configuration, 7:10
Self-healing minefields, sensor networks for, 1:4
Self-orchestrated operation, 6:5
Self-organizing medium access control, 15:16–18
Self-routing smart messages, 13:7
Sensor information networking architecture (SINA), 7:21
Sensor integration architecture, 3–18, 12:3–18
Sensor interface, 2:15
Sensor management protocol (SMP), 10:3
Sensor medium access control, 15:14–16
Sensor networks. *See* Wireless sensor networks
Sensor nodes. *See* Nodes
Sensor protocols for information via negotiation (SPIN), 10:11, 16:5
Sensor querying, 4:7–8, 10:3–4
- script, 6:10

Sensor-specific application programming interface extensions, 14:12
SensorWare, 4:9–11
Service management, 3:5–6, 3:9
Seven-neighbors wireless sensor network, 9:6–9:7
Short-range radio, as communication component, 8:6
Signal strength, positioning by, 11:3–4
Signal-to-noise-and-interference ratio (SNR)
- wireless link, 1:8

Silicon etchants, 5:2
Silicon-on-insulator (SOI) wafers, 5:2
SINA. *See* Sensor information networking architecture
Single level nodes, 7:21
Single node level, 7:21
Six-neighbors wireless sensor network, 9:5–9:6, 9:8
Sleep mode, 1:6, 14:10
Sleep state, 14:3
- power, 14:5
- transition policy, 14:4–5

Sleeping, energy consumption, 1:9
Smart Dust mote, 5:15
- mock up, 5:16

Smart dust networks. *See* Wireless sensor networks
Smart kindergartens, 7:5
Smart messages, 13:5–7, 13:12–14
- directed diffusion, 13:15
- life cycle, 13:6
- mobile code platform, 4:8–9
- multiple, 13:16
- self-routing, 13:7

Smart office spaces, 1:3
Smart refrigerators, sensor networks for, 7:5
Smart roads, control of, 7:6
Smoke sensor nodes, 7:6
SMP. *See* Sensor management protocol
SNR. *See* Signal-to-noise-and-interference ratio
Software, effect of hardware restrictions, 3:9
Software development, 7:20–22
SOI wafers. *See* Silicon-on-insulator wafers
Solar cells, 1:6, 5:12
Solar radiation, as energy source, 5:12
SpawmSM, 13:7
Speed of propagation of information, 1:10
SPIN. *See* Sensor protocols for information via negotiation
Spreadsheet abstraction, 6:5
Static information, 3:21–23
- connection, 3:23
- environment, 3:23
- equipment, 3:22–23
- managed element, 3:22
- managed object classes, 3:22
- network, 3:22
- support object classes, 3:21–22
- system, 3:23
- wireless sensor network goals, 3:23
- wireless sensor network management, 3:23
- wireless sensor network observer, 3:23

Static sensor networks, clustering for, 1:6
Steered agile laser communication, packaging, 5:13
Storage area networks, 2:8
Storage buffer, 2:5
- resource guarantees, 2:8

Stores, tracking of goods in, 1:3
Sun Microsystem, Kilobyte Virtual Machine, 13:8
Sunken vessel location, 7:4
Super-resolution processing, 2:5
Surface-micromachined polysilicon, 5:8
Surface micromachining, 5:2
Surgery, wireless sensor networks for, 1:4
Surveillance, 7:2
- sensors for, 1:4, 2:1, 7:6

Synchronization, 1:11
Synthetic aperture radar (SAR), 2:2–3, 10:11
System faults, 2:10
System-level programmability models, 4:4–6
- active sensor model, 4:4–5
- blackboard-based model, 4:6
- client/server model, 4:6
- database model, 4:4
- distributed programming, 4:4
- embedded programming, 4:4
- frameworks, 4:6–11
- Linda-like model, 4:6
- meeting-oriented model, 4:6
- mobile agent

coordination models, 4:5–6
technology, 4:5–6
System lifetime, of sensor network, 1:5

T

Tag space operations, 13:11–12
Tasking applications, 6:2–3
Tasking language, 4:7–8, 7:21
script, 6:10
Telematics, automotive, wireless sensor networks in, 1:2
Temporal geocast, 11:2
Temporally-ordered routing algorithm (TORA), 1:7
Terrestrial networks, 2:6–7
Terrorist attack, remote sensing, 7:3
Tetramethylammonium hydroxide (TMAH), 5:2
TFLOPS, 2:5, 2:9
Thin-film vanadium oxide battery, 5:11
Three-neighbors wireless sensor network, 9:4
Throughput, 1:5–1:7
polynomial, 1:8
sensor networks, 1:5
Time division multiple access, 17:7
Time synchronization, 8:5
effects of, 15:23
factors influencing, 10:5
protocols, 10:5–7
Timing schedules, 15:15
Timing techniques, 10:6
TinyDR model, 4:7
TMAH. *See* Tetramethylammonium hydroxide
TopDisc. *See* Topology Discovery for Sensor Networks
Topology database, 2:15–16
Topology Discovery for Sensor Networks (TopDisc), 6:9–12
architectural view, 6:11–12
TORA. *See* Temporally-ordered routing algorithm
Toxic substance leaks, tracking of, 7:4
Traditional data networks, sensor networks, contrasted, 4:2
Traffic adaptive medium access protocol (TRAMA), 15:18–20
Traffic distribution, 7:19
Traffic monitoring, sensor networks for, 7:6
TRAMA. *See* Traffic adaptive medium access protocol
Transceivers, 3:8
Transplant organs, monitoring of, 7:5
Transport capacity of sensor networks, 1:5
Transport layer, 3:25
event-to-sink transport, 10:7–8
protocols, 10:7–9
sink-to-sensors transport, 10:8–9
Trapped personnel, detection of, 7:4
Trilateration, 11:4–5
Trillion floating-point operations per second. *See* TFLOPS

U

UCLA Medusa MK-2 node, 8:8–11
Unattended operation, 7:9
Unicast, communication mode, 3:11
Unicast messages, 15:12
Unmanned aerial vehicles, 2:3
Unrechargable energy, wireless sensor networks, 1:4
Urban planning, sensor networks in, 1:3
Urban warfare, sensor networks in, 1:4
Usage standard map, 3:24
Utility meters, sensor use for, 7:5

V

Variable-frequency processors, power savings, 17:4–5
Variable voltage processing, 14:5–6
Variable-voltage processors, power savings, 17:5
Vibration, as energy source, 5:12
Victims, detection of, 7:4
Video camera, node with, 13:12
Virus monitoring, 7:4
Voice communication, mesh networking, 1:1

W

WAN. *See* Wide-area network
Warehouse inventory tracking, 7:6
Water meters, sensor use for, 7:5
Waterproof sensor arrays, 7:4
WBM. *See* Web-based management protocol
Web-based management protocol, 3:24
Web pages, devoted to sensor networks, 1:2
Wide-area network (WAN)
data storage centers in parallel, 2:8
Wide-area network resource managers, local network resource managers, interface, 2:16
Wild land fire detection, 7:4
Wireless Internet access, mesh networking, 1:1
Wireless link, 1:8–9
Wireless reprogramming, 1:11
Wireless sensor networks (WSNs)
Active Badge system, 11:10
admission manager, 13:4
applications, 1:2–4, 6:1–3, 10:2
asynchronous processors, 17:3–4
basic features of, 1:1
Bat system, 11:10
Berkeley Mote Node, 8:7–8
Bluetooth, 15:6–7
BWRC PicoNode, 8:11
capacity, 1:6–7
channel access, 1:7–8
characteristics of, 1:1
classification, 7:6–7
code cache, 13:4
commercial interest in, 1:2
communication network connection, 1:4–11
communication protocols, 10:1–16
computation, 2:8–10, 17:2–5
connectivity, 1:10
cooperative computing, 13:1–20
Cougar system, 4:7
Cricket system, 11:11
data link layer protocols, 10:11–13
data processing techniques, 12:1–20
database mapping, 2:15
DFuse, 4:11

directed diffusion, with smart messages, 13:13–14
directional source-aware routing protocol, 9:13–19
disaster prevention, relief, 7:4
dynamic voltage scaling circuit, 14:6–8
energy efficiency, 1:5–1:6, 5:10–5:12, 8:5, 9:1–20, 14:5–6
environment uses, 7:3–4
evolution of, 12:2
fabrication techniques, micromachines, 5:2–3
fault tolerance, network resource manager, 2:16
functional architecture, 3:17–21
functionalities, 3:8–12
graph generation, 2:12–13
graph search, 2:14–15
hardware-software interaction, 17:5–6
hardware techniques, 7:11–22
high data rates, 12:5–11
high-energy physics, architecture evaluation, 12:18–19
home intelligence, 7:5
identification of mobile targets, 2:2–3
idle power management, 14:2–5, 14:8–9
IEEE 802.11 standard, 15:4–6
imbedding, 8:13–16
in-network processing, directed diffusion with, 4:6–7
information architecture, 3:21–24
integrated circuit sensors, 5:5
integrating architecture, MANNA, 3:15–25
interactive surroundings, 7:5–6
layer protocols, 10:2–4
localization protocols, 10:4–5
location information, privacy protection of, 16:12–15
location tracking, 11:1–14
long-term architecture, 2:3–5
low-power design, 17:1–12
MagnetOS, 4:11
management issues, 2:16–19, 3:1–28
Maté, 4:9
medical care, 7:4–5
medium access control, 15:1–26
metering, in response to security threats, 16:4
metrics object, 2:13–14
microelectromechanical system, 5:2
micropower sources, 5:10–12
middleware, 2:11
military applications, 7:2–3
miniaturization, 5:1–20
minimal generalization principle, 16:11
mobile agent technology, 4:5–6
mobile code, 16:3–4
mobile targets, 2:2–3
modeling, 1:8–10
mote-to-central station communication, 17:9–10
mote-to-mote communication, 17:7–9
multilateration, 11:5–6
multiple shutdown states, 14:2–3
nanosensors, 5:5
National Chiao Tung University system, 11:11–12
neighbors in, number of, 9:1–20
network-based tracking, 11:9–10
network layer protocols, 10:9–11
network resource manager, 2:15–2:16
Nibble system, 11:11
nodes, 1:4–11, 8:4–13, 13:4–5, 17:1–2
number of neighbors in wireless sensor network, 9:1–20
opportunities in, 1:2–4
optical communication, 5:8–10
orientation, 17:10
packaging, 5:12–13
parallel processing, 2:11
pattern matching, 11:6–8
performance metrics, 1:5
physical architecture, 3:24–25
positioning, 11:1–14
power supply, 1:5–1:6, 5:10–5:12, 8:5, 9:1–20, 14:5–6, 15:8–10
preprocessing, 12:12–18
privacy protection, 16:11–15
processing systems, 12:3
processor power modes, 14:9–11
programmability, 4:1–14
programming interface, 13:7–8
protocols, 7:13–20
quality of service, 1:10–11
querying applications, 6:2
RADAR system, 11:11
radio, 8:6–7
radio frequency communication, 5:5–8
real-time distributed network computing, 2:5–6
research increase in, 1:2
resource management, 2:11–16
routing protocols, 1:7
scientific exploration, 7:5
security issues, 1:11, 16:1–18
sensing, 17:2
sensor-centric design, 8:11–13
sensor information networking architecture, 6:4–9
sensor integration, 12:3–18
sensor interface, 2:15
sensor protocols, smart messages, 13:12–13
sensor querying, 4:7–8
sensor-specific application programming, 14:12
SensorWare, 4:9–11
sleep state, 14:3–5
smart messages, 13:5–7
    mobile code platform, 4:8–9
software development, 7:20–22
storage, 8:4–5
storage buffer resources, 2:8
surveillance, 7:6
system architecture, 2:3–5, 3:15–25, 6:1–14, 7:13–20, 8:1–20, 12:18–19
system control, 2:11
tag space, 13:5
tasking applications, 6:2–3
tasking language, 4:7–8
technical challenges, 1:4–11
throughput, 1:6–7
time synchronization protocols, 10:5–7
timing techniques, 10:6
TinyDR model, 4:7
topology database, 2:15–16
topology design, 9:3–8

Topology Discovery for Sensor Networks (TopDisc), 6:9–12
traditional data networks, contrasted, 4:2
transport layer protocols, 10:7–9
trilateration, 11:4–5
UCLA Medusa MK-2 node, 8:8–11
variable-frequency processors, 17:4–5
variable voltage processing, 14:5–6, 17:5
virtual machine, 13:4–5
wireless ATM networks, 15:7–8
Workshops on sensor networks, 1:2
WriteTag, 13:7
WSNs. *See* Wireless sensor networks

Z

Zebranet, 1:3
Zinc-air cell, 5:11
    problems with, 5:11
Zinc electrodes, 5:12